Principles of Animal Learning and Motivation

Roger M. Tarpy
Bucknell University

Lyle E. Bourne, Jr., Consulting Editor
University of Colorado at Boulder

Scott, Foresman and Company

Glenview, Illinois
Dallas, Texas   Oakland, N.J.   Palo Alto, Cal.
Tucker, Ga.   London, England

To Alicia, Elizabeth, and David

Library of Congress Cataloging in Publication Data

Tarpy, Roger M., 1941–
    Principles of animal learning and motivation.

    Bibliography: p. 350
    Includes index.
    1. Learning in animals. I. Bourne, Lyle
Eugene, 1932–   . II. Title.
QL785.T347   156'.35   81–21218
ISBN 0–673–15383–5   AACR2

123456-VHS-868584838281

ACKNOWLEDGMENTS

Cover photograph courtesy of Gen/Rad, Inc./Grason-Stadler Division.

(2–2) Hammond, L. J., *Psychonomic Science,* 1967, *9,* 65–66. Adapted with permission of the author and Psychonomic Society.   *(2–6)* Rescorla, R. A., *Journal of Comparative and Physiological Psychology,* 1968, *66,* 1–5. Copyright 1968 by the American Psychological Association. Adapted by permission.   *(2–7)* Bolles, R. C., Holtz, R., Dunn, T., and Hill, W., *Learning and Motivation,* 1980, *11,* 78–96. Adapted by permission of the authors and Academic Press, Inc.

(3–2) Kalat, J. W., and Rozin, P., *Journal of Comparative and Physiological Psychology,* 1971, *77,* 53–58. Copyright 1971 by the American Psychological Association. Adapted by permission.   *(3–3)* Welker, R. L., and Wheatley, K. L., *Learning and Motivation,* 1977, *8,* 247–262. Adapted by permission of the authors and Academic Press, Inc.   *(3–4)* Fitzgerald, R. D., and Teyler, T. J., *Journal of Comparative and Physiological Psychology,* 1970, *70,* 242–253. Copyright 1970 by the American Psychological Association. Reprinted by permission.   *(3–5)* Nachman, M., and Ashe, J. H., *Physiology and Behavior,* 1973, *10,* 73–78. Adapted with permission of the author and Brain Research Publications.   *(3–6)* From Jenkins, H. M., and Moore, B. A., *Journal of the Experimental Analysis of Behavior,* Vol. 20, pp. 163–181, "Copyright 1973 by the Society for the Experimental Analysis of Behavior, Inc." Reprinted with permission of the author and publisher.   *(3–7)* Rizley, R. C., and Rescorla, R. A., *Journal of Comparative and Physiological Psychology,* 1972, *81,* 1–11. Copyright 1972 by the American Psychological Association. Adapted by permission.   *(3–8)* Reprinted with permission of the author and publisher from: Lynch, J. J. Pavlovian inhibition of delay in cardiac and somatic responses in dogs: schizokinesis. *Psychological Reports,* 1973, *32,* 1339–1346, Figure 1.   *(3–9)* Lubow, R. E., *Journal of Comparative and Physiological Psychology,* 1965, *60,* 454–457. Copyright 1965 by the American Psychological Association. Reprinted by permission.   *(3–10)* Wagner, A. R., A stimulus selection and a

(4–1) Roberts, W. A., *Journal of Comparative and Physiological Psychology,* 1969, *67,* 395–400. Copyright 1969 by the American Psychological Association. Adapted by permission.
(4–2) Franchina, J. J., Escape behavior and shock intensity: Within-subject versus between-subject comparisons. *Journal of Comparative and Physiological Psychology,* 1969, *69,* 241–245. Copyright 1969 by the American Psychological Association. Adapted by permission.   *(4–3)* Fowler, H., and Trapold, M. A., *Journal of Experimental Psychology,* 1962, *63,* 464–467. Copyright 1962 by the American Psychological Association. Adapted by permission.   *(4–5)* Caplan, H. J., Karpicke, J., and Rilling, M., *Animal Learning and Behavior,* 1973, *1,* 293–296. Adapted by permission of the authors and Psychonomic Society.   *(4–6)* From Felton, M., and Lyon, D. O., *Journal of the Experimental Analysis of Behavior,* Vol. *9,* pp. 131–134. "Copyright 1966 by the Society for the Experimental Analysis of Behavior, Inc." Adapted by permission of the authors and publisher.   *(4–7)* Mellgren, R. L., Wrather, D. M., and Dyck, D. G., *Journal of Comparative and Physiological Psychology,* 1972, *80,* 478–83. Copyright 1972 by the American Psychological Association. Adapted by permission.   *(4–8)* Nation, J. R., Wrather, D. M., and Mellgren, R. L., *Journal of Comparative and Physiological Psychology,* 1974, *86,* 69–73. Copyright 1974 by the American Psychological Association. Adapted by permission.   *(4–9)* Kimmel, H. D., Brennan, A. F., McLeod, D. C., Raich, M. S., and Schonfeld, L. I., *Animal Learning and Behavior,* 1979, *7,* 447–451. Reprinted with permission of the authors and Psychonomic Society.

(5–2) Kamin, L. J., *Journal of Comparative and Physiological Psychology,* 1957, *50,* 445–449. Copyright 1957 by the American Psychological Association. Reprinted by permission.   *(5–3)* Bower, G. H., Starr, R., and Lazarovitz, L., *Journal of Comparative and Physiological Psychology,* 1965, *59,* 13–17. Copyright 1965 by the American Psychological Association. Adapted by permission.   *(5–5)* Weisman, R. G., and Litner, J. S., *Journal of Comparative and Physiological Psychology,* 1969, *68,* 597–603. Copyright 1969 by the American Psychological Association. Adapted by permission.   *(5–7)* Rescorla, R. A., *Journal of Comparative and Physiological Psychology,* 1967, *64,* 114–120. Copyright 1967 by the American Psychological Association. Adapted by permission.

(6–1) Shettleworth, S. J., *Journal of Experimental Psychology: Animal Behavior Processes,* 1975, *1,* 56–87. Copyright 1975 by the American Psychological Association. Adapted by permission.
(6–2) Garcia, J., and Koelling, R. A., *Psychonomic Science,* 1966, *4,* 123–124. Adapted with permission of the authors and Psychonomic Society.   *(6–3)* "Illness Induced Aversions in Rat and Quail: Relative Salience of Visual and Gustatory Cues," Wilcoxon, H. C., Dragoin, W. B., and Kral, P. A., *Science,* Vol. *171,* pp. 826–828, Fig. 1, 26 February 1971. Copyright 1971 by the American Association for the Advancement of Science.   *(6–4)* Brener, J., and Goesling, W. J., *Journal of Comparative and Physiological Psychology,* 1970, *70,* 276–280. Copyright 1970 by the American Psychological Association. Adapted by permission.   *(6–5)* "Imprinting," Hess, E. H., *Science,* Vol. *130,* pp. 133–141, Fig. 2, 17 July 1959. Copyright 1959 by the American Association for the Advancement of Science.

(7–1) Burdick, C. K., and James, J. R., Spontaneous recovery of conditioned suppression of licking by rats. *Journal of Comparative and Physiological Psychology,* 1970, *72,* 467–470. Copyright 1970 by the American Psychological Association. Adapted by permission.   *(7–2)* Uhl, C. N., and Garcia, E. E., *Journal of Comparative and Physiological Psychology,* 1969, *69,* 554–562. Copyright 1969 by the American Psychological Association. Adapted by permission.   *(7–3)* Roberts, W. A., *Journal of Comparative and Physiological Psychology,* 1969, *67,* 395–400. Copyright 1969 by the American Psychological Association. Adapted by permission.   *(7–4)* Knouse, S. B., and Campbell, P. E., *Journal of Comparative and Physiological Psychology,* 1971, *75,* 116–119. Copyright 1971 by the American Psychological Association. Adapted by permission.
(7–6) Welker, R. L., and McAuley, K., *Animal Learning and Behavior,* 1978, *6,* 451–457. Adapted with permission of the authors and Psychonomic Society.   *(7–7)* Daly, H. B., *Journal of Experimental Psychology,* 1970, *83,* 89–93. Copyright 1970 by the American Psychological Association. Adapted by permission.   *(7–8)* Ratliff, R. G., and Ratliff, A. R., *Learning and Motivation,* 1971, *2,* 289–295. Adapted by permission of the authors and Academic Press, Inc.   *(7–9)* Dyal, J. A., Sytsma, D., *Journal of Experimental Psychology: Animal Behavior Processes,* 1976, *2,* 370–375. Copyright 1976 by the American Psychological Association. Adapted by permission.

(8–2) Thompson, R., and Dean, W., *Journal of Comparative and Physiological Psychology,* 1955, *48,* 488–491. Copyright 1955 by the American Psychological Association. Reprinted by permission.   *(8–3)* Roberts, W. A., and Grant, D. S., *Learning and Motivation,* 1974, *5,* 393–

PREFACE

For many years, biologists and psychologists have recognized that learning is one of the most important abilities that an animal possesses. Having a large and powerful frame or light but sturdy wings; being armed with sharp and dangerous teeth or a hard, protective shell are all features that may endow the species with enormous advantages within its environmental niche. But having the capacity to learn may also be extremely significant. Learning permits an animal to acquire new behaviors when the old ones are no longer appropriate; it allows the species to cope with a varied and changing world; it permits the animal to remember and to predict future events that may be significant to its survival. If an animal's reactions were fixed and immutable, any environmental change, be it a climatic drought or an invasion by predators, could be devastating.

The mechanisms of animal learning are astonishingly complex. More than sixty years of systematic research have provided us with many valuable clues about learning, but we have barely scratched the surface. Moreover, science does not always progress in a neat and orderly fashion; this certainly has been true for the study of animal learning. There have been years when our research has stagnated in a cumbersome theoretical mire, and, fortunately, there have also been periods of enormous growth.

I believe that the past decade has been a period of growth in animal learning research. Indeed, several important findings and concepts have virtually revolutionized the study of learning. One of the most salient concepts is the notion that learning is, fundamentally, a cognitive or mental activity by which organisms come to predict or expect future events. Years ago such a belief that rats, mice, pigeons, or other such lowly creatures could acquire expectancies was heretical in most camps. In addition, theorists have now begun to recognize that learning functions and is expressed within the constraints of the animal's biological makeup. Previously, many theorists ignored species differences, preferring to believe that such differences were of minor importance in the study of learning. Taken together, however, these new developments suggest that an animal learns to anticipate certain important events in its environment, but its overt actions based on such expectancies reflect its species-specific behavior potentials. If we are ever to understand the mechanisms of learning, then we must surely appreciate these interrelated themes: learning involves cognitive changes and it operates in the context of the animal's natural environment, giving the animal the means for flexible and adaptive behavior within the limits of its biological endowment.

In this book, the recent literature in animal learning and motivation is described and interpreted in light of these two dominant themes—animal cognition (expectancy learning) and evolutionary theory. The material is discussed from a historical perspective so that modern learning research can be seen more clearly as a dynamic process. What we currently believe about learning is merely a collective judgment formulated within the limits of our imagination and available data, rather than an end point in our quest. The material included and the judgments rendered in this text, of course, reflect my own interests and bias; I make no claim that other interpretations are not feasible and I beg indulgence for my errors of omission and commission.

I am deeply grateful to the many teachers, students, and colleagues who, directly or indirectly, have facilitated the writing of this book. In particular, I appreciate the encouragement and help of Robert C. Bolles, Lyle E. Bourne, Jr., and Robert A. Rescorla who read the entire manuscript and provided invaluable criticism and guidance. I also thank Douglas K. Candland who offered insightful suggestions for improving Chapter 1. Mary Brouse, Kay Ocker, Ruth Robenalt, and Alice VanBuskirk helped type the manuscript; their expert skills and cheerful outlook are greatly appreciated. I am also indebted to the editors at Scott, Foresman—Betty Slack and especially Katie Steele for their help and steadfast faith in the project. In addition, I appreciate the courtesy extended to me by the many authors and publishers who allowed me to use their material. Finally, I would like to thank my parents for the years of generous and undying support of my education, and my family for their encouragement and cooperation throughout the many months spent in preparation of this volume.

Roger M. Tarpy
Lewisburg, Pa.

CONTENTS

CHAPTER 1   INTRODUCTION   1
Evolution   3
Expectancy   7
Theoretical Approaches in Learning
    Research   11
Basic Concepts in Modern Learning
    Research   13
Learning Versus Performance / A Theory of
    Stimuli and Responses
Summary   18

CHAPTER 2   BASIC THEMES AND METHODS IN
    ANIMAL LEARNING RESEARCH   20
Introduction   21
Pavlovian Conditioning Tradition   22
Conditioned Excitation / Terms / Aversive
    Conditioning / Measurement Techniques /
    Conditioned Inhibition / Summary of Four
    Types of Pavlovian Conditioning
Thorndike's Conditioning Tradition   32
Reward Conditioning / Terms / Procedures /
    Punishment / Inhibitory Conditioning /
    Summary of Instrumental Learning
    Paradigms
General Laws of Conditioning   40
Stimulus-Outcome Correlation /
    Response-Outcome Correlation / Joint
    Stimulus/Response Expectancies
Summary of Classical and Instrumental
    Procedures   45
Summary   47

CHAPTER 3

STIMULUS LEARNING 49
Introduction 50
Training and Control Procedures 51
Problems in Evaluating Conditioning / Control
  Procedures
**Training Conditions That Influence
  Learning 55**
CS Information / CS-US Interval / CS
  Intensity / CS Quality / US Intensity / US
  Duration / US Predictability / US Quality /
  Intertrial Interval
Basic Pavlovian Phenomena 70
Second-Order Conditioning / Sensory
  Preconditioning / Inhibition of
  Delay / Latent Inhibition / Context
  Effect / Compound CS
  Conditioning
Summary 90

CHAPTER 4

RESPONSE LEARNING 93
Introduction 94
**Training Conditions That Influence
  Learning 95**
Response-Outcome Correlation / CR Type /
  US Intensity / Delay of Reinforcement /
  US Quality / Pattern of Unconditioned
  Stimuli
Schedules of Reinforcement 107
Fixed Interval Schedule / Fixed Ratio
  Schedule / Variable Interval Schedule /
  Variable Ratio Schedule / Comparison of
  Ratio and Interval Schedules / DRL
  Schedules / Complex Schedules /
  Reinforcement Schedules in the Real
  World
Basic Instrumental Phenomena 117
Contrast / Biofeedback / Choice Behavior:
  Matching
Summary 128

CHAPTER 5   STIMULUS-RESPONSE
                 INTERACTIONS   131
                 Introduction   132
                 Simultaneous Stimulus-Response Learning
                 132
                 Two-Factor Theory   136
                 Support for Mowrer's Theory / Challenges to
                   Mowrer's Theory
                 Conditioned Emotion   143
                 Interaction Experiments   146
                 A Matrix of Interaction Studies / Alternative
                   Explanations
                 Summary   155

CHAPTER 6   BIOLOGICALLY SPECIALIZED
                 LEARNING   157
                 Introduction   158
                 Naturalistic Response Patterns   160
                 Biologically Specialized Learning
                   Phenomena   163
                 Acquired Food Preferences / Acquired Taste
                   Aversions / Avoidance Learning
                 Social-Oriented Learning Phenomena   180
                 Imprinting / Animal Language
                 Summary   190

CHAPTER 7   EXTINCTION   192
                 Introduction   193
                 Stimulus Extinction   194
                 Response Extinction   198
                 Reward Training / Omission Training /
                   Avoidance Training
                 Procedures That Affect Extinction   204
                 Magnitude of Reward / Training Level /
                   Delay of Reward / Response Effort
                 General Theory of Extinction   209
                 Stimulus Generalization Decrement /
                   Interference / Counterexpectancy

Partial Reinforcement Effect   218
Stimulus Learning / Conditions Affecting
   Extinction Following Intermittent Reward
   Training / Theory of the Partial
   Reinforcement Effect
Summary   228

CHAPTER 8

MEMORY   230
Introduction   231
General Theories of Memory   233
Consolidation Theory / Retrieval Theory
Short-Term Memory   238
Delayed-Matching-to-Sample / Theories of
   DMTS
Intermediate-Term Memory   244
Long-Term Memory   249
Age Retention Phenomenon / Reinstatement
Summary   253

CHAPTER 9

**GENERALIZATION AND
   DISCRIMINATION   255**
Introduction   256
Generalization Gradients   257
Factors Affecting Generalization   260
Degree of Acquisition / Motivation Level /
   Training-Test Interval / Early Sensory
   Experience / Prior Discrimination
   Training / Inhibitory-Excitatory
   Interactions
Factors Affecting Discrimination   269
Problem Difficulty / Stimulus Information /
   Differential Outcomes / Observational
   Learning
Theories of Discrimination   278
Hull-Spence Theory / Attention Theory
Discrimination Phenomena   282
Overlearning-Reversal Effect /
   Intradimensional and Extradimensional
   Shifts / Learning Sets
Summary   290

CHAPTER 10 MOTIVATION 292
Introduction 293
Historical Perspectives / Drive Theory /
    Incentive Theory
Bio-Regulatory Systems 297
Deprivation and Activity / Deprivation and
    Instrumental Learning / Irrelevant Drives /
    Drive Measurements
Acquired Motivation 305
Fear / Conditioned Appetitive States /
    Conditioned Physiological Reactions
Stimulation Theories 312
Opponent Process Theory / Illustrations of the
    Opponent Process Theory
Nonregulatory Motivational Systems 317
Exploration / Spontaneous Alternation
Summary 320

CHAPTER 11 PUNISHMENT AND LEARNING
    DISORDERS 322
Introduction 323
Avoidance Extinction 324
Response Blocking 326
Factors Influencing Avoidance Extinction /
    Theories of Flooding
Punishment of Avoidance Extinction 331
Theory of Self-Punitive Behavior / Stress and
    Stimulus Predictability / Punishment and
    Aggression
Learned Helplessness 337
Principles of Learned Helplessness / Theories
    of Helplessness
Summary 348
References 350
Author Index 392
Subject Index 400

1

Introduction

Evolution

Expectancy

Theoretical Approaches in Learning Research

Basic Concepts in Modern Learning Research

Summary

Evolution

Have you ever wondered why it is that human beings, or any other animal for that matter, are able to learn? The usual notion is that learning permits us to master our environment, to be successful in shaping our world. Such an idea is right, of course. But consider the alternative: a vast number of organisms on this earth are quite unable to learn. Bacteria float aimlessly about in their environment without ever learning anything. Certain creatures as "complex" as sponges or even insects have existed unchanged for many millions of years, without ever having developed much, if any, capacity to learn. So, again, why is it that the success enjoyed by human beings, on the surface at least, has depended so heavily on our ability to learn and yet these other animals have survived for a much longer period of time without being able to learn?

The answer to that question is not a simple one, nor is it really the primary subject matter of this book. The evolution of learning, however, does bear on how we study and interpret laws of learning. By understanding even a little bit about how or why learning abilities evolved in some species, we can gain a significant advantage in our effort to understand the learning process itself.

Let us start with a familiar argument, the one proposed by the famous nineteenth-century naturalist Charles Darwin. Briefly, he reasoned that animals competed for resources. Food, for instance, had to be found and harvested; mates and territories had to be defended; enemies had to be avoided or conquered. Those individuals that succeeded in securing scarce resources for themselves were, on the average, the ones that survived to raise offspring. And since the world was not infinitely abundant, those individuals that failed to compete successfully were, on the average, more likely to die before successfully raising progeny.

Although an extensive discussion is beyond the scope of this book, it is important to be clear about Darwin's basic arguments. It is a matter of common knowledge that many more organisms are born into the world than there are organisms that live to full reproductive age. Quite simply, some organisms survive while others do not. It is also clear that all individuals of a given species are not identical. Differences may not be dramatic, but variation nonetheless occurs. Darwin combined these two known facts and concluded that those individuals that were born with even a slight advantage relative to the other members of their species would, on the average, be better at securing vital resources. This, in turn, would increase their chances of successfully reproducing, that is, of endowing future generations with their advantage. In short, traits that are adaptive, that facilitate survival, are the ones that are more likely to be passed on to future generations.

Perhaps an example from Darwin's own work would help to clarify this simple but powerful idea. In 1832, Darwin traveled to South America on H.M.S. *Beagle* for the purpose of cataloguing the flora and fauna in that part of the world. One of his most important stops was the Galápagos Islands. Some of the animals on the islands, such as the finches and lizards, resembled forms that he had observed on the mainland about one thousand miles away. Yet there were differences. An iguana, for example, had most of the recognizable features of iguanas found on the mainland, but the island-dwelling animals, unlike their mainland relatives, swam in the ocean and ate seaweed. This extraordinary behavior presumably evolved because the ancestors of those animals, who were trapped on the islands, had a vast resource that they could exploit, namely, the extensive beds of seaweed. The barren and rocky shoreland, in contrast, produced little of the vegetation that iguanas normally require. According to the theory of natural selection, those early individuals that were better at utilizing this new source of food were the ones that, on the average, survived. Other individuals that could not cope with this unfamiliar habitat died before having offspring.

Countless other examples could be cited. All would emphasize the same theme: (a) that individuals vary somewhat in their physical makeup; (b) that the variation, in turn, affects their ability to compete for resources and to secure a measure of safety from predators; and (c) that the more successful forms will, on the average, tend to survive and reproduce while the less successful forms will fail to be perpetuated. The notion of selection of the fittest, of course, applies to life forms at all levels: from plants in the forest that must compete for sunlight and nutrients, to insects, to blue-ribbon poultry, to human beings.

Up to now, we have emphasized the evolution of morphological characteristics, that is, the physical traits which characterize the bodies of animals. Fish possess a streamlined shape because those individuals that first had this characteristic were better able to navigate in water; accordingly, they were selected for. Similarly, birds display specialized bone structures (their wings are light yet sturdy) because early ancestors of the modern bird species, who happened to have such a trait, were selected for during evolution. Indeed, there are very few characteristics which can't be shown to have represented a valuable asset to the animal in its evolutionary development. The important concept to which all this discussion leads, however, is that evolution has acted upon characteristics far less obvious than limb structure or body shape. *An animal's behavior potential, including its ability to learn, has also evolved in specialized ways.*

Quite simply, the capacity to learn has evolved in many animals because it has allowed those animals to behave more adaptively than they would otherwise. Recall the astonishing changes that our world has undergone throughout evolutionary history. When changes in the environment did occur, whether those changes were slow or rapid in developing, pressure was put on individuals to cope. For example, new sources of food had to be found because existing supplies were uncertain. This put pressure on the individuals that already were utilizing those food sources. The result was that competition for the resources arose, and defense became necessary. All in all, survival was more problematic in a world that was constantly shifting, constantly running out of supplies, and continually providing new threats to survival.

The ability to learn, almost by definition, endows the individual with flexibility. An animal that is able to learn can cope with new contingencies or conditions in the environment, can locate and utilize new food supplies, can alter its routine when faced with direct or indirect threats; *it can remember.* Overall, then, learning abilities have been selected for during evolution because they, like certain morphological characteristics, endow the individual with a capacity to behave adaptively.

But *how* has learning ability been passed from one generation to the next? Consider the following analogy: animals are in some ways like computers. They start life with certain behavior potentials encoded in their genes in essentially the same way that computers possess performance potentials embodied in their electronic network. These behavior potentials are formless until the environment molds and shapes their development; similarly, computers are unable to perform even the simplest task until they are programmed to do so. For an animal, the ability to become modified, the capacity to be shaped by environmental influences, is inherent in its genetic makeup. The better the potential, the more effective the environmental input is in producing appropriate behaviors. And, for computers, the better the hardware, the more effective is the programming, and the more solutions the computer can derive.

The point is that natural selection operates on those genetic codes that underlie the *capacity* to learn. If a species possesses genetically coded behavior potentials, then the potentials will be passed on to future offspring provided that the animal survives and reproduces. Learning helps the animal to do just that. By acting in a more adaptive fashion, the animal has a greater probability of surviving and of raising offspring. The crux of the argument then is simple: physical characteristics, that is, genetic configurations that are responsible for the ability to learn, will be passed on from one generation to the next because those individuals that have these characteristics are better equipped to survive. It makes no difference if the adaptation is via a physical trait or a behavioral strategy. Just as the configuration of genes which is responsible for the shape of a bird's wing is selected for (because more efficient wings are adaptive), so too are the genetic configurations that permit an organism to learn about its environment (the resulting ability to learn is also adaptive).

Does this mean that we, as a species, are programmed at birth to follow a predetermined course, to be constrained by a genetic intelligence level, and to be blindly controlled by an environmental program? The answer is an emphatic no. As a species, our potential for learning is simply that—a potential to be modified by the environment. Indeed, learning potential has liberated us from the bonds of simple genetic programming because the very essence of learning is flexibility.

An example may help to show how certain learning potentials were selected for. Imagine the ancestors of some modern mammal. One of the pressing demands placed upon these early creatures would have been the regulation of diet, that is, the ingestion of nutritive substances and the avoidance of harmful foods. Food items, of course, varied enormously in nutritive value; similarly, watering holes surely became contaminated with minerals and bacteria from time to time. The immediate problem for these animals, therefore, was one of identifying the good and bad food items in a world that contained a multiplicity of food types. But how could the animals meet this challenge? It might be possible for them to evolve instinctual mechanisms for recognizing each of the many eatable substances available to them. But the animals would not be able to cope effectively with new substances. Their recognition of food items would be rigidly fixed at birth.

It would seem obvious that learning could provide the needed flexibility. An animal that could learn to associate the quality of the food (for example, its taste, smell, or sight) with the consequences of ingesting that food (that is, whether it induced illness or provided nutrients) would be at a selective advantage relative to its competitors that did not have this potential encoded in their genes. In other words, animals that could not recall which food had caused their illness were more likely to eat that same food a second time, thus increasing their chances of being fatally poisoned and of leaving no offspring. In contrast, animals that *could* recall the features of the food that had made them ill would be less likely to consume the food a second time. Because of this ability, they would have a greater likelihood of passing their learning potential on to future generations. It is not that they would survive because they had better taste (or vision or smell). Rather, they would survive because they possessed a greater capacity to associate the sensory qualities of food with the consequence of ingesting that food.

Let us review the important thesis developed in this section. Genetic codes are, on the average, passed from one generation to the next when the properties that they determine enhance the animal's ability to survive in an uncertain and changing world. Some of these codes relate simply to morphological traits like wing structure. Others, however, provide the individual with a potential to learn and the ability to behave adaptively and flexibly. Learning, therefore, is a process that heightens adaptation; it is designed to serve the specific biological needs of the organism within its natural environment. This ultimately means that our study of the learning process, as in the subsequent chapters of this book, must reconcile the basic laws of learning, as we observe them in the laboratory, to the basic principles of evolutionary development. Learning processes, viewed in the absence of evolutionary principles, would merely be a solution in search of a problem.

Expectancy

In our discussion of evolution, we emphasized that natural selection has produced species whose genetic configuration allows them to learn. But what is the fundamental character of this learning potential? What, in fact, do animals learn when they exercise this potential? In this section, we will begin the discussion of this question and thereby set the stage for a detailed analysis in later chapters.

Consider the simple model depicted in Figure 1–1. The model suggests that environmental stimuli impinge upon an organism which contains two simple functions: learning mechanisms and species-typical (genetic) behavior potentials. As a result of these elements, the organism is able to perform adaptive responses. Learning, therefore, is a process that helps to mediate or connect environmental inputs and behavior.

Let us be more specific about this simple model. What would learning have to "do" or "be like" in order to help translate environmental inputs into adaptive behavior outputs? To answer the question, consider for a moment what learning is like in an everyday sense. You observe a dark rain

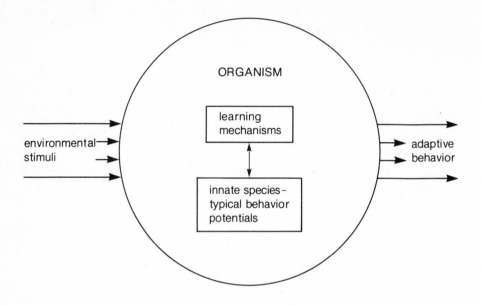

Figure 1-1. A simple model indicating that an organism's adaptive behavior in response to environmental stimuli is due to an integration of its learning mechanisms and innate species-typical behavior potentials.

cloud and proceed to cancel a picnic you had planned. You cancel the picnic because you have "learned" that dark clouds mean rain is imminent. Take another example. Many zoo animals, such as bears, become highly active —almost agitated—when they observe their caretakers approach at feeding time; presumably the bears have learned to recognize these people and have learned to associate them with food.

We could give many more examples but each of them would illustrate the same message: that learning is fundamentally a process whereby the organism comes to anticipate a future event based upon the "information" or stimuli that it receives from the environment. Through learning, an organism is able to order its environment, utilize signals, and anticipate changes.

This view of learning puts a special emphasis on stimuli in the environment because stimuli are the things that any species must cope with. Specifically, the animal must be able to anticipate patterns in the stimuli inputs, to resolve matters of sequencing, and to anticipate various relationships. In short, the fundamental task for any organism, in terms of learning, is to be able to predict changes in its environment, to anticipate the future.

This is a fundamental point. Learning involves the acquisition of expectancies about the environment. As stimuli impinge upon an organism, various patterns of stimulation can be perceived (for example, dark clouds are nearly always followed by rain). If patterns can be perceived, then certain events can be anticipated. The simple model in Figure 1–1 therefore depicts an organism with two major components: species-typical behavior potentials coupled with learning mechanisms. The behavior potentials limit the range of responses that the organism can perform (all species have various limitations on their behavior by virtue of their physical characteristics). The learning mechanisms, on the other hand, process and interpret the stimuli from the environment. They do so by detecting patterns and recurring sequences which allow the organism to anticipate future events. And anticipation of future events is precisely what permits an organism to behave adaptively, for example, to remember the location of new resources and to avoid trouble.

Another important principle, however, complicates this model. As shown in Figure 1–2, the behavior of an organism *changes* the environment in which it lives. The behavior feeds back and affects the original stimulus inputs. In other words, the stimulus configuration that impinges upon an animal produces behavior (integration of learning mechanisms and species-typical behavior potentials) which, in turn, alters the original input itself.

Consider this very simple example: an animal smells the presence of a predator and runs away. The original stimulus input, the smell of the predator, impinges upon the organism. Because the organism has experienced this same stimulus in conjunction with the predator in the past, the smell has a particular meaning; based upon the odor that it experiences, the organism can anticipate encountering the predator. The organism's species-typical behavior therefore comes into action—in this case, it runs away. Other species, of course, may have executed different sorts of reactions such as flying, swimming, screeching, playing possum, or even sounding alarms. Whatever the behavior, it would be executed because the animal anticipated danger on the basis of a stimulus it received. But the important point is that the behavior *alters* the stimulus input (running away removes the odor). The environment of the animal now is quite different.

Let us consider another simple example of this feedback model. An animal is hungry; it searches for and later consumes food. Again the environmental input is the first step in the process (in this case, the stimulus is the internally generated stimulus of hunger). The learning mechanisms in the organism allow the animal to remember, to anticipate, the location of food sources. They do so because, in the past, the sight or smell or sound of various locations have been consistently associated with food. In the second stage of this behavioral sequence the species-typical behavior poten-

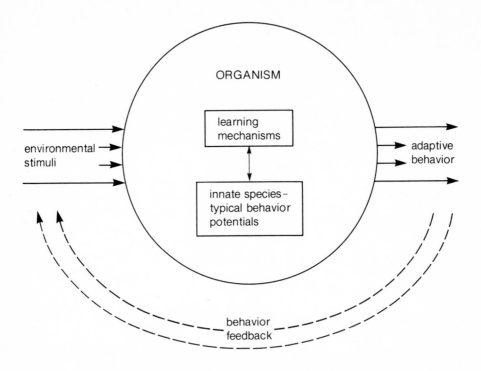

Figure 1-2. A simple model of an organism indicating that the organism's behavior feeds back and affects the original stimulus input from the environment.

tials are translated into responses (searching and consuming food). Finally, the original stimulus input, the feeling of hunger, is eliminated because the behavior has fed back and affected it; the animal has eaten. In summary, an organism comes to anticipate certain events based upon the consistencies it has perceived in the environment. In addition, an organism's behavior feeds back and affects those stimuli.

This fundamental point is the basic theme in learning research. Much of this research suggests that learning is the acquisition, or confirmation, of expectancies which interact with the organism's innate response potentials to produce adaptive behavior. The behavior itself may, in turn, influence the stimuli on which the expectancies are based. Regardless, the basic task for any organism's learning, is to predict, that is, to expect future outcomes and to order the patterns and configurations of inputs in such a way that future events can be anticipated. If there is a single message that you should derive from this book, let it be this: learning is the acquisition of expectancies, based upon patterns of stimulus inputs and response feedback, which allow an animal to behave in an adaptive fashion. An animal can learn about the stimuli in its world as they occur naturally (recurring stimulus patterns), and it can learn about its world in terms of how its behavior affects those stimuli. In either event, the animal is learning to anticipate future inputs. To deny that learning involves the acquisition of expectations would be to fail to grasp the fundamental truth about the learning process: that it has evolved to help animals deal with uncertainties, to allow them to perform more adaptively in a changing world, to permit them to exert some order over a complex and changeable environment.

Theoretical Approaches in Learning Research

The previous section emphasized the idea that learning involves the acquisition of expectancies. An animal comes to expect certain events based upon consistencies in the environment itself or based upon feedback from the animal's own behavior. This notion, however, implies very little about the strategies that we have adopted in our attempt to understand how these expectancies come into being. The purpose of this section, therefore, is to provide a brief and selective account of scientific strategies so that the research on basic learning discussed in subsequent chapters will be more meaningful.

To facilitate the discussion of scientific strategy, consider the distinction between the object of study and the level of analysis. The object of study refers to the kind of behavior on which one focuses, the sort of questions one would ask about the learning process. A scientist, for instance, could choose to study memory or how learning improves with age or how fast an organism performs as a function of its state of hunger or the effect of social interactions on learning. The level of analysis, on the other hand, refers to the type and amount of detail one uses in explaining the behavior. For example, one could use personality constructs such as the id to explain a person's actions; alternatively, one could appeal to mathematical functions or physiological mechanisms; or one could explain behavior using concepts such as hunger, fear, and frustration. Each of these examples represents different levels of analysis.

The choice of level of analysis has been regarded as a serious issue in psychology. While an extensive discussion is not appropriate here, two opposing positions can be described. The first, espoused most clearly by B. F. Skinner and his associates, could be termed the functional or "black box" approach. Skinner claims that it is useless to try to explain the internal workings of organisms in terms of theoretical constructs (see Skinner, 1950, 1963). Rather, behavioral scientists should concentrate on describing the rules of behavior with reference merely to the stimulus inputs that precede the responding. Put differently, Skinner argues that we should not attempt to develop explanatory theories of learning by appealing to inner mechanisms or mentalistic concepts. Skinner objects to the unscientific nature of these mentalistic concepts. Furthermore, he suggests that they don't really help us understand the behavior because any explanation we could offer would itself have to be explained.

Consider this example. Imagine that you saw a tape recorder for the first time. You could try to understand its "behavior" in a variety of ways. You could manipulate the various knobs, hoping that something you did accidentally would produce a recognizable response. Or you could take the tape recorder apart and try to explain its behavior by analyzing the internal components. If you did this, however, according to Skinner, you would be faced with even more problems than you started with. Upon discovering that the tape recorder contained tubes, transistors, wires, and the like, you would now have to explain those elements before you could ever go back and explain the behavior of the tape recorder as a whole. In effect, the regression is never ending. By digging deeper into the "black box," you merely generate more questions. Each new component or mechanism or subunit needs to be explained before you can use it as the explanation for the original behavior. According to Skinner's strategy, then, one ought to avoid using such inner mechanisms to explain behavior; one should simply concentrate on describing the relationship between the inputs to the system and the behavioral outputs.

There is a great deal of value in this approach. After all, the only things about which we can be objective are the environmental stimuli that impinge upon an organism (the inputs) and the organism's behavior (the output). And the approach may be efficient. If one does not have to explain behavior and in turn explain one's explanations, then greater effort can be put into discovering the relationships between environmental conditions and behavior.

However, let us consider one alternative approach to understanding behavior. In this method, we open the black box and try to identify the neural, chemical, or behavioral mechanisms that are causing the behavior. Consider the tape recorder analogy once again. If we investigated the wires, tubes, transistors, and other elements of the tape recorder, or at least the amplifiers, recording heads, and motor drives, we might appreciate how the tape recorder was constructed and therefore how it behaves. Moreover, we might be able to improve the tape recorder by adding extra transistors, replacing defective tubes, or making other adjustments.

This mechanistic strategy is an attractive one because it promises to explain behavior in terms of its "real" causes. It is appealing to think that we might be able to point to a physical cause for something that is as complex and puzzling as behavior. After all, we are used to thinking in a biological framework. We explain diabetes in terms of a lack of insulin or a malfunction of the pancreas. We explain our good health in terms of an adequate supply of vitamins and nutrients. Similarly, the proposition that we might be able to explain our behavior in terms of hormones, neurons, and chemicals is exceedingly attractive. By investigating an organism's physiology we may be getting at the source of the organism's behavior. And because there are important continuities and similarities between species, especially higher organisms, some of the biological principles that determine behavior in one type of animal are very likely to be important in other types.

These are only two juxtaposing views or strategies. The essential point is that most objects of study, most questions about behavior or learning, can be investigated on several different levels of analysis. There is no singular correct approach. For example, a topic like memory can be studied either in terms of physiological or biochemical substrates, or in terms of the environmental inputs that affect later memory. Neither of the two extreme strategies are entirely inappropriate; both have merit. Too often our scientific enterprise fails to recognize this fact. Different investigators often appear like the blind men who inspected the elephant in Saxe's famous poem. Each had a vastly different perspective and belief about what constituted an elephant, and each claimed that his was the true perspective. Of course, all of the men were right to a limited degree, but it wasn't until the pieces were put together that the true qualities of the elephant as a whole were appreciated. It is similar in science. We attempt to investigate many different objects of study; we try to solve different questions about behavior, without realizing that each object or question can be analyzed on several different levels. Fortunately, integration between levels is achieved periodically and the picture becomes clearer and more complete.

Basic Concepts in Modern Learning Research

In this section, we will review several basic concepts in learning research. These are fundamental to much of the discussion that follows in later chapters.

Learning Versus Performance

The distinction between learning and performance is so fundamental that it pervades our thinking at every turn. Learning, the acquisition of expectancies, is an inferred, internal change. It is the creation of a new behavior potential or response capacity. It reflects a capability not present before the learning conditions were in effect. It is the acquisition of an expectancy system that we cannot observe directly. Performance, on the other hand, refers to the overt behavior itself. Performance is the end result, the result of behavior potentials being translated into real action. Performance is the responding that we observe objectively: rats or monkeys pressing levers, pigeons pecking disks to get food, and so forth.

This distinction is important because we often forget that we cannot study learning directly. We must use the animal's performance or overt behavior as a way of assessing the inner learning process. Yet we must not think that we are *always* observing learning when we look at performance. Performance can occur without any learning having taken place. Countless examples could be cited. Assume, for instance, that a hungry rat is placed in an alley containing a start and goal chamber and is given food in the goal box. Normally the rat easily learns to run down the alleyway to get the food reward in the goal box. The psychologist observing this would surely claim that the animal had learned; the animal behaved as if it expected to receive food in the goal box. But suppose that the psychologist then gave the rat a large meal just prior to testing it. Most likely the rat would sit in the start box and never venture toward the goal. Would the psychologist conclude that learning had been disrupted? Did feeding the animal prevent it from remembering the location of food? Clearly not. The elimination of hunger merely changed performance without affecting learning. In summary, performance may or may not reflect learning. We usually try to structure the experiment in such a way that it does reflect the learning process, but often we are unable to do this in a convincing fashion.

Just as performance can take place without learning, it is equally true that learning can take place without performance. It is impossible to observe this directly because learning is an internal process and therefore must be inferred from the overt behavior of the animal. Nevertheless, under the proper conditions we can confirm that learning may take place without performance being affected. Tolman and Honzik (1930) conducted one such experiment. They tested three groups of rats in a complex maze containing fourteen choice points. For one of the groups, food was available in the goal box. Animals in this group could run through the maze, discover which turns were correct, and receive food at the end. This is not a particularly difficult task for a rat and these animals showed typical improvement in performance—their speed of running in the maze increased over trials, thus indicating greater and greater efficiency in negotiating the various turns.

A second group was allowed to traverse the maze, but the animals in this group were not given any food in the goal box. Rather, they were taken out and later given their daily ration of food in their home cage. The speed of these animals during successive training trials never changed very much. They continued to make errors and to run slowly. In terms of their performance, therefore, they appeared to have learned nothing at all.

The third group of rats was the crucial one. Like Group 2, these animals were not given food in the goal box for the first 10 trials, and, like the Group 2 animals, their performance was consistently poor. However, starting on Trial 11, these animals were shifted to the other condition, that is, they were fed when they reached the goal box. The resulting change in their behavior was dramatic: following the first rewarded trial, Trial 11, the performance of the animals improved almost immediately. In fact, the improvement in the performance was so rapid that Tolman and Honzik were forced to conclude that learning had been taking place all along, even during the first 10 nonrewarded trials. If learning had not taken place during those early trials, then the animals should have shown a very gradual improvement in performance once they began receiving reward in the goal box. After all, the animals that had been rewarded right from the start improved only gradually. But the improvement in the performance of the shifted subjects was not gradual; it was sudden, suggesting that they had learned about the maze all along but simply had not demonstrated that fact in their overt behavior. Later, however, when food was provided, their performance accurately reflected their previous learning.

This study and others like it illustrate that learning can take place even though the animal does not perform in a way that demonstrates learning is taking place. Surely we must use performance later to verify the presence of learning. But if we do not keep in mind the distinction between learning and performance, we might erroneously conclude that certain behaviors were due to learning when, in fact, they were not. Learning *is* always inferred from performance, but there may be principles unique to each process.

A Theory of Stimuli and Responses

There are two qualities or characteristics that are possessed by any stimulus or response. First, stimuli and responses vary along a dimension of reflexivity; they vary in the degree to which they are innate or learned. Second, stimuli and responses vary along a hedonic dimension; they differ in the degree to which they are pleasant or aversive.

Given the emphasis on the expression of learning variables in the context of evolutionary theory, it is not surprising that a dimension of reflexivity would be stressed. By reflexivity, we mean a continuum ranging from biologically determined to environmentally acquired; from exclusively instinctual to entirely learned; from nature to nurture. As depicted in Figure 1–3, there are highly reflexive behaviors and stimuli which are instinctual in nature. Lower organisms, of course, are most typically associated with these sorts of behaviors and stimuli. The elaborate burying response of squirrels is, presumably, instinctual; no learning is involved. The red underbelly of a stickleback fish is a stimulus that innately induces aggression. On the other side of the spectrum are learned behaviors or stimuli. New rela-

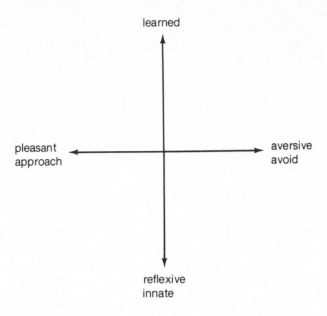

learned

pleasant
approach

aversive
avoid

reflexive
innate

Figure 1-3. Matrix indicating that stimuli (environmental events) and responses (behavioral reactions) can be characterized as falling along two independent dimensions.

tionships between stimuli may be learned such that one stimulus takes on meaning due to its association with another stimulus. Similarly, new behaviors become more probable through learning; none was preprogrammed or instinctual.

The main point of this reflexivity dimension is this: while the ends of the continuum are conveniently labeled instinct versus learning or innate versus acquired or biologically determined versus environmentally conditioned, there is no sharp dividing line between the extremes. Rather, *all* behaviors and *all* stimuli, to some degree, reflect a mixture of both instinctual and acquired characteristics. Although some behaviors and stimuli may be largely innate, while others may be almost totally learned, most reflect a mixture of the two. The important problem, therefore, is determining not whether a behavior is innate or learned but rather how and to what degree the reflexive and learned components interact.

Behaviors reflect a mixture of learning and reflexiveness primarily because the two are not mutually exclusive properties or independent dimensions of animal behavior. In one sense, learning abilities are a special case of instinctual behavior patterns. Although they are not programmed like innate behaviors, learned responses have evolved as a capacity to serve the organism's innate behavioral potentials. Animals still are constrained to perform behaviors which their nerves and muscles and genes allow them to perform. But learning is an ability to perceive new relationships and to make novel behavioral adjustments that are consistent with the species-typical behavioral capacities. Thus, for most animals, behaviors are usually in the middle of the reflexive-learning continuum, exemplifying the dynamics of learning principles while at the same time reflecting the inevitable biological constraints of the species.

The same sort of argument can be applied to environmental stimuli. Very few are wholly innate and, similarly, few are exclusively learned. Consider the animal's recognition of food items. There is a strong element of reflexivity involved in that animals are adept at choosing appropriate items. And yet there is a significant amount of sampling and exploration; many animals are quite capable of learning about new food items.

There is a second dimension in addition to the reflexiveness continuum. This dimension, as shown in Figure 1–3, is a hedonic scale, a pleasure-pain dimension. The end points of the hedonistic dimension would be harmful versus helpful, aversive versus pleasurable, avoidance versus approach, negative versus positive, useful versus deleterious. Both stimuli and responses may be characterized in terms of this scale. In the case of stimuli the terms aversive and pleasant apply; for behaviors, verbs such as avoid and approach are more appropriate.

The two independent dimensions shown in Figure 1–3 indicate that behaviors and stimuli may be classified according to two dimensions simultaneously. A stimulus or response is more or less learned (versus more or less innate) and, *at the same time,* more or less aversive (as opposed to pleasant). In short, both stimuli and responses can be located somewhere in this matrix depending upon these two dimensions.

Consider a few simple examples. The most fearsome predators to some small songbirds are large birds of prey like hawks. If such a songbird were foraging for seeds and a hawk flew into the vicinity, the songbird might show signs of alarm and flee from the area. In this example, the stimulus "hawk" would be highly innate and highly aversive. Similarly, the avoidance of the predator would represent an aversive and innate behavior. The stimuli and responses in this example would be located in the lower right-hand quadrant of Figure 1–3. Both are highly reflexive in nature and, presumably, aversive since the behavior is one of flight or avoidance.

There are many examples of innate stimuli or responses that are essentially pleasant in nature. Sexual responses are essentially approach behaviors; various food items, for some animals, appear to be instinctively recognized and are approached; and many social behaviors, and the stimuli eliciting them, fall on the pleasant side of the continuum as well. These examples would be located in the lower left quadrant of Figure 1–3.

Examples that would be located in the two remaining quadrants of Figure 1–3 are abundant; indeed, the bulk of this book is devoted to explaining such examples. Without going into detail here, we can state clearly that animals can learn to perform responses which are either approach or avoidance behaviors. Similarly, stimuli can come to signify aversive or pleasant consequences.

In summary, behaviors that an animal performs and stimuli that impinge upon the animal can be classified according to two independent dimensions. Usually the behaviors and stimuli will be some mixture of reflexive tendencies and learning; at the same time, they will reflect various degrees of aversion and pleasure. The model depicted in Figure 1–3 is useful because it illustrates the theoretical boundaries and relationships that govern an animal's response repertoire. In addition, the model gives us an overview, a framework, for relating stimuli and behaviors to each other. Without such a system, it would appear on the surface that stimuli and responses are either innate or wholly learned or that stimuli and responses are either pleasant or aversive. In reality, however, stimuli and responses reflect a mixture.

One additional point should be made concerning the hedonistic dimension. We cannot infer anything about the qualities of the experience elicited by a stimulus or underlying a response even though we may classify that stimulus along the pleasant-aversive dimension. When a songbird sees a hawk, it becomes agitated and flies away. We infer, therefore, that the hawk is an aversive stimulus. In fact, our definition of aversion is an operational one based on the behavior of the animal. Similarly, we claim that a behavior is pleasant or that a stimulus elicits pleasure when we observe the animal approaching that stimulus. We know nothing about the nature of the experience, but it is certainly reasonable to infer that such an experience takes place. In other words, it is convenient to classify stimuli in terms of pleasure-pain because adaptation to our environment necessarily requires approaching some objects and avoiding others. But we classify stimuli in this manner not in terms of any presumed insight into the nature of the experience itself but rather in terms of the animal's behavior.

Summary

Learning abilities, like morphological traits, have evolved in those animals that demonstrate learning. They have been selected for because learning permits the organism to behave adaptively and flexibly in a changing world. Thus, animals that possessed certain learning capacities were more likely, on the average, to survive and pass these genetically coded potentials on to their offspring.

Learning is fundamentally a process whereby the animal comes to expect a future event based upon the patterns of stimuli in the environment or upon its own behavior. Such expectancies interact with the species-typical behavior potentials to produce adaptive responding.

There are two opposing approaches or strategies in learning research. One approach advocates the study of the rules of learning by specifying the relationship between stimulus inputs and response outputs. The other approach emphasizes the investigation of internal mechanisms for learning, such as the physiological substrates for learned behaviors. It would appear that both strategies are not only appropriate but are, in fact, complementary.

Two important ideas are crucial to the understanding of learning research. First, we must recognize the distinction between learning and performance. Learning is an internal change in behavior potential or response capacity which we infer has taken place; it is not directly observable. Performance, on the other hand, is the actual, objectively observed behavior —the learning potential translated into action. Although we use performance as an index of learning, the two processes are independent.

The second important concept in learning research is the classification of stimuli and responses along two dimensions: innate-learned and pleasant-aversive. Stimuli and responses vary not only according to how innate (reflexive) or learned they are but also according to how pleasant or aversive they are.

2

Basic Themes in and Methods Animal Learning Research

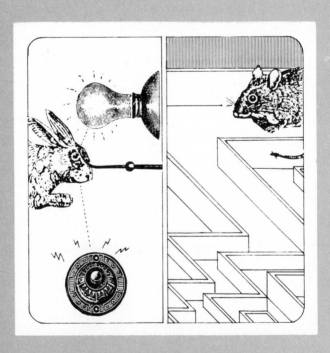

Introduction

Pavlovian Conditioning Tradition

Thorndike's Conditioning Tradition

General Laws of Conditioning

Summary of Classical and Instrumental Procedures

Summary

Introduction

In Chapter 1, we claimed that learning involved the acquisition of expectancies. An animal lives in a sensory world; it communicates with the world through its senses. Some sensory inputs are exceedingly important from a biological viewpoint, for example, stimuli such as food or warning signals from predators are life-sustaining commodities. It therefore makes sense that animals have evolved various capacities to deal with biologically important stimuli, different means that allow them to anticipate events that affect survival.

If an animal's main task is to understand its environment, to predict biologically significant events which promote its survival, *and* if, at the same time, an animal can change the world through its behavior, then two different sorts of expectancies should exist (see Bolles, 1972). The first is a stimulus expectancy. An animal can come to expect the delivery of important stimuli *based upon* prior environmental cues or signals. The environment may contain stimuli that are presented to an animal in such a way that one stimulus becomes a signal for the presentation of the other stimulus. Imagine the situation in which you anticipate freshly baked bread simply by the smell emanating from the kitchen. This is an example of an expectancy based upon a stimulus—the odor has always been associated with the bread in the past.

21

The second kind of expectancy is a response expectancy. Animals can learn to expect the delivery of prominent events based upon their behavior. A human behavior that exemplifies this is the expectation that a car will stop if the driver steps on the brake. Here, the expectancy is based on a response.

In summary, an organism's world consists of inputs from the environment, some of which can be modified by the organism's behavior. An animal can come to predict events in the environment (for example, biologically important stimuli) based upon a prior stimulus or signal (a stimulus expectancy) or based upon its own behavior which produces or modifies that event (response expectancy). Note that these are not expectancies *of* stimuli or responses, but rather expectancies about future events *based upon* the occurrence of a stimulus or response.

Now, let us consider learning. The two major traditions in learning research that have developed over the last seventy-five years deal with these two types of expectancy. One, the classical conditioning tradition, beginning with the work of Pavlov, concerns itself with stimulus expectancies. The second tradition or approach, instrumental conditioning, began with the work of Thorndike; this approach deals with response expectancies. Our purpose here is merely to introduce these two research paradigms. A more detailed discussion of stimulus and response expectancy learning will be given in Chapters 3 and 4 respectively.

Pavlovian Conditioning Tradition

Ivan Pavlov (1849–1936), a Russian physiologist, won the 1904 Nobel Prize in physiology and medicine for his studies on the physiology of the digestive processes in dogs. In the course of his experiments, Pavlov quite accidentally discovered classical conditioning. He routinely measured the salivary reflex as a means of studying the digestive processes; he observed, however, that the dogs salivated when he entered the room before the actual experiment had begun. Pavlov labeled these responses "psychic secretions" because they occurred in the absence of any biological stimulus that, presumably, was needed to elicit salivation. Pavlov was perceptive enough to see that these psychic secretions were of great significance, and, fortunately for science, he devoted the rest of his life to their study.

Conditioned Excitation

Pavlov's original experiment is well known. The basic study went as follows (see Pavlov, 1927, for more details): the subject, a dog, was prepared for the experiment by implanting a small glass tube in the salivary duct so that the saliva could be directed to a flask. The dog was then restrained in a harness and presented with two stimuli. First, Pavlov sounded a metronome; several seconds later he presented food powder which made the dogs salivate copiously. After about 10 presentations of the metronome-food combination, Pavlov noticed that the metronome *by itself* induced salivation. Admittedly, the salivation was only a few drops, but after another 20 pairings of the metronome and food, the dog was observed to salivate profusely immediately upon the presentation of the metronome.

This simple experiment illustrates the acquisition of a stimulus expectancy: the animal came to expect the food powder whenever the metronome was sounded. The temporal consistency of these two inputs, that is, the fact that they were repeatedly given in close temporal proximity, allowed the animal to form such an expectancy or, in more traditional language, to develop an association between the two stimuli.

One important characteristic of Pavlovian conditioning—in fact, the defining characteristic—is the stipulation that *the animal's behavior and the presentation of the two stimuli are thoroughly independent.* It was Pavlov who determined beforehand which stimuli would be used and when they would be presented; the fact that the animal salivated was quite irrelevant to that decision. This is an essential point. The association or expectancy that is acquired by an animal is derived from the consistent temporal pattern of the two stimuli independent of its behavior. We can be sure we are dealing with a stimulus expectancy *only* when we can show that the presentation of the two stimuli is not influenced by the animal's behavior.

Terms

There are four major terms in learning research that are especially pertinent to classical conditioning. All of these are exemplified in Pavlov's original experiment.

The unconditioned stimulus (US) is defined as a biologically potent stimulus, one which reliably evokes a massive, reflexivelike reaction. In Pavlov's experiment the US was the food powder. Unconditioned stimuli have predictable effects on an animal; by definition they elicit a reflexivelike behavior over which the subject has little control. Imagine, for example, that someone placed a drop of lemon juice on your tongue. Since you would be quite unable to avoid salivating to that stimulus, we would label it as a US.

The definition of a US in terms of its potency does present some problems. While it is plausible to consider food as a strong biological stimulus for a hungry animal, it is not at all clear where we should draw the line between strong and weak stimuli. In order to solve this dilemma, conventions have been established that are really based on common sense.

Food, water, shock, or other painful stimuli are often used because they have been shown to have a powerful and *unambiguous* effect on the subject. Other stimuli may also qualify as USs if they too elicit strong reactions. Thus, we may not be able to give a precise definition of the term *potent,* but we certainly can identify many stimuli that do produce large, reliable, reflexivelike, and persistent reactions.

It is important to appreciate the fact that biologically powerful stimuli (USs) are precisely those events that an animal must be able to predict if it is to behave adaptively and survive in its environment. Weak stimuli do not have serious consequences, but USs do. For example, an animal must be able to predict when and where food will be available if it is to remain healthy; similarly, it must be able to predict when and where painful threats will occur if it is to avoid its predators. Presumably, it is by means of the classical or Pavlovian conditioning paradigm that animals may learn to predict the occurrence of these important stimuli.

The unconditioned response (UR) is the second major term in Pavlov's study. The UR is the regular and measurable *unlearned* response elicited by the US. In Pavlov's original experiment the UR was salivation to the food powder. The important point is that *the UR is determined by the US;* it is essentially a reflex that is biologically linked to the US. A US may elicit more than one UR, however. Food powder, for example, might elicit a variety of reactions including salivation, changes in heart rate or blood pressure, pupil dilation, and gross motor behavior.

The third important term is the conditioned stimulus (CS). This is a cue which is essentially neutral at the start of the experiment. Pavlov used the sound of a metronome, but virtually any innocuous stimulus could be used, provided that the animal can perceive it. Tactile stimuli such as vibrations to the skin, as well as visual, auditory, olfactory, or gustatory stimuli are all appropriate in Pavlovian experiments.

We have the same problem in defining the CS as a weak stimulus as we did in defining the US as a strong stimulus. What do we really mean by *weak?* First, we mean that the CS does not elicit a strong and *lasting* reaction. To be sure, the CS may provoke a measurable response by the animal, such as the turning of the head or limbs (Pavlov called these orienting reactions and they demonstrate that the animal does indeed perceive the stimulus), but these reactions subside rather quickly. Second, and more important, CSs are not in any clear fashion linked to the biological well-being of the animal; they are simply innocuous signals.

The final term in Pavlovian conditioning is the conditioned response (CR). This is the *learned* reaction, the new behavior that indicates that an expectancy has been formed. In Pavlov's experiment, the CR was salivation to the metronome alone. We know that the salivation response was learned or acquired because the metronome was unable initially to elicit it. Through its consistent pairing with the US, however, the metronome had gained a predictive value (the animal could use it to anticipate the presentation of food powder) and came to elicit the response. Salivation was the outward manifestation that an expectancy between the CS and US had been formed. In short, the CR represents an association between a weak and a strong stimulus: given one stimulus (the CS), the animal learns to expect the second one (the US), the visible CR indicating that such an expectancy had developed.

A simple diagram of this classical conditioning process is given in Figure 2–1. During conditioning the CS and US are presented in a sequential fashion, the US eliciting a UR. After conditioning, however, the CS itself is able to elicit a reaction, namely, the CR.

Aversive Conditioning

As suggested above, Pavlovian or classical conditioning is not limited to situations involving a food US. Many other strong stimuli could be used provided they elicit a reliable UR. One important class of stimuli are aversive USs. Refer back to Figure 1–3 in Chapter 1. There we noted that stimuli vary along some hedonistic dimension. Some are strongly aversive and some are strongly pleasurable. Many, of course, are in between those extremes. Note that USs, by definition, tend to be those stimuli toward the extremes of this dimension while CSs tend to be in the middle of the scale.

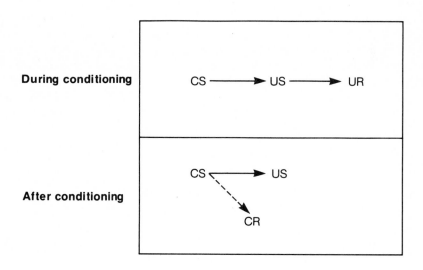

Figure 2-1. Schematic diagram indicating the basic structure and sequence in a Pavlovian conditioning experiment.

Pavlov used a pleasurable US (food) in his research. However, numerous experiments similar to Pavlov's work have been done using USs that are strongly aversive.

One of the first of these experiments was by Bekhterev (see Bekhterev, 1913). A neutral stimulus (a tone CS) was sounded and, a few seconds later, an electric shock was applied to the forepaw of the dog. Initially, the tone had no effect upon the dog, but the shock elicited a strong and persistent reaction—the reflex of withdrawing the paw. After repeated pairings of the tone and shock, however, the tone itself became a more powerful stimulus: it could elicit the paw withdrawal in the absence of the shock. Obviously, the only major difference between Pavlov's and Bekhterev's experiment was the fact that Pavlov used an appetitive or pleasurable US while Bekhterev used an aversive US. To be sure, this difference is an important one in terms of the overt behavioral reactions. But the difference is irrelevant to the fact that what was learned in both experiments was an association between an innocuous stimulus (a CS) and a more powerful one (a US).

There are many aversive stimuli besides electric shock that are commonly used in research on Pavlovian conditioning. For example, an animal or human may be given an air puff to the cornea of the eye (this produces a reflexive blinking reaction). It is rather easy to condition the eye blink response by presenting the CS (a tone, for example) prior to the air puff. After sufficient training, the tone itself will elicit the eye blink reaction.

Aversive USs do not necessarily have to be applied to the external surface of the animal. It is possible to use internally applied stimuli too (see Chapters 6 and 10 for more details). Consider the following example: if an animal were to consume a poison it would, at the very least, become sick and nauseous at some later time. According to our definition, the poison would certainly be a US because it elicits a strong and persistent reaction over which the animal has no control. If a poison were paired with a neutral CS, usually a taste or smell, conditioning is observed to take place. Specifically, the animal shows signs of nausea (Garcia, Hawkins, & Rusiniak, 1974) and avoids consuming the flavored CS in the future.

Measurement Techniques

In a conventional classical conditioning study, there are actually two ways of measuring the CR. The first technique is called the anticipation method. In this method, the CS is given for a long enough period of time to allow the subject to anticipate the US onset; the CR occurs prior to the US which is always presented. In the second technique, the US is occasionally omitted (so that it may be shown that the CR occurs prior to the CS alone). This is called the test trial technique.

Both of these methods have advantages and disadvantages. The advantage of the anticipation method is that every trial utilizes both the CS and

US presentation. But the problem with this method is that in order to demonstrate conditioning, one may have to pair the CS and US in a less-than-optimal fashion. The galvanic skin response (GSR), for example, doesn't become a full reaction until about 2 seconds after the onset of the CS. The optimum CS-US interval, however, is about .5 seconds (White & Schlosberg, 1952). Therefore, in using the anticipation technique, one may be utilizing a CS-US interval that is not ideal for strong conditioning. The disadvantage of the test trial technique lies in the fact that the US is occasionally omitted. If the US is omitted every so often for the purpose of "testing" for the CR, conditioning proceeds more slowly. After all, conditioning depends upon the consistency of the CS-US pairings and omission of the US partially destroys that consistency (see Sadler, 1968).

The strength of Pavlovian CSs may be measured in quite a different way. Imagine that you were performing some sort of routine task, say, reading. Suddenly, a meaningful stimulus occurs, for example, a friend shouting "fire!" More than likely you would stop your reading momentarily at least to assess whether your friend was joking. We can assume that the stimulus "fire" does indeed have strength because it disrupts your behavior; if the word *fire* did not have meaning, you would ignore it and continue to read at the same pace. In fact, the extent of the disruption in your reading rate would constitute an indirect assessment of the strength of the stimulus.

Many experiments in Pavlovian conditioning use precisely this technique. It is called the conditioned emotional response (CER) measure. Rats, for example, are taught to perform a routine behavior, such as pressing a lever for food pellets. The animals do this quite readily and they maintain consistent levels of performance for long periods of time. Then, quite independently of their behavior (recall that this is the defining operation of a Pavlovian experiment), a Pavlovian conditioning study is conducted. A CS, say, a tone, is paired with the US, a mild shock. Later, when the tone is sounded while the animal is pressing the lever, one observes disruption of the lever pressing. The more meaningful the tone (that is, the stronger the conditioned reaction to the tone), the greater the disruption. In other words, classical conditioning in this context is not measured directly as in the usual heart rate, salivation, or paw flex studies. Rather, the strength of the Pavlovian CS is assessed *indirectly* in terms of the degree to which it disrupts an ongoing response.

The CER technique is a popular method in learning research and we shall be discussing many studies that use the technique. Therefore, we will here clarify what the results of a CER study actually indicate.

Conditioning strength is measured in terms of the degree to which the CS disrupts some simple ongoing behavior like lever pressing. The measure of disruption is the CER suppression ratio. This is calculated in the following way. The number of lever presses during the CS presentation *and* during an equal period of time just prior to the CS presentation are counted. With these two numbers, a ratio is then calculated. Specifically, the CER ratio is the number of responses during the CS divided by the total number of responses both prior to and during the CS. If, for example, the animal pressed 25 times in the minute prior to the CS and 25 times during the 1-minute CS, then the ratio would be .5 (25 divided by 50). This ratio would indicate that no disruption took place; the CS had essentially no power (note that the animal performed the same number of responses prior to and during the CS). In contrast if the subject presses 25 times prior to the CS presentation, but only 10 times during the CS, then the ratio would be .29. This ratio indicates that considerable response suppression was caused by the CS. The rate went from 25 per minute prior to the CS to only 10 responses per minute during the CS. In summary, a CER ratio of .5 indicates no suppression; the CS has virtually no power. However, a CER ratio of less than .5 indicates that conditioning indeed has taken place; substantial disruption of the ongoing behavior has occurred during the CS presentation.

Conditioned Inhibition

We have discussed the idea that the CS comes to predict the US because it is consistently paired with it. The CS, more simply, conveys information about the US presentation. But what if the outcome of the CS were not a US; what if the CS predicted "no US"? Wouldn't conditioning still occur because the CS conveys information; it reliably signals an outcome?

The answer to this last question is yes. In fact, many experiments have shown that conditioning will take place when the CS reliably predicts "no US." The nature of this conditioning, however, is different from what we have been considering (see Rescorla, 1969); it is inhibitory rather than excitatory.

Up to now, we have been discussing excitatory conditioning. Inhibitory conditioning procedures differ from excitatory procedures in one important respect: the CS is explicitly paired with no US. The CS (noted as CS— in contrast to the CS+, which is an excitatory Pavlovian cue) does not remain neutral but rather acquires associative strength; it comes to elicit an expectation of an event, namely, the nonoccurrence of the US just as the CS+ comes to elicit an expectation of the US presentation. In every respect, then, the conditioned inhibition reaction is the mirror image of a normal excitatory CR. The response tendencies are simply opposite in nature. Whereas the CS+ tends to elicit behavior like salivation or paw flex, the CS— tends to inhibit the appearance of these reactions. In summary, conditioned inhibition requires the explicit pairing of the CS and no US during acquisition; the CS gains associative strength and does elicit a conditioned reaction, namely, an inhibitory response which is the antithesis of an excitatory CR.

Obviously there is a problem in measuring an inhibitory CR. One can observe positive instances of behavior like salivation and paw withdrawal, but not negative instances of behavior such as the absence of salivation or paw flex. Fortunately, psychologists have devised two indirect ways for measuring inhibition, the summation method and the retardation-of-acquisition technique.

The summation technique involves the presentation of two stimuli, the CS+ and the CS−, together. The CS+, of course, is the component that elicits a measurable reaction (excitatory CR) while the CS− is presumed to elicit a counter or inhibitory CR. If this, indeed, were the case, that is, if the CS− were to elicit a reaction opposite to that evoked by the CS+, then the *overall* CR strength should be lower when the two stimuli are combined than when the CS+ is given by itself. In other words, the summation test assumes that inhibitory tendencies subtract from or cancel the excitatory tendencies. The net behavioral effect when both stimuli are presented indicates whether the CS− does, indeed, have inhibitory powers. If it does not, the joint presentation will produce a response just as strong as the behavior produced by the CS+ alone. But if the CS− does have inhibitory strength, the joint presentation should produce a weaker conditioned reaction.

A study by Hammond (1967), using the CER technique, illustrates the summation procedure. First, the rats learned to press a lever for food. This provided a stable response which later could be disrupted by Pavlovian cues (recall the CER technique discussed earlier). In a different phase of the study, the Pavlovian stimuli were given independent of the subjects' behavior. Specifically, subjects in the inhibitory group (Group I) were given a tone CS+ that was followed by a shock US. This constituted normal excitatory conditioning. They were also given a light CS− that was explicitly paired with no shock. The light, therefore, was expected to become a conditioned inhibitor. In short, subjects in Group I experienced both excitatory and inhibitory conditioning.

A control group (Group R—random group) received the CS+ and shock pairings (excitatory conditioning), but for them, the CS− was presented randomly with respect to shock. In other words, this group received excitatory conditioning to the tone, but no inhibitory conditioning; the CS− could develop no predictive power because it was not consistently paired with "no US." In summary, these two groups were equivalent in terms of excitatory conditioning, but differed in terms of inhibitory conditioning: Group I received inhibitory training because the CS− was explicitly paired with no US; Group R, on the other hand, did not receive inhibitory conditioning because their CS− (the light) was not consistently paired with any particular event.

The results demonstrated the concept of conditioned inhibition. Figure 2–2 illustrates the findings of the summation test. The ordinate of the graph

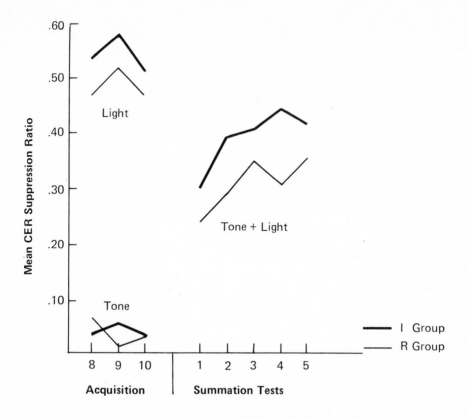

Figure 2-2. Mean CER suppression ratio to the tone CS+ and the light CS− during acquisition (left) and during the summation test for conditioned inhibition where the S+ and S− were given together (right) for the inhibitory (I) and random (R) groups. (From Hammond, 1967)

is the CER ratio: the number of lever presses emitted during the CS is divided by the number of presses performed prior to the CS and during the CS. The result is the CER suppression ratio. Recall that a ratio of .50 means that a stimulus had no disruptive effect on the behavior, but lower ratios signify disruption.

Group R showed a substantial degree of disruption, ranging from about .25 to .45 over the 5 summation tests. Group I showed less disruption to the tone-light combination. (Remember that on a CER test, *less* disruption means a *higher* CER ratio; *greater* disruption means *stronger* conditioning and a *lower* CER ratio.) The reason for the higher ratio for Group I was not that the tone CS+ was weaker; the same number of tone-shock pairings had been given to each group. But the subjects in Group I experienced an inhibitory CS− along with the CS+, and the presence of the CS− subtracted from the excitatory strength of the tone CS+. Thus, the *combination* of CS+ and CS− produced less disruption of lever pressing (less excitation) in Group I than in Group R because the excitatory effect of the CS+ was reduced by the presence of the inhibitory cue.

The second technique for demonstrating conditioned inhibition, the retardation-of-acquisition technique, involves a two-phased experiment. In the first phase of the experiment, a CS is used as a conditioned inhibitory stimulus; the CS− signals "no US." Naturally, we do not see any overt behavior during this phase because the CS− is presumably eliciting a "negative" CR.

In the second phase of the study, however, we can observe indirectly the effect of having used the conditioned inhibition procedure in Phase 1. During the second phase the former CS− is used in a "normal" excitatory conditioning experiment. Quite simply, the CS− is now paired with a US. The prediction is that a cue should be hard to convert into an excitatory CS+ if it already is a conditioned inhibitor. In other words, if a cue had become a conditioned inhibitor during the first phase of the experiment, then considerably more excitatory training trials should be needed to convert that cue into a conditioned exciter during Phase 2 relative to the number of training trials needed to establish any other neutral cue as a conditioned exciter.

A good example of this is an experiment by Marchant and Moore (1974). They studied the conditioning of the eye blink response in rabbits using light and tone CSs and a mild shock to the eyelid as the US. One group of subjects was given 50 excitatory trials in which a CS+ (the light) was followed by a brief shock US. Intermixed with these trials were 50 inhibitory trials in which the light CS+ and a tone CS− were followed by no shock. (Note that although the CS+ was presented on these inhibitory trials, the critical feature, that is, the single element that predicted no US presentation, was the CS−. Therefore, inhibitory conditioning to the tone CS− should occur.) The authors observed that conditioning took place in a normal fashion on the excitatory trials and that the subjects did not respond on the inhibitory trials. The question was whether the CS− was developing conditioned inhibitory strength during this phase.

To test this possibility, Marchant and Moore completed a second phase during which the tone CS was used for the first time as an excitatory stimulus. In effect, the authors tried to convert the CS− into a CS+, a change from conditioned inhibition to conditioned excitation, by presenting the CS followed by the US. They discovered that normal excitatory conditioning took place more slowly than it did in a naive group of subjects who had not been previously subjected to the inhibitory conditioning phase. The mean number of trials required to produce excitatory conditioning to the tone CS was 135.4 for the experimental subjects, but only 81.6 for the control group. This outcome clearly demonstrates that conditioned inhibition had indeed occurred during the first phase of the study. Had the tone *not* become a conditioned inhibitor in Phase 1, no retardation would have been evident in Phase 2; experimental and control subjects would have acquired the excitatory CR at the same rate.

You should note that inhibitory conditioning is by no means limited to aversive USs. In the two experiments just described, the US was a negative, or aversive, one, but the phenomenon also occurs with appetitive, or positive, USs.

Summary of Four Types of Pavlovian Conditioning

Let us review the four principal types of Pavlovian conditioning experiments. These are depicted in Figure 2–3. The upper two cells of the figure refer to excitatory conditioning: those experiments in which the CS is explicitly paired with a US. On the left, the US, which the CS comes to predict, is a positive or pleasurable one like food. On the right, the US is an aversive or negative one, for example, shock.

The same arrangement is evident in the bottom two cells of Figure 2–3. These refer to inhibitory conditioning studies where the CS− is explicitly paired with no US. The left cell represents inhibitory conditioning in the context of a positive US. Here, the CS+ would signal a pleasurable US whereas the CS− would signal "no US." Clearly, if a CS− can ever have meaning as a signal for "no US," then a US must be presented at other times during the study. After all, how could an animal learn to expect "no US" following a certain cue if it never received USs at other times?

The bottom right cell of Figure 2–3 illustrates the two experiments just described: in the context of excitatory conditioning using an aversive US, inhibitory conditioning occurred to a cue that signaled no US.

Thorndike's Conditioning Tradition

E. L. Thorndike (1874–1949) was an American psychologist whose important work began at about the turn of the century while he was a student of William James at Harvard. Thorndike's early experiments were not new; in a sense, they had been performed thousands of times before by animal trainers. But the conclusions that Thorndike drew and the theory that he developed from his observations were important contributions to science. In retrospect, we consider Thorndike an innovator because his experiments provided a new and systematic framework for investigating the learning process.

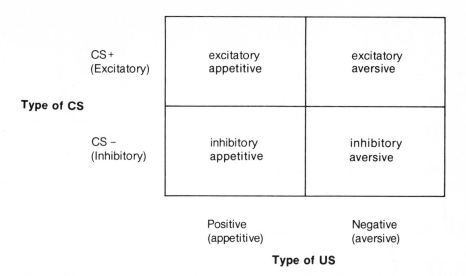

	Positive (appetitive)	Negative (aversive)
CS+ (Excitatory)	excitatory appetitive	excitatory aversive
CS− (Inhibitory)	inhibitory appetitive	inhibitory aversive

Type of CS (row label)

Type of US (column label)

Figure 2-3. Matrix showing the four principal types of Pavlovian conditioning experiments.

Reward Conditioning

In Thorndike's (1898) original work, cats were placed in a small box containing a latch. When the cats hit the latch, the door would spring open allowing the cats access to the food placed nearby. When Thorndike put the cats in the cage for the first time, they thrashed about in an apparent random fashion, as if they were seeking a way to escape. Sooner or later, of course, the cats hit the latch and were allowed access to the food. When Thorndike repeated this procedure a number of times, he noticed that the cats became more and more efficient; once they were put into the box, they required less time before they hit the latch. What was learned, therefore, was a new behavior, namely, hitting the latch in an efficient manner. Since the behavior was instrumental in securing food, this type of conditioning is called instrumental conditioning.

Thorndike's basic procedure serves as a prototype for virtually all experiments in this field of research. Unlike Pavlov's dogs, Thorndike's cats "controlled" the presentation of the food. In other words, the stimuli presented to the cats, such as the food, were *not* independent of their behavior; rather, food was contingent upon hitting the latch. This is an all-important and fundamental point: in contrast to classical conditioning, *instrumental conditioning stipulates that the US is contingent upon the CR;* the animal, not the experimenter, is the one who "decides" when the US will be administered.

Terms

The language used to discuss instrumental conditioning is generally the same as that used for classical conditioning. The important biological stimulus, the US, is evident in instrumental conditioning studies, but it is often called the reinforcer. Reinforcers are normally potent stimuli like food or shock, although other strong stimuli are used too, including access to a sexual partner, heat or cold, sweet-tasting substances, access to activity devices, or changes in sensory stimulation. In general, one tests a reinforcer, that is, the strength of a US, by observing how the animal reacts to it. If a stimulus is sought by the animal, or, at least, if it is never avoided, then we call the stimulus a positive reinforcer. An aversive US, in contrast, is one that the animal will never seek but rather will avoid. In other words, if the stimulus is powerful enough to serve as a reinforcer for an instrumental response, then the subject will invariably demonstrate approach behaviors (if it is positive) or avoidance behaviors (if it is aversive). Note that this definition of reinforcer is precisely the behavioral criterion we used to define stimulus strength in Chapter 1 (see Figure 1–3).

Although the second term in instrumental conditioning, like classical conditioning, is the unconditioned response, the UR is relatively unimportant. We assume that it occurs. For example, in Thorndike's experiment, we assumed that the presentation of food, once the cats had escaped the box, made them salivate. But the UR is rarely a point of focus in instrumental conditioning studies.

Instrumental conditioning studies may or may not involve a salient CS. In Thorndike's original study, no obvious or deliberate stimulus was presented. The cats were, of course, subjected to the stimuli in the environment including the sight of the latch, but otherwise no special lights or tones were employed as is typical in Pavlovian studies.

Deliberate signals, however, may be used. Such cues in instrumental conditioning studies are called discriminative stimuli (S+). A discriminative stimulus is a neutral cue that signals to the animal when reinforcement is available. It informs the subject when to respond. To illustrate, Thorndike could have changed his experiment to include a S+: after the cats were placed in the box, he might have sounded a buzzer to signal when the lever response should be executed. The buzzer, in a sense, would help the cat to know when the appropriate time for the response had occurred. In short, the discriminative stimulus is like the CS in Pavlovian conditioning, but its specific function is to signal reinforcement availability (pending the response execution), whereas in classical conditioning its function is to signal the immediate presentation of the US (without regard to any response that may or may not occur). (See Chapter 5 for a more thorough discussion of this point.)

The final term in instrumental conditioning, the conditioned response, is quite unlike the UR. This is the most startling difference between the two forms of conditioning. In classical conditioning, the CR is essentially like the UR, whereas in instrumental conditioning, the CR is a relatively arbitrary behavior (see later chapters for qualifications however). Thorndike could have required his cats to ring a bell, or pull a string, or run from one side of the box to the other before being given food. The choice of the learned reaction, was, in that sense, arbitrarily made by the experimenter. Most often, the CR is a gross motor behavior such as lever pressing or running down a maze, but other less arbitrary behaviors, such as internal visceral reactions, are also possible to condition (see Chapter 4).

Procedures

To understand how the typical instrumental experiment is constructed, consider Figure 2–4. Before conditioning, the animal is faced with a myriad of stimuli in its environment, one of which may be a salient stimulus, the S+. In addition, the animal is capable of performing many behaviors, only one of which is designated as the "correct" response by the experimenter. In Figure 2–4, the response of R-lever is designated as correct.

During conditioning, whenever the designated response occurs after the S+ onset, reinforcement is presented to the animal. This is shown in the middle portion of Figure 2–4.

Finally, after conditioning has taken place, the extraneous behaviors are no longer performed; the animal's performance is now efficient. What has been learned is an expectancy for the reinforcer based upon the execution of the response during the S+. The performance of the animal, the increase in efficiency, indicates unambiguously that such an expectation has been formed.

Punishment

The Thorndike experiment that we have reviewed is a typical example of reward conditioning; the response was followed by a pleasant appetitive US. Punishment training is the mirror image of this study; it occurs when the CR is followed by an aversive stimulus. This is analogous to Bekhterev's aversive experiment in classical conditioning. Bekhterev paired a CS with

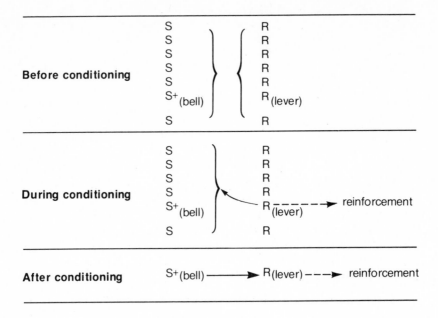

Figure 2-4. Schematic diagram indicating the basic structure and sequencing in a typical instrumental learning experiment.

an aversive US, but in punishment training it is the behavior, rather than a stimulus, that signals the presentation of the aversive US.

We could cite many examples of punishment training (see Chapters 4 and 5). The notable fact is that punishment leads to the suppression of the CR. For example, if a rat were given a mild shock after pressing a lever, it would quickly cease to respond and would begin to engage in a variety of other behaviors. In short, the punishment procedure is just like reward training (the US is contingent upon the CR) except for the fact that an aversive US is involved. The effect of punishment is suppression of responding rather than excitation.

Inhibitory Conditioning

The two types of conditioning discussed above are analogous to the excitatory conditioning paradigms of classical conditioning. For both reward and punishment training, the response leads to the *presentation* of a US; a positive US in the case of reward training and an aversive one in the case of punishment. There are two additional instrumental training procedures which are analogous to the inhibitory procedures of Pavlovian conditioning. In both of these cases the response leads to the *nonoccurrence* or termination of the US.

In the first type, termed omission, a response leads to the nonoccurrence of a positive US; that is, the response prevents the occurrence of, say, food. Failing to respond during omission training usually results in reinforcement, while performing a response results in no reward. The overall consequence of omission training is the suppression of the CR. Not only is the criterion behavior reduced in strength because it results in no reward, but alternative behaviors, specifically any response other than the criterion CR, are strengthened (see Chapter 7).

The second major form of instrumental conditioning in which the response leads to the nonoccurrence of a US is termed avoidance conditioning. In the typical avoidance study, a rat is placed in a shuttlebox containing two compartments. A discriminative stimulus, like a tone, is then presented which signals to the animal that responding is appropriate. If the animal makes the response, namely, jumping from one side of the box to the other, shock is omitted for that trial. If the animal fails to respond within the specified time, however, the shock is presented. When this happens, the animal can terminate the shock by running to the other side. In avoidance learning, therefore, a response leads to the nonoccurrence of an aversive US.

Actually, there is a procedure called escape learning that is very similar to avoidance. Here, the response leads to the termination of the aversive US, rather than to the avoidance of it. That is, the noxious US comes on (the experimenter determines when) and then goes off after the animal has made the appropriate response. This is analogous to a training procedure in classical conditioning in which a shock US is presented and then, just before the shock ceases, a tone CS is sounded. The tone is a signal for the *termination* of the US. It is the same in escape conditioning except that it is a response, not a stimulus, that signals the termination of the shock. Both avoidance and escape conditioning procedures involve a correlation between a response and the absence of an aversive US. Hence, they are analogous to the inhibitory procedures in classical conditioning in which an aversive US is used. In escape training, however, the US is actually *terminated* by the response, but in avoidance, the US is *eliminated* altogether by the response.

Avoidance and escape responses often occur in the same experiment. Consider a simple study designed to teach an animal to avoid a shock US. Early in training the animal does not know that shock can be avoided, that the shock will not occur if it responds appropriately. Therefore, the animal can only escape the shock after it comes on. Executing the escape response and causing the shock to be terminated, of course, is reinforced; the painful US ceases. Later, when the animal has learned to expect the termination of shock after its response, it will begin to execute the avoidance behavior. That is, the animal will begin to respond *before* the shock is actually presented. This is the avoidance behavior; once it appears, it continues to be performed at a stable level because it is reinforced by the nonoccurrence of the painful shock. In other words, a typical avoidance experiment has two parts. During the first few trials, before the subject has discovered that responding will prevent the presentation of the shock, the animal merely escapes or terminates the noxious US. Later, however, the animal discovers that a response will prevent the US altogether; it begins to make avoidance reactions.

Avoidance learning is not really the mirror image of reward training because during avoidance the animals are under a time constraint; they have a limited amount of time during which they must either execute the avoidance response or else be shocked. In reward training, on the other hand, no such time restriction is placed on the subject. The animal has unlimited time in which to execute and receive the reward (but there are some techniques in which this is not true and these will be discussed in Chapter 4). Despite this significant difference in methodology, avoidance training, like reward training, leads to the strengthening of the criterion behavior.

There are other paradigms for studying avoidance besides the two-compartment shuttlebox technique described above. One variation is called Sidman avoidance. Here, the animals must press a lever in order to postpone shocks. Failure to do so means that shock will be delivered shortly, perhaps within 10 or 15 seconds. When the animal presses the lever, however, the clock is reset and there is a shock-free interval. Moreover, if an animal presses at a sufficiently high rate, it can avoid shocks entirely. Unlike the shuttlebox experiment previously described, Sidman avoidance usually does not employ a discriminative stimulus. The animal must anticipate the pending shock merely from the temporal spacing of shocks rather than from any deliberate external signal.

A third form of avoidance conditioning is called passive avoidance. Here, the animal receives shock or other aversive stimuli at a particular place in the apparatus, say, on the black side of a two-compartment box. Given the opportunity, the animal will, of course, escape the shock by running into the safe side of the box. Later the animal will perform a passive avoidance response: it will remain passively on the safe side of the box and avoid entering the previous shock compartment.

Summary of Instrumental Learning Paradigms

The four major types of instrumental conditioning procedures that we have discussed are depicted in Figure 2–5. In the top two cells are the reward and punishment techniques in which a US is produced by or follows the CR. For reward training, the US is an appetitive or pleasurable one; for punishment, the US is aversive. The two additional types of instrumental conditioning which involve the nonoccurrence or termi-

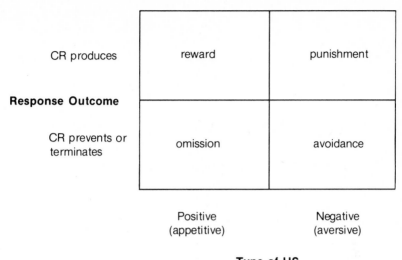

	Positive (appetitive)	Negative (aversive)
CR produces	reward	punishment
CR prevents or terminates	omission	avoidance

Response Outcome

Positive (appetitive)   Negative (aversive)

Type of US

Figure 2-5. Matrix showing the four principal types of instrumental conditioning.

nation of the US following the response are shown in the lower two cells. For omission training, a positive or appetitive US is omitted following the CR, while for avoidance learning, the CR terminates or avoids an aversive US.

It is important to note that the types of instrumental learning procedures illustrated in Figure 2–5 are virtually the same as those used in classical conditioning. Compare Figures 2–5 and 2–3. In both cases the presence or absence of an appetitive or aversive US is predicted by some prior event. In the case of classical conditioning the prior event which predicts the presence or absence of a US is a stimulus. In contrast, for instrumental conditioning, the event that predicts the presence or absence of a subsequent US is a response. In other words, a Pavlovian CS+ is comparable to an instrumental CR in the reward or punishment situation; both signal a US presentation. Similarly, a Pavlovian CS− corresponds to an instrumental response in the omission and avoidance situation; both signal the absence of the US. Thus, we see that classical conditioning pertains to stimulus-US relationships whereas instrumental conditioning pertains to response-US relationships. In both classical and instrumental conditioning the four possible outcomes are identical: presence versus absence of the US and positive versus negative US. The only difference is in terms of what predicts these outcomes: a stimulus in the case of classical conditioning or a response in instrumental conditioning.

General Laws of Conditioning

The previous sections have presented the procedures that are followed in normal conditioning experiments. Let us now formulate the general laws that depict how these training techniques operate.

Stimulus-Outcome Correlation

The traditional theory of Pavlovian conditioning emphasized stimulus contiguity. The close occurrence of the CS and US somehow allowed the two stimuli to become "bonded" or associated. The CS was thought to become a substitute or surrogate US, and it did so because of the contiguity. In a very important article, Rescorla (1967b) reexamined this notion and provided a new perspective on conditioning. Rescorla's claim was straightforward: a CS acquires information value (association or predictive power) concerning whatever event follows it in time. If the outcome of the CS is a US, then the CS provides information about the occurrence of the US. The animals can use the CS to predict the US presentation and excitatory conditioning occurs. If, on the other hand, the CS is reliably followed by "no US," as is true in inhibitory conditioning, then the CS provides information about that outcome. Conditioning still occurs, but the CR is inhibitory in nature. In both cases, then, conditioning is based upon the consistency of the stimulus-outcome relationship or correlation, not just on the number of CS-US pairings.

Rescorla's information theory was an exceedingly important contribution because it provided a new framework for understanding how conditioning operates. In a 1968 article, Rescorla provided strong support for his theory. According to his theory, no conditioning should take place if the CS and US are presented randomly with respect to one another because the predictive relationship would be absent. That is, when the US has as much chance of occurring during the CS as it does at other times, then the CS should develop no predictive powers at all. Conditioning should be absent because the correlation between the CS and US presentations would be essentially zero. However, as the CS becomes more highly correlated to the US occurrence, excitatory conditioning should increase.

Rescorla used the CER technique. The CS was a 2-minute tone. During the tone, the probability that a subject would receive shock was .4. In other words, 40 percent of the tones were accompanied by shock. The probability of receiving shock at other times (that is, in the absence of the tone CS) varied among groups. For one group, the probability of receiving these extra shocks was .4; no conditioning was anticipated in this group because the probability of receiving a shock during the CS was equal to the probability of receiving a shock during no CS. The CS certainly could not become a reliable predictor of shock under such conditions since it wouldn't be consistently paired with shock. For the other groups, however, the probability of receiving extra shocks during other parts of the trial diminished. For one group it was .2; for another it was .1; and for the fourth group, Group 0, no extra shocks were given. These groups, therefore, reflect a decrease in the extent to which shock occurred at times other than during the CS. As the probability of receiving these extra shocks decreased, the predictive value of the tone would increase. Conversely, as the probability of receiving shock at times other than during the CS increased, the correlation between tone and shock would become weaker, and conditioning should diminish.

The results are illustrated in Figure 2–6. Actual testing involved the presentation of the CS without the US. The disruption caused by the CS therefore should extinguish over the test sessions. Subjects in Group .4, for whom shock was as likely to occur during the CS as during other times, showed very little conditioning. The CS presentation did not disrupt their lever-pressing behavior very much. In everyday language, these animals could not use the CS to predict when shock would be presented since shock was as likely to occur at other times as it was during the CS.

Conditioning was evident in the other groups, however. Subjects in

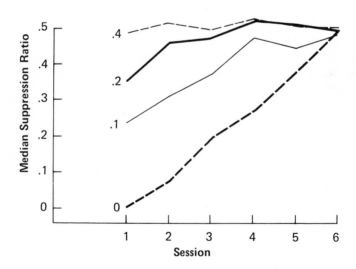

Figure 2-6. Median CER suppression ratio for the four groups over the six test sessions. All groups had a .4 probability of receiving the US during the CS, but each group differed in terms of the probability of receiving extra USs in the absence of the CS. (From Rescorla, 1968)

Group 0, who were not given any extra shocks, showed the greatest suppression (that is, the strongest conditioning). For those animals, the CS was most exclusively a signal for the US. Subjects in the other two groups showed intermediate suppression levels reflecting the fact that the CS and shock were only moderately correlated.

It is very important to remember that all the subjects in Rescorla's experiment had the same *number* of CS-US pairings. Conditioning strength, however, did not reflect this fact. Rather, the strength of the CS depended upon the predictive validity of the CS. The more the CS-US correlation was degraded by the presentation of extra shocks during the no CS periods, the more the conditioning was diminished.

In summary, Rescorla's work suggests that the fundamental law of Pavlovian conditioning is the CS-outcome correlation. It is this ingredient which is critical to the formation of a stimulus expectancy. When the correlation between the CS and a particular outcome is weak, conditioning fails to occur even though an appreciable number of CS-US pairings may be involved.

Response-Outcome Correlation

As noted earlier, the important relationship in instrumental conditioning is between the response and the US (or absence of US as in avoidance and omission training). The corresponding law, therefore, ought to involve the response-outcome correlation. The higher the correlation between these two events, the greater the response expectancy and the stronger the instrumental conditioning.

Although there is little research on the topic of response expectancies, it would appear that such expectancies do develop (see Bolles, 1972). Indeed, it is difficult to imagine why an animal would ever make a response, at least a voluntary response like lever pressing, if it did not expect some particular outcome like food. Recently, there has been a number of studies which, in a rather direct fashion, support the notion that response expectancies develop during instrumental training. One such study was done by Bolles, Holtz, Dunn, and Hill (1980). They trained rats to execute two different responses using the same specialized lever. That is, the rat could both press the lever and pull it out from the wall. Both of these behaviors were learned quite readily, and the animals achieved a fairly high and stable rate of response. Then, during two punishment sessions, one of the behaviors was followed immediately by a shock whereas the other was not. The authors argued that any kind of generalized suppression of behavior, that is, general disruption caused by the sight of the lever, would reflect stimulus learning. In contrast, the *differential* suppression of one response relative to the other would indicate that a response expectancy had developed. Specifically, the subject should continue to perform the unpunished behavior (either pressing or pulling, whichever was unpunished), but cease to execute the punished behavior. Here, suppression of the *specific* behavior would reflect response learning; the animal would have no generalized fear of the lever but would be reluctant to execute the punished response because it would expect to receive shock following the response.

As shown in Figure 2–7, the performance rate for both responses fell during the first punishment session indicating a general disruption due to shock. The unpunished reaction, however, began to recover during the last part of this punishment session. Then during the second punishment session, differential suppression was observed. Clearly, the animals were not reluctant to manipulate the lever (the stimulus, that is, the sight of the lever, was not influencing their behavior much), but they were reluctant to perform the particular behavior which led to the punishment. This differential reaction could have occurred only if the punished response itself induced the expectancy for punishment but the unpunished response did not.

Joint Stimulus/Response Expectancies

Other experiments have demonstrated the joint occurrence of response and stimulus expectancies. For example, in a study by Church, Wooten, and

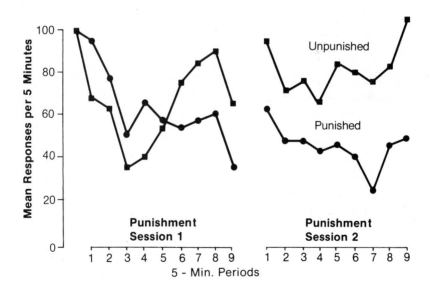

Figure 2-7. Mean number of responses per 5 minutes as a function of the nine 5-minute segments in both punishment sessions. (From Bolles, Holtz, Dunn, & Hill, 1980)

Matthews (1970), one group of rats was punished following a lever press; presumably these animals developed a response expectancy. A second group of animals was punished following a stimulus presentation; a stimulus expectancy presumably developed. The authors found that a differential effect was produced: suppression was far greater in the response-punishment group than it was in the stimulus-punishment group.

In a second experiment, the authors reasoned that suppression of the lever press response would not be as severe for the response-punishment group if those animals were allowed to execute a different (nonpunished) CR. After all, if the suppression were due to the specific response-shock correlation as discussed in the previous section, then such a contingent relationship would be absent if the response were changed. More specifically, if the subject were given shock immediately after a lever press, then there is no reason to believe that a second response, such as chain pulling, would be suppressed. This prediction was confirmed. Suppression was substantial for a response that was immediately followed by shock, but other behaviors that were not punished directly were not suppressed (see also Williams, 1978).

Given that animals may learn to expect future events based upon *either* a stimulus or a response, how do these types of expectancies interact? The problem is a complex one and few studies on the subject have been completed. Two recent papers showing the interaction of stimulus and response expectancies were done by Pearce and Hall (1978) and St. Claire-Smith (1979). The basic procedure in both of these studies was as follows. Subjects were allowed to press a lever in order to receive food. The food, however, was not given for every lever press, but only after several had been performed. In terms of response learning, the lever press was not perfectly correlated with food because occasionally it did not produce the food reward. Whenever the food was delivered, however, it was delayed for a half second during which time some of the animals received a stimulus (a light or a tone). Because the light *always* preceded food, stimulus learning was expected to be rather strong. Which element, the stimulus (the light or tone) or the response (the lever press), would be the better predictor of the food reward? The answer is the stimulus: it was perfectly correlated with food presentation (it always preceded food delivery) whereas the response predicted food only some of the time. The actual results showed that the group that received the stimulus prior to reward performed the lever press response at a lower level than did the group that got no stimulus. The stimulus, because it was a better predictor of reward, became the stronger element. The response, of course, predicted food to a degree but when a better predictor, the light, was available, performance of the response tended to suffer. These experiments are exceedingly important not only because they support the notion that expectancies develop from both stimuli and responses, but also because they illustrate that stimulus and response learning interact in predictable ways. (See Chapter 5.)

Summary of Classical and Instrumental Procedures

Let us review the instrumental and classical procedures so that the similarities between the two may become apparent. Consider Figure 2–8. Here, we have two independent dimensions which are labeled excitatory versus inhibitory and aversive versus appetitive. The four types of Pavlovian procedures and the four major forms of instrumental conditioning can be classified in terms of these two dimensions. For instance, in the upper left-hand quadrant of Figure 2–8 are located the usual Pavlovian appetitive experiment (a CS signals the presentation of a food US) and standard reward training (a response leads to the presentation of an appetitive US). The Pavlovian and instrumental procedures are conceptually the same; the only difference is that a stimulus signals a US in the case of Pavlovian conditioning, whereas a response signals the US in the case of instrumental conditioning.

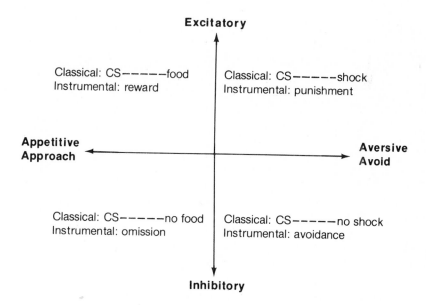

Figure 2-8. Matrix summarizing the four types of classical and instrumental conditioning experiments.

The essential similarity between Pavlovian and instrumental operations is evident in the other quadrants of Figure 2–8. In the upper right-hand portion we have a classical aversive experiment, as shown originally by Bekhterev, and we also have a typical punishment procedure. These two are forms of excitatory conditioning because the predictive elements, the CS in the case of classical conditioning and the response in the case of instrumental conditioning, predict the occurrence of a US. At the same time, these two forms of conditioning employ aversive USs.

The inhibitory forms of conditioning are shown in the bottom two quadrants of Figure 2–8. On the right are cases that employ an aversive US. For classical conditioning a CS that reliably predicts, say, no shock becomes an inhibitory stimulus. Similarly, avoidance conditioning takes place when the response predicts or leads to the nonoccurrence of an aversive US. We don't normally think of avoidance conditioning as being an inhibitory phenomenon because the behavior, the CR, actually increases. This happens because the nonoccurrence of the aversive US is rewarding and, therefore, it strengthens the behavior that leads to it. In terms of the animal's expectancy, however, avoidance conditioning and classical inhibitory (aversive US) conditioning are identical. The animal comes to expect the nonoccurrence of an aversive US based upon a stimulus (classical CS) or a response (instrumental avoidance reaction).

The lower left-hand quadrant in Figure 2–8 contains, on the one hand, omission training (instrumental) and, on the other hand, appetitive inhibitory (classical) conditioning. Again, both forms of conditioning are inhibitory in the sense that the predictive element leads to the nonoccurrence of the US. The US that is experienced at other times, however, is an appetitive one. Formally speaking, then, the Pavlovian and instrumental procedures are analogous except for the fact that the expectation for no food reward is based upon a stimulus presentation in the case of classical conditioning but is based upon the execution of a response in the case of instrumental conditioning.

Before we conclude this introductory chapter, we should consider one final point. Most students come to study psychology out of their interest in human behavior. This is natural and proper since as humans we are concerned with understanding ourselves more thoroughly. Given this interest in human behavior, you might ask why study animals?

The answer to this question takes several forms. First of all, it often makes good sense from a methodological point of view to study animals as well as humans. Behavior is really what psychology is all about. And we know that behavior is, to a large degree, a function of past experiences. It is therefore important to specify and often control experience. This can be done with animals but not with humans. We can make sure that our white rats and mice have had certain experiences but not others. We can house them under specified and restricted circumstances; we can control their genetic makeup; we can make sure that each of our subjects has undergone the same experience. This is not possible in humans both from a practical as well as an ethical point of view.

Animals such as white rats and monkeys may not be appropriate substitutes if one is interested in studying human behavior, but there may be sufficient continuities across the species that warrant studying animals. This brings us to our second major reason for using animals in learning research. Humans and other animals may have certain basic learning strategies in common; that is, they may share on a very molar level at least, general patterns or modes of learning. Each species, of course, has its own behavior potentials (as discussed in Chapter 1). But because some of the life-sustaining tasks faced by many species, such as remembering the location of food sources, are similar, the underlying principles by which the different species approach these situations might also be similar. The specific mechanisms for behavior, therefore, may be species-typical, and yet the broader learning strategies, which bring these mechanisms into action, may be shared by many diverse species, including humans. The point is not that humans are like rats or pigeons; they are, emphatically, not (although in terms of some biological parameters, humans and rats are not too dissimilar). The point is rather that humans and rats have evolved from common ancestors (both are mammals) and share many physiological and neurological traits which may be important for behavior. In other words, certain principles of learning may be basic to many higher forms of animal life. What is true of the learning in lower animals may, *in principle,* also be true of humans. Certainly differences between species exist and it is important to investigate these differences. And yet, important continuities may be present too. In conclusion, it is quite possible to have the understanding of human behavior as the ultimate goal and yet investigate animals as one means of assessing the fundamentals of human learning. Our current understanding may be highly erroneous (the way we understand animal learning may have very little to do with the way humans actually learn), but final judgment must be reserved for the future.

Summary

There are two major traditions in learning research: classical conditioning and instrumental conditioning. A Pavlovian experiment is one in which two stimuli are presented to the animal independent of its behavior. The stimuli normally are a weak or innocuous one, the CS, followed by a biologically potent one, the US, which elicits a reflexive UR. The learned reaction, the CR, is later elicited by the CS alone. This may be demonstrated by showing that the CS will elicit the CR on a test trial or the CR will occur prior to the US presentation.

Conditioned inhibition occurs in Pavlovian conditioning when the CS, in the context of normal excitatory conditioning, is consistently followed by no US. The inhibitory CR that develops must be measured using special techniques. Specifically, the inhibitory CS— will cause an excitatory CR to be reduced in strength; conditioned excitation and inhibition thus summate. Normal excitatory conditioning using a former CS— is difficult to achieve; this result, too, indicates that conditioning occurs when a CS is paired with no US.

Instrumental conditioning is the second major model in animal learning research. In instrumental conditioning, a response is followed by an outcome. The US may be a pleasant one like food in reward conditioning, or it may be an aversive US like shock in punishment training. The critical feature is that the US, or its absence, is contingent upon the response.

The terms used in instrumental conditioning are essentially the same as those used in Pavlovian conditioning except that in instrumental conditioning the CS is usually called a discriminative stimulus. It signals when reinforcement is available.

Like Pavlovian conditioning, instrumental conditioning may involve inhibitory procedures in which the response may lead to the nonoccurrence of a pleasant US (omission) or the nonoccurrence or termination of an aversive US (avoidance, escape). Similarly, an instrumental CR may produce a positive US like food (reward) or an aversive US (punishment) training.

From the study of the Pavlovian and instrumental conditioning models, we can formulate the general laws of conditioning. In the case of classical conditioning, the fundamental law of learning is the stimulus-outcome correlation: stimuli that consistently signal an important outcome, such as a US in the case of excitatory conditioning, become conditioned. The underlying law in instrumental conditioning is the response-outcome correlation: responses that predict important outcomes are learned and the animals come to expect the outcome based on the execution of the response.

3

Stimulus Learning

Pavlov

Introduction

Training and Control Procedures

Training Conditions That Influence Learning

Basic Pavlovian Phenomena

Summary

Introduction

In the previous chapter we outlined the essential methodology and theory for Pavlovian, or stimulus, conditioning. Recall that the stimuli are given independent of the animals' behavior; this is the operational definition of a Pavlovian trial. What is learned is an expectancy: the CS comes to predict the US (or no US in the case of inhibitory conditioning), that is, the animal learns to expect the US presentation based upon the CS. It was primarily the work of Rescorla that showed that conditioning depends on the predictive value of the CS. According to Rescorla's theory, conditioning is based on the correlation between CS and US, not simply on the number of times they are paired. The higher the correlation, the greater the predictive value and the stronger the conditioning.

Also recall the important idea relating classical conditioning to evolutionary theory. Unconditioned stimuli are strong biological stimuli that, under normal circumstances, would have an important consequence on the animal's survival. Such USs are precisely those events which the animal must learn to predict if it is to behave adaptively. Pavlovian conditioning, then, is a mechanism that allows an animal to form these predictions.

Chapter 2 barely scratched the surface of the research on Pavlovian conditioning. Therefore, in this chapter we shall discuss more fully the details and theories pertaining to stimulus learning.

Training and Control Procedures

If the CS is to be used by the animal as a signal for the US (or any other outcome), then it is only logical that the CS onset should precede the US. Pavlov himself recognized this notion: "Further, it is not enough that there should be overlapping between the two stimuli; it is also an equally necessary that the conditioned stimulus should begin to operate before the unconditioned stimulus comes into action" (1927, p. 27).

As shown in Figure 3–1, there are actually two basic modes of forward conditioning. The first and most common is termed delay conditioning. In this mode, the CS lasts at least until the US is presented; often there is some degree of overlap. In any event, the CS remains on at least until the US occurs.

The second mode of forward conditioning is termed trace conditioning (see Figure 3–1). This is similar to the delay conditioning technique except that the CS offset occurs *before* the US is presented. There is an empty interval between the CS offset and the US onset. The technique is called the trace procedure because it is really the decaying neural memory trace of the

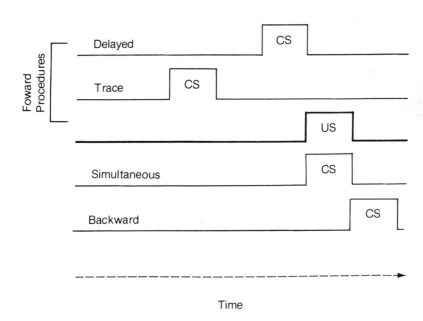

Figure 3-1. The four major classical conditioning procedures which differ with respect to the temporal relationship between the CS and the US.

CS that is contiguous with US onset. In other words, during the empty interval the memory of the CS (the neural excitation) decays gradually; the vestige of this neural memory trace, then, persists in time and is present when the US ultimately occurs.

A number of studies have compared these two forward procedures, but there is disagreement about which produces superior conditioning. Ross and Ross (1971) and Wilson (1969) found that trace and delayed procedures were equally effective in eye blink and heart rate conditioning. Other studies (Black, Carlson, & Solomon, 1962; Manning, Schneiderman, & Lordahl, 1969; Schneiderman, 1966), however, have shown that the delay procedure produces a stronger CR than the trace conditioning technique. The difference between trace and delay procedures may be linked to differences in the total CS onset-US onset interval. Manning et al. (1969), for example, found that the two techniques were essentially the same when the CS-US interval was short; with longer intervals, when the empty period was relatively lengthy and the neural "reverberation" had to persist for some time, the trace design was significantly inferior to the delay conditioning procedure.

One of the things that makes trace conditioning somewhat complex is that it produces not only excitatory CRs but inhibitory CRs as well. It has long been known that if the empty interval is sufficiently long, "inhibition of delay" is generated (see the section on this topic later in this chapter). Hinson and Siegel (1980) have recently demonstrated that trace conditioning may also lead to conditioned inhibition. They conditioned the rabbit's eyelid reaction using a trace procedure in which the empty interval was 10 seconds. They discovered that the CS later was inhibitory: it subtracted from an excitatory CR on a summation test and it was difficult to "convert" it into an excitatory CS on a retardation-of-conditioning test. In summary, the two forward methods of conditioning, although often equal in their effectiveness, may differ dramatically. Specifically, the trace technique may involve both excitatory and inhibitory CRs.

Problems in Evaluating Conditioning

Although it seems clear that the forward procedures lead to excitatory conditioning, there still are problems in terms of specifying the exact strength of the conditioning level. More precisely, we may be fooled into thinking that we have demonstrated true conditioning when, in fact, the behavior we observe is not an example of conditioning at all. Many investigators have noted that CRs may be given without the forward conditioning procedures being in effect. Such responses are called pseudoconditioned CRs (false conditioning) and they are caused by an inflation in the CR strength due to factors other than the association between the CS and US. For example, Domjan (1977) found that giving an illness-inducing drug to rats later made the animals fearful of consuming new flavors. The flavors, the novel CSs, had never been associated with any US, much less an illness-inducing drug. Yet when they were presented to the animal after it had been poisoned, an aversion (actually a pseudoconditioned reaction) was evident.

Domjan observed a pseudoconditioned response because the animal had been sensitized by the previous US administration. Sensitization is a motivational state that enhances an animal's natural reaction to stimuli. It is well known that rats are neophobic to novel flavors, that is, they show some natural avoidance of them (see Chapter 6); administration of the illness-inducing drug, therefore, seems to have enhanced this reaction. In short, Domjan observed what could have been mistaken as a CR, but in fact was only a pseudo-CR based upon sensitization, that is, potentiation of the natural neophobic reaction to the flavor.

Quite a different picture was evident in an important study by Sheafor (1975). While studying the reflexive jaw movements in the rabbit, he observed pseudoconditioned responses. Specifically, he found that administration of the US by itself was sufficient to produce pseudo-CRs. Further investigation revealed the reason for this finding. Sheafor discovered that the exposure of the animal to the US did involve conditioning, not to the CS but to the environment or background stimuli (the apparatus, etc.). The tone CS merely triggered this conditioned reaction to the background stimuli (see the last section of this chapter for more information on this point).

In summary, sensitization or enhanced neophobia (as observed by Domjan) or background conditioning (as in Sheafor's study) are but two possible mechanisms for a pseudoconditioning effect (which is an *apparent* CR but not a *real* CR based on the appropriate CS-US pairings). Pseudo-conditioning is a pervasive problem; it is therefore imperative that we evaluate the strength of conditioning only with the proper control procedures.

Control Procedures

The way to deal with pseudoconditioned responses, in fact the only way to evaluate properly the strength of classical conditioning, is to compare the conditioned group (trace or delay procedure) with a proper control group. Presumably, the control group and the conditioned group have been treated identically with one exception: the control group lacks the essential ingredient responsible for true conditioning.

One control procedure that has been cited by some as being effective (Beecroft, 1966) is the simultaneous presentation of the CS and US (see Figure 3–1). If the CS is to become a signal for the US, then it would appear that the simultaneous procedure would not lead to conditioning; the CS can't be used to anticipate the US presentation. Therefore, simultaneous conditioning ought to provide a "no conditioning" control technique.

There are really two problems with this argument, however. First is the fact that the US may overshadow the CS. That is, the potent biological stimulus may simply overpower the innocuous CS so that any conditioning, were it in fact taking place, would be obscured and undetected. Second, and even more important, recent research has shown that conditioning does, indeed, take place with a simultaneous procedure. For example, Burkhardt and Ayres (1978) paired a shock US and a noise CS in a simultaneous fashion. Their study confirmed that simultaneous conditioning did take place under these conditions by showing that the CS later caused the rats to delay drinking. The animals were reluctant to drink from a water spout when the CS was given. This would not have resulted had the CS remained essentially neutral and innocuous. Other experiments have also demonstrated in various fashions that simultaneous conditioning leads to learning (for example, Heth & Rescorla, 1973; Mahoney & Ayres, 1976; Sherman, 1978; Sherman & Maier, 1978).

A second technique for establishing a control baseline is backward conditioning (see Figure 3–1). In this procedure the US precedes the CS. Traditionally this procedure was believed to result in no conditioning because the CS did not precede the US and therefore it could not come to serve as a signal for the US (see Gormezano & Moore, 1969; Plotkin & Oakley, 1975).

The issue, however, is more complex than that. First of all, the backward procedure is known to produce inhibitory conditioning (see Chapter 2). Inhibitory associations develop because the CS predicts no US. But there is a second reason why the backward paradigm is not a good control technique. Paradoxically, it may produce excitatory conditioning. We shall explore this idea in more depth at the end of this chapter after we have developed additional concepts. However, one study in which excitatory conditioning was found deserves mention here.

Keith-Lucas and Guttman (1975) shocked rats while they were eating in a novel box and then immediately exposed them to a complex, seminaturalistic stimulus. The backward CS that was dangled in front of the rats was a toy hedgehog, a small hemispherical-shaped toy with soft rubber spikes radiating from its body. The next day the authors tested to see whether the rats avoided the forward CS (the location where they were shocked) or the backward CS (the hedgehog). They discovered that the backward conditioned CS exerted a more powerful effect upon the animals than did the forward CS; the animals avoided the hedgehog more than the novel box, indicating that backward conditioning was stronger than forward conditioning. The interpretation of the study was based upon certain evolutionary arguments. Naturalistic stimuli, especially those that conceivably might represent a predator, are more easily conditioned than highly artificial stimuli like lights, tones, and stripped walls. So it is that backward CSs, even though they are not optimally associated with USs, may come to develop excitatory conditioning.

Finally, we come to the random control procedure. As discussed in Chapter 2, this control procedure is based upon Rescorla's theory that the correlation between the CS and US is essential to conditioning. A random procedure, in which the CS and US presentations are independent, eliminates such a correlation and therefore should eliminate conditioning. Despite this claim there are studies which indicate that conditioning can take place even with a random control procedure (Benedict & Ayres, 1972; Keller, Ayres, & Mahoney, 1977; Kremer, 1971, 1974; Kremer & Kamin, 1971). The excitatory conditioning, however, appears to be due to the chance CS-US pairings that occur early in the session. When an animal is given random CS-US presentations, the two stimuli occasionally occur close together; if this happens *early* in the training session, conditioning results. But if the training is extended, excitatory conditioning decreases (Keller et al., 1977; Rescorla, 1972). In summary, the random control procedure *does* appear to be the most appropriate technique; with enough training the animals do learn that the CS and US presentations are unrelated. There are some genuine excitatory reactions that are conditioned early in the session or that are conditioned to background stimuli (Kremer, 1974), but these exceptions do not invalidate the more general notion that the random control procedure is the most appropriate one because it precludes any predictive relationship between the CS and US.

Training Conditions That Influence Learning

In the following section we shall discuss many of the important parameters of classical conditioning. These parameters represent variations in training procedures that affect the strength of conditioning. The parameters, or variables, are divided into three major groupings: those related to the CS, those related to the US, and other factors.

CS Information

In both Chapter 2 and the preceding section we discussed the all-important notion that classical conditioning is based upon the correlation between the CS and the outcome. Rescorla's work in particular illustrates that it is the correlation, not merely the number of CS-US pairings, that is important. Predictive value or information, then, is the crucial parameter. The importance of information value of a stimulus has been demonstrated in several interesting ways. In one study, Egger and Miller (1962) conducted the following experiment: rats were initially trained to press a lever for food. Then, during a second phase of the study, all of the subjects were given Pavlovian conditioning using two CSs (a tone and a flashing light) and a food US. For some of the animals, the first CS was followed .5 second later by the second CS; both of the stimuli then continued for an additional 1.5 seconds until food delivery. According to the traditional theory, the second CS should have developed greater strength because it occurred closer to the US presentation. From an informational point of view, however, the second stimulus should have been weaker because it was redundant; in order to be able to predict food an animal merely had to pay attention to the first CS. To determine which factor, information value or contiguity, was more important, Egger and Miller tested the strength of the two CSs in a third phase of the study. They did this by first extinguishing the lever press response and then beginning a reacquisition phase during which a lever press resulted in the presentation of either the first or the second CS. The idea was that the stronger the conditioning, the faster the reacquisition; that is, animals should press the lever in order to receive a stimulus that had been paired with food, and the stronger the stimulus, the greater the lever pressing. Egger and Miller found that the more informative cue was the stronger: the mean number of lever presses during a 10-minute period was 115.1 for the subjects who were presented with the first CS following a lever press; in contrast, the mean number of responses was 65.8 for the other subjects who were given the second CS.

To strengthen the claim that information value is more critical to Pavlovian conditioning than mere CS-US contiguity, Egger and Miller tested additional subjects. They were treated in essentially the same manner as before but this time one group of subjects occasionally experienced the first CS during the intertrial interval. This procedure had the effect of rendering the first stimulus unreliable; the animals could not depend upon that stimulus as a signal for food because it often occurred without being followed by the US. If the first CS was unreliable, then the second stimulus should become the important informative cue for the animals. This is precisely what Egger and Miller found. The number of responses to the first stimulus now dropped to a mean of 76.1, whereas the formerly weak stimulus, the second CS, gained in strength: animals presented with this stimulus after a lever press response performed an average of 82.6 lever presses during the 10-minute test. In summary, the Egger and Miller study was an important precursor to Rescorla's theory because it highlighted the notion that Pavlovian conditioning involves the acquisition of information; cues that are reliable and predictive are the ones that gain strength. (For additional experiments on CS predictive value see Ayres, 1966; Burstein & Moeser, 1971; Libby & Church, 1975; Nageishi & Imada, 1974; Seger & Scheuer, 1977; and Seligman, 1966.)

CS-US Interval

Although the correlation between a CS and US is most essential to conditioning, the contiguity, the temporal closeness of the two stimuli, is also a factor of great importance. A stimulus may be perfectly informative in that it reliably signals a pending US. At the same time, however, the US may be delayed sufficiently so that conditioning is retarded; when the two stimuli are not contiguous, conditioning suffers.

There appears to be three general divisions or categories of responding relevant to the CS-US interval. The first category is motor or skeletal responses, such as the conditioned eyelid reaction. Ross and Ross (1971) used both trace and delay procedures and found that the optimum interval for eyelid conditioning in college students was about .5 second. Conditioned responding decreased sharply with intervals shorter than .5 second, but somewhat more gradually with intervals ranging up to 1.4 seconds.

The optimum CS-US interval for eyelid conditioning in rabbits appears to be similar. Smith, Coleman, and Gormezano (1969) and others (Coleman & Gormezano, 1971; Smith, 1968) have suggested that conditioning is best with intervals of .2 to .4 second. Little conditioning is evident with a CS-US interval of .06 second or less (Salafia, Lambert, Host, Chiaia, & Ramirez, 1980). Levinthal (1973) has shown that the optimum interval may depend in part upon the number of trials given to the subject per session.

The conditioning of another skeletal response, the jaw movement in rabbits, differs somewhat from eye blink conditioning in terms of the optimum CS-US interval. Gormezano (1972), for instance, has shown that conditioning was maximum for an interval of 1 to 4 seconds; conditioning was relatively poor when using shorter intervals like .5 second that were best for eyelid conditioning.

The second category of classical CRs is the autonomic or visceral responses. Much research has attempted to assess the optimum CS-US interval for these behaviors as well. Generally, the interval which is optimum for visceral conditioning is longer than that found optimal for skeletal conditioning. For example, heart rate in rats is conditioned with a CS-US interval of 6 seconds (Fitzgerald & Martin, 1971; Fitzgerald & Teyler, 1970), 2 to 6 seconds in rabbits (Deane, 1965), 2.5 to 10 seconds in dogs (Black et al., 1962), and 13 seconds in humans (Hastings & Obrist, 1967). In salivary conditioning, also an autonomically controlled behavior, Ellison (1964) demonstrated effective conditioning when the interval between the CS offset and the US onset was 8 seconds; some conditioning was even found when the interval was 16 seconds. Although most of these studies indicate that visceral conditioning is different from skeletal conditioning in terms of the optimum CS-US interval, GSR conditioning, which is also an autonomic response, paradoxically appears to be best with a .5 second interval (Prokasy, Hall, & Fawcett, 1962; but see Badia & Defran, 1970).

Finally, the third major category of classical conditioning that differs from the other forms in terms of the CS-US interval is taste aversion learning. Recall that in taste aversion learning an animal is given a flavor CS followed by an illness-inducing drug. A form of conditioned nausea may take place (Garcia, Hawkins, & Rusiniak, 1974; Zahorik, 1972). In addition, the animals later demonstrate a strong aversion to the flavor, indicating that the originally "neutral" flavor has indeed become aversive. The remarkable nature of taste aversion is that the maximum CS-US interval which still allows for learning is vastly different from the conditioned reactions previously discussed. In particular, many studies have demonstrated that the maximum possible interval is several hours rather than seconds (see, for example, Andrews & Braveman, 1975; Garcia, Ervin, & Koelling, 1966; Kalat & Rozin, 1971; and see Rozin & Kalat, 1971; Riley & Clarke, 1977 for reviews). In the study by Kalat and Rozin (1971), for instance, different groups of rats were given sucrose water followed by an injection of lithium chloride, a drug which induces nausea in the animals. The two stimuli were separated by either .5, 1, 1.5, 3, 6, or 24 hours. Two days after this training session, each subject was tested for an aversion to the sucrose solution. As shown in Figure 3–2, Kalat and Rozin found that the percentage preference for sucrose (relative to water) increased as a function of the CS-US interval. Although the strongest aversion was in the .5 group, learning was possible with US delays even up to 3 hours. In fact,

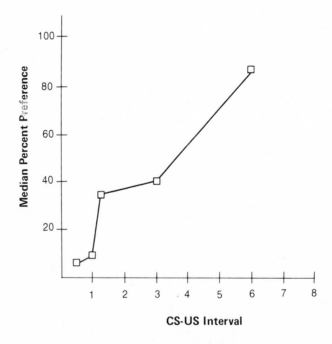

Figure 3-2. Median percent preference for the sucrose solution as a function of the CS-US interval (in hours). (From Kalat & Rozin, 1971)

other studies (Andrews & Braveman, 1975) have found that stimuli may become aversive even when the US is delayed as long as 12 to 15 hours. These experiments, of course, confirm the importance of CS-US contiguity: conditioning is stronger when the two stimuli occur close together in time. However, unlike visceral and skeletal response conditioning, taste aversions may be learned even when the US is delayed for many hours.

Why should there be such a vast discrepancy between conventional CRs and taste aversion reactions in terms of the CS-US interval? The answer appears to be related to evolutionary principles. Classical conditioning in general involves the prediction of important biological events. Food substances that are potentially harmful are biologically important and therefore should be predicted by the animal if it is to avoid consuming those poisons in the future. But since the effects of poisoning rarely occur immediately, it is adaptive for animals to be able to associate the flavor of a food with its consequences over a long period of time. Thus, animals who could bridge a long interval between the CS and US and still associate the flavor with illness were at a selective advantage during evolution. They were more likely to pass their ability on to future generations.

CS Intensity

The third major factor or characteristic of the CS which affects conditioning is the strength of the conditioned stimulus (see Gray, 1965; Grice, 1968). The general finding has been that conditioning increases as a function of CS intensity. Both Beck (1963) and Moore (1964) have shown that human eyelid conditioning is enhanced by more intense CSs. Similar effects using the CER technique have been shown by Kamin (1965) and Kamin and Schaub (1963).

The notion that conditioning should be related to CS intensity makes intuitive sense. Animals live in a stimulus environment and are constantly bombarded by CSs. The more salient or intense a specific CS is relative to what the animal has become accustomed to, the more it will demand attention, be better isolated, and thus be better conditioned. In other words, CSs that are more intense are more easily discriminated from background stimulation; if they are better discriminated, then they are isolated phenomena that have a greater chance of becoming connected to the US.

Such a theory about CS intensity has been supported by a number of experiments. Kamin (1965), for instance, provided an 80-db white noise to several groups of rats; this noise constituted a background level of stimulation. Using a CER technique, Kamin then presented a discrete CS followed by a brief shock. Specifically, the CS was a reduction in the background noise to 70, 60, 50, 45, or 0 db. The degree of lever press suppression, that is, the standard CER index, showed that the larger changes resulted in greater conditioning. Minimal suppression was observed when the CS was a change in noise intensity from 80 to 70 db, but almost complete suppression was observed when the CS was a reduction from 80 to 0 db. In summary, then, CS intensity has a dramatic effect on conditioning; it appears to affect the subject's discrimination of the CS from the background cues. (See the section on context conditioning later in this chapter for more information about this phenomenon.)

When we consider a more biologically oriented learning phenomenon, namely, taste aversion learning, CS intensity has quite a different meaning. Does one mean by CS intensity the quantity of solution that is provided during training or the concentration of the flavor solution? Both factors have been investigated, and both seem to affect conditioning strength. In the case of concentration, the point is rather clear: taste aversion conditioning is a direct function of flavor concentration (Barker, 1976; Dragoin, 1971). The Barker study, for example, showed that only a few seconds exposure to a very concentrated flavor solution was needed in order to produce maximum conditioning.

In the case of CS amount, the picture is somewhat less clear. A number of studies have shown convincingly that taste aversion conditioning is a positive function of the amount of flavor consumed at training (Barker, 1976; Bond & Digiusto, 1975). In contrast, other investigators have found that CS amount is either unimportant (Kalat, 1976) or that larger amounts actually produce less conditioning—at least when relatively long CS-US intervals are used (Deutsch, 1978). It would appear that CS amount does indeed relate to conditioning level and, in this sense, the amount of flavor functions in a manner similar to the concentration of the flavor. However, with relatively long CS-US intervals, conditioning is weakened with greater CS amounts.

CS Quality

Another extremely important parameter affecting conditioning is the very quality of the CS. Let us first consider the quality of the CS in taste aversion learning. Kalat and Rozin (1970) discovered that not all flavors are equivalently associable to a poison US; paired with an illness-inducing drug, some flavors "naturally" seem to become more aversive than other flavors. The researchers term this a difference in salience; the quality of the CS, or its salience, determined, in part, how strong conditioning would be quite independent of the flavor's other parameters such as its concentration or quantity.

The salience effect has been explained in several ways. First, Kalat (1974) claimed that the novelty of the flavor was really the inherent dimension of salience. In his study, animals that were raised with a particular flavor showed conditioning to weaker concentrations of that flavor; clearly conditioning in that case was a function of the novelty of the CS rather than its concentration. In summary, Kalat's position, therefore, is that salient flavors are those that are more novel to the subject; they are relatively unfamiliar compared with the flavors that the animal has experienced. Flavor CSs that are low in salience, on the other hand, tend to be like those that the animal has had in the past.

Salience may also be due to other dimensions of the CS flavor. Best, Best, and Lindsey (1976) showed, for instance, that many taste stimuli actually contained an olfactory component. The animals not only tasted the stimulus but also smelled it. When the subjects were deprived of their olfactory sense, that is, when they could no longer smell the stimulus because of a special operation performed on their olfactory nerves, the salience of many CSs was reduced. This study strongly suggests that salience is, in part, due to the olfactory component of a taste stimulus.

The quality of the CS has an important bearing on conditioning in conventional Pavlovian studies as well. Holland (1977) put rats in a stabilimeter and gave them Pavlovian conditioning using either a tone or a light CS and a food US. What he observed was changes in both gross activity and other specific motor movements during the CS. Holland demonstrated that the tone elicited qualitatively different CRs than did the light: the tone tended to elicit head jerks when it was sounded prior to food, whereas the light elicited more rearing postures and investigations of the food tray. In other words, the activity of the animal depended upon which CS was used.

Some studies have suggested that the way a CS operates during conditioning is related to the instinctual or naturalistic tendencies of the animal. For example, Allison, Larson and Jensen (1967) noted that rats are nocturnal animals and therefore prefer dark areas. The Allison et al. study demonstrated that the rat's preference for black (for example, the black side of a two-compartment box) was enhanced following a shock. The authors claimed that the subjects are naturally prepared to learn that white is associated with aversive states and black with safety.

This finding, the enhancement of a natural preference for dark areas following shock, was amplified in a study by Welker and Wheatley (1977). Their experiment was a typical CER study using two different qualities of a CS. One group received an increment in the illumination level; for these subjects the light was made brighter. The other group received a decrement in illumination level; for them the background lighting was dimmed. As shown in Figure 3–3, Welker and Wheatley found that there was greater suppression of lever pressing when the CS was an increment in luminance than there was when the CS was a decrement in light level. Interestingly, these two stimuli did not differ when the US was food; the difference was observed only when the cues were used to signal an aversive US. Those authors came to the same conclusion as Allison et al.: since rats are nocturnal, they are quite ready to learn that a bright CS signals an aversive US, but the animals are less capable of learning that a dim or black stimulus is

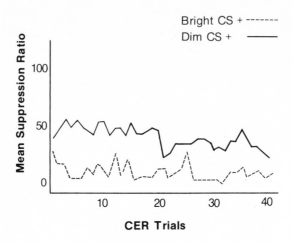

Figure 3-3. Mean suppression ratio as a function of successive CER trials for the bright CS and dim CS groups. (From Welker & Wheatley, 1977)

a signal for an aversive state. For them, in nature, darkness has always been associated with safety. The lesson that we derive from these studies is not simply that light CSs of different intensities differ in terms of their ease of conditioning in an aversive situation. We also derive a more general lesson, namely, that the quality of the CS, relative to the animal's evolutionary history, is an important dimension in Pavlovian conditioning.

The studies cited above suggested that the success of the CS in conditioning reflects innate dispositions of the animal. In particular, some CSs may have a special "affinity" for certain USs. (See Chapter 6 for an elaboration of these arguments.) But there is another point of view to be considered. Associations between USs and CSs may develop in accordance with the temporal patterning, intensity patterns, and location of the two stimuli. For instance, Testa (1974, 1975) claimed that animals have an inclination to associate stimuli when they appear in the same location or when they display the same temporal or intensity patterns. In one CER study, Testa (1975) employed four kinds of light CSs. For two of the groups, the light was located at the bottom of the cage underneath the grid floor. For the other two groups of animals, the light was located at the top of the testing chamber. One group in each of the above conditions had a constant light, whereas the other group in each condition had a pulsated light. Testa used a pulsating shock as the US.

Testa demonstrated that conditioning was stronger when the light was located at the bottom of the cage and when it pulsated. These procedures led to stronger conditioning because there was a match or similarity between the CS qualities and the US qualities: both were located "near the floor" and both pulsated. The implication is that an animal is better able to associate the US and CS because the two are similar along those important dimensions. What Testa's experiment demonstrated was not that stimuli have a particular affinity for one another; animals are not necessarily "prewired" to associate one stimulus with another. Rather, Testa showed that animals *are* inclined to associate stimuli when their location and temporal-intensity patterns are similar.

Finally, in discussing CS quality let us review the previously mentioned experiment by Keith-Lucas and Guttman (1975). Those authors gave rats a novel stimulus panel followed by shock, and, a few seconds later, a backward CS. The backward CS was a rubber toy in the shape of a hedgehog. Keith-Lucas and Guttman found that the backward CS gained greater strength than the forward CS (the novel stimulus panel). That is to say, a naturalistic stimulus object was conditioned more strongly than an unnatural arbitrary stimulus.

US Intensity

One of the most reliable effects in Pavlovian conditioning is the relationship between the CR and the intensity of the US. The fact that conditioning is stronger with more intense USs appears to hold over a wide variety of responses and species. Let us consider four principal forms of classical conditioning. First, the intensity of shock affected the rate of eyelid conditioning in rabbits (Smith, 1968) and the conditioned jaw movement responses (Sheafor & Gormezano, 1972). Eyelid conditioning in humans is affected, too; the CR level increased as a function of the strength of the air puff to the eyelid (Spence & Platt, 1966).

Second, similar relationships have been observed for conditioned visceral reactions. For example, the strength of salivary conditioning is directly related to the magnitude of the food US (Wagner, Siegel, Thomas, & Ellison, 1964) and to the concentration of an acid US (Ost & Lauer, 1965). Conditioned heart rate is a visceral response that also fluctuates with US intensity. Fitzgerald and Teyler (1970) have found that the amplitude of the change in heart rate increased as a function of shock intensity. As shown in Figure 3–4, mean change in a conditioned heart rate CR increased as a function of the US strength using both trace and delay procedures (Fitzgerald & Teyler, 1970). Although there may be important interactions between the quality of the CS and the intensity of the US, (see, for example, de Toledo, 1971), the generalization that conditioning is stronger with more intense USs is not in doubt.

It has long been recognized that shock intensity is an important parameter in CER conditioning. Annau and Kamin (1961), for example, showed that suppression of lever pressing increased with shock intensity.

Finally, the fourth type of conditioned reaction that shows a regular

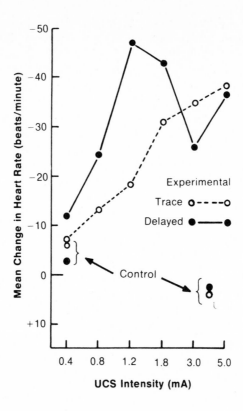

Figure 3-4. Mean change in heart rate as a function of the US intensity for both trace and delayed procedures. (From Fitzgerald & Teyler, 1970)

relationship with US intensity is taste aversion learning. A number of studies have demonstrated that more intense USs, that is, more concentrated illness-inducing drugs, lead to greater aversions. This appears to be the case for a variety of poisoning agents (see Andrews & Braveman, 1975; Nachman & Ashe, 1973). As shown in Figure 3–5, mean intake decreases as the concentration of the US increases.

The preceding information indicates that US intensity is important to conditioning. But consider a trial on which the animal responds to the CS prior to the US presentation. How is it that the CR level can reflect the intensity of the US which is yet to be delivered? The answer appears to be that animals have a memory for the US which is based, in part, upon the strength of the US. If that memory is degraded or enhanced, the conditioned reaction changes appropriately. In other words, if the animal experiences some procedure which changes its estimation of the US intensity, its memory for the US will change accordingly; later, when the CR is tested, the level of the responding will also have changed because of the change in the memory for the US.

Let us consider an experiment in which the responding decreased following a degradation of the US memory. Rescorla (1973a) used a CER technique in which the CS was a flashing light and the US was an intense noise. During acquisition, suppression of the lever pressing increased, indicating that conditioning took place in a normal fashion. One group of subjects was then given a habituation procedure: during 3 sessions, 24 unsignaled presentations of the noise US were given while the animal pressed the lever for food. This procedure was designed to habituate the animal to noise and thus to diminish the subjective intensity of the noise.

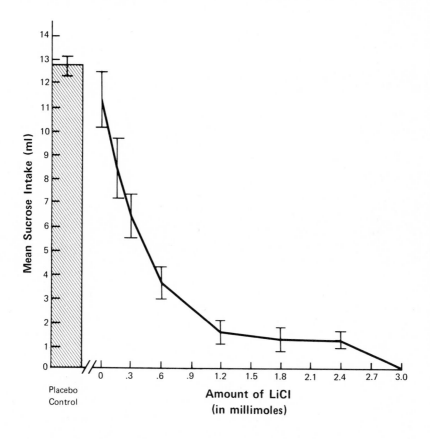

Figure 3-5.　Mean sucrose consumption as a function of the amount of lithium chloride US present in the injection (expressed in millimoles). Concentrations varied from .15 M to .65 M and injection volumes varied from 0 to 20 ml/kg. (From Nachman & Ashe, 1973)

(This procedure would be similar to subjecting oneself to an unpleasant stimulus like a cold-water shower in order to "get used to it.") A second group did not receive this habituation training. In a third phase of the study, Rescorla measured the conditioned reaction to the light CS in all animals. A clear difference was observed. Suppression remained almost complete in the control group, but lever pressing recovered rather quickly in the habituation group. The conclusion was that habituation of the US caused the animal to adjust its memory of the US intensity; this, in turn, changed the conditioned responding to the CS. In other words, conditioning does not depend merely on the physical intensity of the US but also on the animal's memory of the US experience (see also Holland & Rescorla, 1975b).

Rescorla (1974) has also demonstrated the opposite finding, that if the subject's memory for the US intensity is enhanced, then the conditioned reaction is strengthened. In this study, Rescorla used a CER technique in which the US was a shock of medium intensity. In some groups of animals, a more intense shock was given either before or after the conditioning session. This procedure had the effect of increasing suppression on the CER test because it altered the subject's memory of the US. In other words, conditioned responding to a CS based upon a US of moderate intensity was enhanced when the animal experienced an even stronger shock later on in a different situation.

Finally, a factor which influences the US intensity is the degree to which it is associated with stimuli that are opposite to its nature. Generally, unpleasant stimuli like shocks tend to be reduced in intensity when they are associated with food (see Dearing & Dickinson, 1979; Dickinson & Pearce, 1976, 1977). In one study, Pearce and Dickinson (1975) showed that the aversiveness of shock is reduced when it is paired with food; moreover, the shock US is unable to function as effectively in a CER experiment after this counterconditioning procedure. Again, the principle that we learn from these studies is that US intensity is a function of more than just the physical parameters of the stimulus; USs also may be modified when they are associated with stimuli opposite to their nature.

US Duration

A second major factor concerning the unconditioned stimulus is its duration. Many studies have shown the effect of US duration on Pavlovian conditioning. For instance, Frey and Butler (1973) trained an eye blink response in rabbits using shock durations of .05, .1, and .2 second. All the animals acquired the conditioned eyelid reaction although subjects in the .05 group were slower to develop the CR.

Similar results were obtained by Riess and Farrar (1973). Using a CER technique, these authors paired a light CS with a shock US, the shock lasting either 0, .5, 1, 2, or 3 seconds. Suppression of lever pressing indicated that US duration exerts a strong effect on conditioning. Subjects in the .05 group showed almost no suppression of lever pressing; disruption of the lever press increased as a function of duration so that the 1- and 3-second subjects were maximally suppressed after about 5 sessions of training. Recent evidence, therefore, appears to support the generalization that the strength of a Pavlovian CR is a positive function of the US duration.

US Predictability

A third important factor is US predictability. It is not surprising that the occasional omission of the US during conditioning weakens the CR because omission reduces the correlation between the CS and US, thus degrading the essential ingredient for conditioning. This principle seems to be true in human eyelid conditioning (Ross & Hartman, 1965) as well as in salivary conditioning (Fitzgerald, 1963; Sadler, 1968; Wagner et al., 1964). Thus, the acquisition of a classically conditioned response is a function of the correlation between CS and US. The greater the number of US omissions, the lower that correlation (but see Thomas & Wagner, 1964; Vardaris & Fitzgerald, 1969; and Willis, 1969).

US Quality

Finally, an important US factor that affects conditioning is the quality of the US, or, in some cases, the degree to which the US is a naturalistic stimulus. In a 1979b study by Holland a tone or a light CS was paired with several kinds of USs. Some animals received a food pellet delivered to a recessed trough; other subjects were given a food pellet in a tray inside the cage; and a third group was presented with a sucrose US inside this tray. Holland discovered that the type of CR depended, to a large degree, upon which US was used. In fact, the type of CR that was learned resembled the form of the UR that was elicited by the particular US. For example, analysis indicated that CRs resembled startle reactions when the US was a food pellet delivered to the recessed trough. The CRs, however, involved grooming behavior when the US was sucrose. In the group that received a food pellet in the open food cup as a US, the CR involved orientation behaviors to the food cup.

These results are, perhaps, not very surprising. It has long been known that the US essentially determines the nature of the conditioned reaction. We therefore have come to expect different reactions when using different USs—shocks lead to CRs that are quite unlike the conditioned reactions produced by food. Despite this long-standing principle, Holland's experiment is important because it illustrates that even subtle differences between USs can produce substantially different behaviors.

The importance of US quality was demonstrated in a different way by Jenkins and Moore (1973). A number of studies in the last decade have shown that pigeon key pecking is a Pavlovian reaction. When a plastic disk is illuminated from behind and food is presented soon thereafter, pigeons reflexively begin to peck at the disk. This Pavlovian phenomenon is similar to conditioned salivation in its methodology and theory (see Chapter 6 for more details). Jenkins and Moore showed that the form of the CR depended upon the quality of the US. In one study, the illumination of a disk (CS) was followed by a water US; other subjects in the study, who were deprived of food, received a food US. All subjects learned to peck the disk (the classical CR), but the form of the peck response varied according to the type of US used. As shown on the left of Figure 3–6, the subjects that received the food US made sharp and vigorous pecks at the disk; the response resembled the animals' natural reaction in securing food. In contrast, the thirsty subjects, shown on the right of Figure 3–6, that received the water US, made much slower reactions; their pecks were more sustained. Their beaks were often open and the pecks were accompanied by drinkinglike movements of swallowing and licking. If one were to look superficially at the two groups of animals, one might conclude that the CR was equivalent

Figure 3-6. Illustrations showing the form of the pigeon's key peck when the US is food (left) and when the US is water (right). These CRs resemble the animal's natural URs to those USs. (From Jenkins & Moore, 1973)

even though different USs had been employed. However, upon precise examination, it is clear that qualitatively different USs lead to qualitatively different CRs. In this respect, the principle concerning US quality appears to be the same as the CS quality: the form of the CR depends to a large degree upon the nature of the US.

Intertrial Interval

The intertrial interval is a situational variable that affects classical conditioning. Most of the experiments on this topic have involved eyelid conditioning either in human beings or rabbits. It appears that short intertrial intervals produce inferior conditioning in humans (Spence & Norris, 1950) and rabbits (Salafia, Mis, Terry, Bartosiak, & Daston, 1973) although the decrement appears to occur only with very short intervals. Specifically, intertrial intervals ranging from about 5 to 20 seconds produce inferior conditioning in humans (Prokasy & Whaley, 1963) and intervals of less than 15 seconds produce inferior conditioning in rabbits (Salafia et al., 1973). Studies on the variability of the intertrial interval have produced conflicting evidence. Generally it is thought that variability has little effect on conditioning (Prokasy & Chambliss, 1960; Prokasy & Whaley, 1961) although recently Salafia et al. (1973) have shown a slight superiority for a constant intertrial interval relative to a variable intertrial interval.

The intertrial interval has been investigated recently in the context of the pigeon's key peck response. Recall that pigeons peck at an illuminated plastic disk whenever the light behind the disk is followed by food reward; the simple pairing of these two events causes pecking behavior. The rate at which a pigeon begins such pecking varies as a function of many parameters; one parameter is the intertrial interval. Terrace, Gibbon, Farrell, and Baldock (1975), for example, showed that when the intertrial interval was about 10 to 18 seconds, about 70 to 130 light-food presentations were required before the birds would make their first peck. However, with intervals ranging from 150 to 400 seconds, only 10 to 20 presentations, on the average, were necessary. The rate of key peck acquisition is an inverse function of the intertrial interval. While this inverse relationship does appear to hold in general, the duration of the presentation, specifically the length of the light CS, is also involved (see, for example, Gibbon, Baldock, Locurto, Gold, & Terrace, 1977). In particular, the intertrial interval length *relative to* the trial duration determines the speed of acquisition. Pigeons begin to peck sooner with longer intertrial intervals, but when the trial duration is extended, then proportionately longer intertrial intervals are needed to maintain the same learning rate. In other words, it is not just the time between trials that is important, but rather the time between trials relative to the CS length that is crucial. Different time values may be used but the rate of learning will be constant provided the ratio of values is constant.

Basic Pavlovian Phenomena

Up to now, we have discussed the most general notions concerning classical conditioning. Actually years of research have brought to light a number of interesting phenomena beyond the mere fact that conditioning takes place when a CS is correlated with a US. Some of the examples that follow do not conform exactly to the principles outlined in previous sections; they are either extensions of those principles or exceptions to the rule.

Second-Order Conditioning

One of the most important of the Pavlovian phenomena is second-order, or higher-order, conditioning. Second-order conditioning occurs whenever a new CS is paired with an already powerful CS. The two stimuli become associated so that the new CS acquires the power to elicit the original CR. In other words, novel CSs may gain excitatory strength not because they are paired with a biologically potent US, but because they are consistently correlated with a "surrogate" US, an already powerful CS that was previously conditioned to a US. In one sense, this phenomenon is an exception to the rule because it is an instance of conditioning without the use of a biologically powerful US.

Second-order conditioning is an extremely important learning ability. Think back to the past 24 hours. How many unconditioned stimuli have you experienced? Surely you have had some food, and perhaps you have experienced some unpleasant events. But often these USs are not paired with any one CS, at least they are not so paired repeatedly. That is to say, strong biological events do not pervade our lives so extensively that they can account for all the associations we develop. Let's state the argument in the opposite direction. There are so many signals, so many meaningful CSs, in our lives that it is hard to conceive how they could have developed their strength through association with USs. Many of them are far removed or distant from USs. And many of these CSs must have gained their strength through second-order conditioning. In summary, then, second-order conditioning is important because it provides a mechanism for the establishment of associations that do not depend upon a US presentation.

Various techniques have been used to illustrate second-order conditioning. One approach is to use aversive conditioning such as the CER technique (see Bond & Harland, 1975; Kamil, 1968, 1969; Rescorla, 1973b; and Szakmary, 1979). Consider the CER study by Rizley and Rescorla (1972). Rats were first taught to press a lever in order to receive food. Once the rate of lever pressing had stabilized at a high level, the authors conducted a Pavlovian conditioning experiment in which lights and tones served as the CSs and a mild shock as the US. There were a number of groups. During the first phase of the experiment, the second-order conditioning group (Group E) received light-shock pairings (CS–US pairings); in the second phase, the tone CS was paired with the now potent light CS (CS_2–CS_1 pairings). Later, as shown in Figure 3–7, the presentation of the tone (the second-order cue) during lever pressing caused suppression of the lever press behavior, indicating that the tone had, indeed, acquired the ability to elicit the CR.

Several important control groups were run to confirm that the tone and light had become associated during Phase 2. One control group (Group C1) received the original CS_1–US pairings in Phase 1 of the study but random presentations of the tone and light in Phase 2. The purpose of this group was to demonstrate that although the light had become a powerful CS during Phase 1, the acquisition of strength to the tone depended upon the correlation between the tone and light in Phase 2. If the tone were able to elicit suppression of lever pressing *without* being paired explicitly with the light during Phase 2, then one would be forced to conclude that the effect was not true conditioning.

A second control group (Group C2) received random presentations of the CS_1–US in Phase 1 but CS_2–CS_1 pairings in Phase 2. Under these

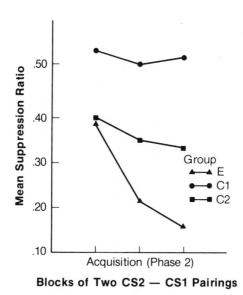

Figure 3-7.   Mean CER suppression ratio during three blocks of second-order conditioning trials. (From Rizley & Rescorla, 1972)

conditions, the CS should fail to develop sufficient strength during Phase 1 because it is not consistently paired with the US. If the CS_1 is not powerful, then it cannot "transfer" its power to a new CS during Phase 2.

As shown in Figure 3–7, no conditioning was evident in either of these control groups. It would therefore appear that second-order conditioning is a "true" conditioning phenomenon: the development of a CR to a second-order stimulus requires not only that the CS_2 be paired with the first-order CS_1 (Phase 2 of the study), but also that the original CS_1 be paired with a US (Phase 1). A deficiency in *either* of those phases, as shown in the control groups, results in a lack of second-order conditioning.

There are other examples of Pavlovian second-order conditioning using an appetitive US. Holland and Rescorla (1975a) performed a study that was essentially like the one just described, except that an appetitive US was used during Phase 1 and gross locomotor activity rather than suppression of lever pressing was measured. The results of the study confirmed the generality of second-order conditioning. Subjects who received the light-food pairings in Phase 1 followed by clicker noise-light pairings in Phase 2 showed a conditioned reaction to the clicker CS. Control groups who received explicitly unpaired presentations of the two stimuli during either of the phases failed to show second-order conditioning.

It is also interesting to note that second-order conditioning is not confined to excitatory CSs. Rescorla (1976), for instance, demonstrated that an inhibitory CS_1 can become associated with a novel CS_2 such that the new CS_2 will demonstrate second-order conditioned inhibition. Judging from these results, it would seem that virtually any conditioned stimulus, either excitatory or inhibitory, is capable of being used to support second-order conditioning.

Surprisingly little is known about the conditions that influence second-order conditioning. It is presumed that the training conditions that affect first-order conditioning influence second-order conditioning in much the same way. Maisiak and Frey (1977) have shown that second-order conditioning is stronger when the CS_2 and CS_1 overlap. Rescorla and Furrow (1977) found that second-order conditioning was stronger when the two CSs were similar (same sensory modality) than when they were dissimilar. That is, second-order conditioning to a tone CS_2 was much stronger when the CS_1 was also a tone; similarly, conditioning was stronger when the two stimuli were lights. Much weaker effects were observed when the two stimuli differed.

Perhaps the most important finding concerning second-order conditioning involves the relationship between the strength of the CS_2, the second-order cue, and the continued strength of the CS_1. Several studies have demonstrated that the CS_2 strength is unaffected when the original CS_1 is weakened. For example, in the Holland and Rescorla (1975a) study, one group received standard second-order conditioning, that is, conditioning of the CS_1 in Phase 1 followed by second-order conditioning in Phase 2. Another group received the same treatment except there was an additional Phase 3: in this phase, the CS_1 was extinguished. Subsequent measurement of the CS_2 strength indicated that extinction of the first-order CS_1 had no effect on the strength of CS_2. In other words, second-order conditioned stimuli are impervious to changes in the CS_1 strength, once the second-order conditioning phase has been completed.

This generalization is dramatically different from what is found in first-order conditioning. For example, consider the 1975b study by Holland and Rescorla. In this study, the animals received first- and second-order conditioning in a normal fashion. One group then was satiated following second-order conditioning. This reduction in hunger affected the strength of the first-order stimulus: animals no longer performed the CR in the presence of that stimulus. Interestingly, however, satiation did not substantially change the strength of the second-order cue; even though the animals were no longer hungry, they performed the CR when the second-order cue was presented. It appears, therefore, that a reduction in the strength of the US (food is not a very potent stimulus to an animal who is not hungry) affects the strength of the first-order conditioned stimulus, but not the CS_2 (see also Rescorla, 1973a).

The reason for such a dramatic divergence between first- and second-order conditioning is not entirely clear. Rescorla believes that first-order conditioning is an association between two stimuli, the CS and US. Second-order conditioning, on the other hand, is an association between a CS and the conditioned reaction. Although such a dichotomy seems possible, it would appear that *all* associations could involve associations between stimuli as well as associations between a stimulus and a reaction. The separation observed by Rescorla and his colleagues may, in fact, be due to their particular testing procedure rather than to any basic difference between the two forms of learning.

Sensory Preconditioning

An important phenomenon that is closely related to second-order conditioning is sensory preconditioning (see Seidel, 1959; Thompson, 1972). A typical sensory preconditioning study is conducted as follows: two CSs are paired, then one of the CSs undergoes conventional classical conditioning using a US; in a third phase one observes that the other CS, the one that had not been directly paired with the US, also has the ability to elicit the CR. It should be noted that sensory preconditioning is almost identical to the second-order conditioning studies just discussed. The principle difference is in the reversal of the first two stages. In sensory preconditioning, the two neutral CSs are first paired and then in Stage 2, one of them becomes associated with a US. In second-order conditioning, however, Stage 1 involves the association of a CS and US, and Stage 2 entails the pairing of the two CSs.

The fundamental importance of sensory preconditioning lies in showing that *any* two stimuli, even innocuous CSs, can become associated. That is to say, Pavlovian conditioning normally involves a strong biological stimulus, but when special precautions are taken it is possible to observe that associations develop between two CSs as well.

A good example of sensory preconditioning was shown by Rizley and Rescorla (1972). They used a CER technique to investigate this phenomenon. Three groups of rats were used. One group was the standard sensory preconditioning group; animals in this group received CS_2–CS_1 pairings in Phase 1, CS_1–US trials in Phase 2, and then the CS_2 presentations in Phase 3 (the test of sensory preconditioning). Subjects in one control group received random presentations of the CS_1 and CS_2 during Phase 1, but otherwise the same treatment as the experimentals. This group, obviously, was designed to show that any sensory preconditioning that occurred in the first group required CS_2–CS_1 pairings in Phase 1. Animals in the other control group were treated like the experimentals, except that this third group received random CS_1–US presentations during Phase 2.

The results of the study indicated sensory preconditioning: the CS_2 elicited the conditioned reaction in Phase 3 of the study. It did so even though the CS_2 had never been paired directly with the US. In one sense, the explanation for sensory preconditioning is rather simple: it merely reflects the combination of two separate Pavlovian experiments into a single test. In the first "experiment" two innocuous stimuli are associated. In the second "experiment," the CS_1 is associated with the US. Not only does the CS_1 come to elicit the expectation of the US presentation, but a visible conditioned reaction develops. Finally, in the third phase of the study, the CS_2 elicits the expectation of the CS_1 (the memory of that stimulus) which is sufficient to trigger the expectation of the US. That is, the CS_1 is merely a mediating link between the CS_2 and the US. In summary, sensory preconditioning operates by establishing two expectancies; the first is an expectancy for another innocuous CS and the second expectancy is for the US. These expectancies can be "chained" together so that the presentation of the first stimulus, the CS_2, will elicit the overt behavioral CR that initially was given only to the CS_1.

Although sensory preconditioning seems to be a "simple" case of stimulus learning, nevertheless some investigators claim that it has unique properties. First of all, it has been shown that the optimum CS_2–CS_1 interval is 4 seconds (see Lavin, 1976; Wynne & Brogden, 1962). With shorter *or* longer intervals, the strength of sensory preconditioning decreases. This is unlike regular conditioning which shows an orderly gradient in which conditioning declines as a positive function of the interval. A second finding is that sensory preconditioning reaches its maximum strength within relatively few trials. For example, Prewitt (1967) discovered that four pairings of the two CSs were sufficient to produce a maximum effect; additional pairings were superfluous. Finally, Adamec and Melzack (1970) have shown that sensory preconditioning is stronger if the subjects are motivated during Phase 1 of the study than if they are satiated.

Perhaps the most important characteristic of sensory preconditioning, at least relative to second-order conditioning, is the fact that the CS_2 strength depends entirely upon the continuing strength of the CS_1. Consider the study by Rizley and Rescorla once again. In one of their treatments, subjects were given sensory preconditioning (CS_2–CS_1 pairings), followed by normal conditioning (CS_1–US pairings), and then extinction of the CS_1 strength prior to testing for sensory preconditioning (the elicitation of the CR by CS_2). A control group received the identical treatment except for the fact that no CS_1 extinction was given prior to the test. Recall our previous discussion that when such an extinction procedure was given following second-order conditioning, the strength of the second-order CS was unaffected. In sensory preconditioning, however, extinction of the CS_1 strength eliminates altogether the strength of the CS_2. This curious fact indicates that sensory preconditioning operates in a substantially different way from second-order conditioning. That is, CS_2–CS_1 associations develop differently depending upon whether the CS_1 already has acquired conditioned strength (the case of second-order conditioning) or whether the CS_1 is still weak (as in sensory preconditioning). If the CS_1 is already strong, then the strength is transferred to the CS_2. It is then possible to extinguish the CS_1 without affecting the strength of CS_2. In contrast, if the CS_1 is not strong when it is paired with the CS_2 but only becomes so in the future, then the CS_2 strength depends entirely upon the status of the CS_1; if the CS_1 is extinguished, the CS_2 suffers.

Inhibition of Delay

As we have discussed, a CS that reliably predicts the presentation of the US normally leads to excitatory conditioning. There are exceptions, however. Recall that Hinson and Siegel (1980) found that trace conditioning, the procedure that involves an empty interval or gap between the offset of the CS and the onset of the US, may lead, paradoxically, to conditioned inhibition. Inhibition of delay is another form of inhibitory conditioning that occurs when the trace procedure is used. More specifically, if the US onset is delayed for a substantial period of time following the CS termination, then over the course of training, the subject gradually comes to withhold or inhibit the CR for longer and longer periods of time. Ultimately, the excitatory CR is executed only just prior to the US presentation. Thus we have an example of both excitatory *and* inhibitory conditioning taking place with the same training conditions. The excitation stems from the fact that the CS and US are paired. The inhibition arises from the particular configuration: the US presentation is delayed. Although inhibition of delay is normally accomplished using a trace procedure, it can also occur with delay conditioning.

A recent study by Lynch (1973) illustrates this phenomenon. Dogs were trained to make a paw flex response; the CS was a tone and the US was a shock. Over the 30 days of training, Lynch measured the latency of the CR, that is, the time it took for the animal to flex its paw following the onset of the CS. For each trial, Lynch noted whether the CR latency fell within one of four categories: the CR could be given after 1 to 10 seconds, 11 to 20 seconds, 21 to 30 seconds, or 31 to 40 seconds. For each block of 50 trials, Lynch computed the percentage of CRs that fell within each category. If inhibition of delay were observed, then a greater percentage of CRs should occur in the later categories after extensive training. Correspondingly, the percentage of CRs that would be given immediately upon CS presentation, that is, in the 1 to 10- or 11 to 20-second categories, should decline over training if inhibition of delay were developing. This is precisely what Lynch discovered. As shown in Figure 3–8, during the first 50 trials only about 15 percent of the CRs were executed with a latency of 31 to 40 seconds. This figure increased with training so that on the last 50 trials, the dogs were performing about 75 percent of their CRs during this 31 to 40-second category. In short, the dogs continued to execute an excitatory response, but they gradually began to delay it until just prior to the US onset.

We know that inhibition of delay represents an inhibitory process, a temporary suppression of the excitatory CR, because when we distract the subject we interfere with that ongoing inhibitory process and observe, immediately, a full-blown excitatory CR. This procedure, called disinhibition (see Pavlov, 1927, p. 93), is the process of disrupting the inhibition and observing excitation. An example of this is found in Kimmel (1965). Kimmel conditioned the GSR response in humans using a red light CS and a shock US and found that inhibition of delay developed. The latency of the GSR response increased from about 2.7 seconds following the CS onset to about 4.1 seconds after 50 training trials. After 50 trials, he presented his subjects with a novel distracting cue (a tone) and observed that the latency of the response on the next 4 trials dropped from about 4.1 seconds to about 2.4 seconds. In other words, the inhibitory process that was causing the response to be delayed was disrupted by the tone; the excitatory CR, then, was performed immediately.

Why is it that animals show inhibition of delay? This phenomenon, which appears to be paradoxical on the surface because it mixes both excitatory and inhibitory processes, actually makes good sense from an evolutionary perspective. Recall that USs are strong biological stimuli that, in one sense, need to be predicted and dealt with by the subject. The gradual

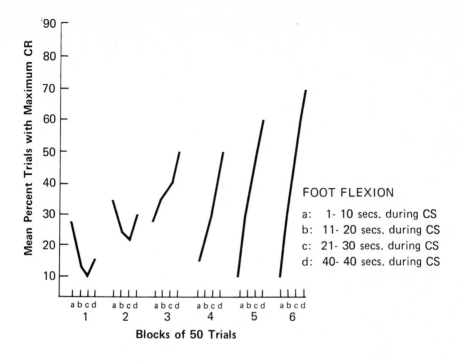

Figure 3-8. Mean percent maximum paw flex CRs executed with latencies of 1–10 (a), 11–20 (b), 21–30 (c), or 31–40 (d) seconds as a function of blocks of fifty training trials. (From Lynch, 1973)

delaying of the CR is a way of coping with the US. That is, by delaying the CR the animal may help to alter the US to its advantage. For example, the impact of a negative or aversive US might be lessened if the animal readied itself *just prior* to the US presentation rather than well before the US. Similarly, behavioral adjustments, including activity and salivation, might conceivably enhance the benefit from a positive or appetitive US. This is not to say that the animals have much control over the US; by definition, in Pavlovian conditioning, they do not. And we take great pains to develop methods that eliminate such control over the US. But there still could be some benefit derived by responding at progressively longer intervals.

Latent Inhibition

Another important inhibitory phenomenon that has been studied extensively in the last decade is latent inhibition. It is defined as the retardation of conditioning because the CS was preexposed to the subject prior to conditioning. That is, exposures of the CS alone will make it more difficult for that CS to become conditioned when it is later paired with a US.

Latent inhibition has been demonstrated in a wide variety of organisms using many different kinds of stimuli (see Lubow, 1973; Siegel, 1972b for reviews). For instance, Lubow (1965) used three groups of sheep and goats as subjects. One group was given normal Pavlovian conditioning using a light CS and a shock US. The leg flex CR increased over the 80 trials in a normal fashion. The other two groups of subjects, however, were preexposed to the light CS prior to conditioning. One group received 20 preexposures and the other group received 40. As shown in Figure 3–9, it was later observed that these preexposed groups conditioned rather poorly. The number of CRs exhibited during the 80 training trials was lower than in the nonpreexposed group. Prior exposure to the CS, therefore, made the subjects relatively insensitive to excitatory conditioning procedures later on.

Another important example of latent inhibition comes from the taste aversion literature. Numerous investigators (for example, Siegel, 1974) have shown that experience with a flavor makes it difficult for that flavor to be used in a taste aversion experiment; animals later fail to learn that the flavor signals poison and malaise because they have been preexposed to that flavor before conditioning. The interesting thing about this example is that latent inhibition, the retardation of conditioning, can occur following a single preexposure to the flavor. This contrasts with the latent inhibition results discussed above. There, approximately 15 to 20 CS preexposures were required before retardation was observed (Lubow, 1973). Regardless of the number of CS preexposure trials required to produce the retardation effect, latent inhibition lasts for quite a while. For example, Crowell and Anderson (1972) found that a noise CS that had been preexposed prior to conditioning retained its latent inhibitory properties for a week. In a study with flavors, Siegel (1974) showed that latent inhibition can last for 24 days.

One of the more interesting characteristics of latent inhibition is that it is positively related to the CS intensity. The stronger the CS, the greater the retardation effect. In the study by Crowell and Anderson (1972), for instance, groups were first preexposed to a noise CS that differed in intensity (either 70 or 100 db). A control group was not given any preexposure to the noise. Then, in Phase 2 of their study, the noise CS and a shock US were paired in a separate box. Under these conditions, the CS normally acquires aversive properties and can serve to suppress ongoing behavior when superimposed on that behavior (this is the typical CER procedure). Finally, in Phase 3, the CS was given while the animals were licking a water tube. Suppression of licking was observed, but the amount of suppression was much less for the preexposed groups; this, of course, shows that latent inhibition had occurred. Interestingly, greater suppression was observed in the group that received the more intense CSs. In other words, the stronger the preexposed CS, the more learning is retarded to that CS later on.

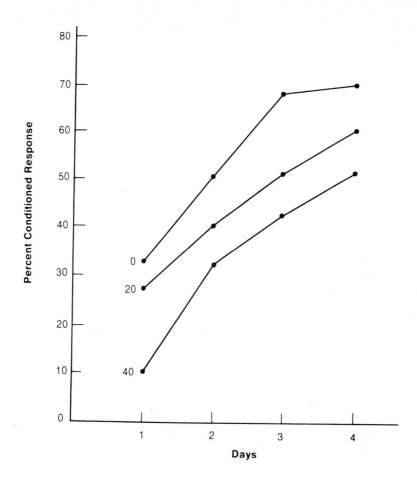

Figure 3-9. Percentage of conditioned responses executed on four acquisition days as a function of the number of prior CS exposures. (From Lubow, 1965)

Some recent studies on latent inhibition include procedures that counteract the effect (Best, Gemberling, & Johnson, 1979; Lubow, Schnur, & Rifkin, 1976). One of these techniques involves presenting the animal with another stimulus right after the CS is presented during the preexposure phase. That is, rather than merely preexposing the CS alone, the experimenter follows presentation of the CS with a second stimulus. When this is done, and the CS is later used in a conditioning experiment, latent inhibition fails to occur. In other words, the mere presentation of a second stimulus during the preexposure phase is sufficient to eliminate the latent inhibition effect. This disruption of latent inhibition occurs whether the CSs are conventional exteroceptive cues like lights and tones or whether they are flavors. The second cue must be novel and it is most effective if it is presented immediately following the preexposed CS. All of this suggests that the function of this added cue is to maintain the subjects' attention to the stimulus.

How does latent inhibition work? What theory best accounts for the fact that mere exposure to a CS makes it difficult for that CS to become conditioned later on? First of all we should note that latent inhibition is specific to the preexposed CS; the animal does not suffer a general loss in learning ability but only a relative inability to learn about that specific preexposed CS (Schnur & Ksir, 1969). Given this fact it may be that the CS comes to serve as a conditioned inhibitory stimulus during the preexposure phase. Later, during the test, excitatory conditioning would naturally be retarded if the CS were already a conditioned inhibitor. Such a theory (Lubow, 1965) assumes that some kind of inhibitory conditioning is taking place during preexposure; the CS is, quite simply, becoming a conditioned inhibitor because it is not followed by a US. Then, when an attempt is made to make that stimulus into a conditioned excitatory cue, progress is very difficult because, according to the theory, the cue already is a conditioned inhibitor. The typical control group, the one that received no preexposure, begins its excitatory conditioning with an essentially neutral cue; conditioning is therefore easier for subjects in the control group than for the preexposed subjects for whom the cue is a conditioned inhibitor.

Does preexposure, in fact, lead to conditioned inhibition? The answer appears to be no. Conditioned inhibition and latent inhibition are similar to the extent that excitatory conditioning is more difficult following both those procedures. Nevertheless, a latent inhibitor is not a conditioned inhibitor. Retardation occurs in both cases but it is due to different mechanisms. This has been demonstrated in several important studies (Best, 1975; Halgren, 1974; Reiss & Wagner, 1972; and Rescorla, 1971a). Each of these investigators argued that if preexposure to a CS did result in conditioned inhibition, then it ought to be *easier* for the stimulus to become a conditioned inhibitory cue. In other words, conditioned inhibitory learning should take place more readily to a cue that is already partly a conditioned inhibitor (due to the preexposure phase) than to an entirely novel cue that was not preexposed. In those studies, however, conditioning of *any* sort, either excitatory *or* inhibitory, was more difficult after CS preexposure. Inhibitory conditioning was retarded just as much as excitatory conditioning by the preexposure procedure. Therefore, we can conclude that simple CS preexposure does not result in conditioned inhibition; CS preexposure results in the retardation of *both* excitatory and inhibitory conditioning.

If retardation of learning is not due to the fact that the CS had become a conditioned inhibitor during preexposure (as the research cited immediately above indicates), then what does cause the retardation in later conditioning? The most widely accepted theory postulates that animals have a difficult time learning about a stimulus after it has been preexposed because they are not very attentive to that stimulus. The stimulus becomes less salient during preexposure. This would account for the fact that conditioning of any sort, excitatory or inhibitory, is later retarded because the subject is simply not paying sufficient attention to the CS.

Loss of attention to a CS that was preexposed could be only one of the ways in which latent inhibition operates. Other investigators have favored yet a third theory: what the subjects are actually learning during the preexposure phase is that the CS is irrelevant. It is not a matter of losing attention; there is an active learning process in which the animals are acquiring the expectancy that the CS is irrelevant. This learned irrelevance, then, makes it more difficult for the subjects to adopt a new expectancy later on when the stimulus is, in fact, a relevant predictor of the US. Evidence for the learned irrelevance position comes from a study by Dexter and Merrill (1969). They performed a typical latent inhibition study except that some of the subjects who were preexposed to the CS were later tested in a different apparatus. The authors found that latent inhibition did not develop for these subjects. A change in the environment eliminated the latent inhibition effect altogether. Conditioning took place at a normal rate in this new apparatus even though the subjects had received CS preexposure. This finding suggests that the subjects had learned to ignore the CS in the former environment where it was irrelevant, but that they readily learned the significance of the CS in a new environment. Had the subjects simply been inattentive to the stimulus, that is, if latent inhibition were merely a function of a loss in attention, then retardation of conditioning should have been observed in both testing environments, even the new one (see also Baker & MacKintosh, 1977; 1979; MacKintosh, 1975a). As these various studies indicate, there is no single explanation for the way that latent inhibition operates. Retardation of learning after CS preexposure may be the result of a loss of attention *and* learned irrelevance. Both processes or mechanisms may be operating at the same time.

Context Effect

The experiments by Dexter and Merrill (1969) cited above suggest an important principle concerning latent inhibition: latent inhibition depends entirely on the context in which it is tested. If excitatory conditioning is conducted in a different environment from the preexposure environment, then retardation of learning is not observed (see also Rudy, Rosenberg, & Sandell, 1977).

This general importance of the experimental context has been demonstrated in a paper by Lubow, Rifkin, and Alek (1976). The investigators used two testing boxes. One was a circular box with ridges on the floor and plastic grids on the ceiling. The second apparatus was a rectangular cage with a layer of sand on the floor and a flat metal ceiling. Attached to each cage were two plastic cylinders each of which was fitted with an automatic feeder. In addition, a cotton ball containing an odor (extract of lemon or wintergreen) could be located in one of these plastic cylinders. The experiment began with a preexposure phase: rats were housed in one of the two boxes for a period of 14 days during which time they experienced one of the odors. That is, each rat was preexposed to a specific stimulus, the odor, as well as to a general environment, the apparatus. Testing was then begun. The animals had to learn to go to the cylinder containing the "correct" odor; upon entering the cylinder and touching its nose near the source of the odor, the rat would receive a food pellet. The question was: How readily would these stimuli, the odors, acquire strength by virtue of their association with the food reward?

The test was conducted in a particular way for four different groups. For two of the groups, training was carried out in the same cage to which they had been preexposed. The environment for these groups was the "old" familiar one. For the other two groups, testing was carried out in a new environment, that is, in the novel apparatus. One group in each of the above conditions was presented with the old smell, the one the group had experienced during preexposure; the other group in the above conditions received the novel odor. In summary, one group received a novel stimulus in a novel environment; a second group received the preexposed odor in the familiar environment; a third group received the novel stimulus in the old environment; and a fourth group received the old familiar stimulus in the new environment.

Let us first consider the effect of stimulus preexposure (that is, latent inhibition). On the surface, it would appear that preexposure of the odor would retard conditioning to that stimulus. The work of Dexter and Merrill, however, suggests that latent inhibition occurs only when the subject is tested in the same environment. This later finding is precisely what Lubow, Rifkin, and Alek demonstrated: preexposure of the odor led to retardation of learning later on only when the subject was tested in the familiar environment. Learning took place normally if the subject was trained in the novel environment.

Lubow and his colleagues also showed that the novelty of the odor was not sufficient for conditioning to take place. Subjects given the novel stimulus, the new odor, learned only if the odor was presented in the old preexposed environment. In other words, when both the stimulus and the environment were novel, learning did not take place; when one of them was novel and the other familiar, then conditioning proceeded as usual. It appears, therefore, that stimuli may acquire strength in conditioning only to the extent that they *contrast* with the background. The context of conditioning makes all the difference. If both the CS and the background are familiar then learning is retarded; this is the latent inhibition effect. Similarly, a novel cue will not be learned if it fails to be contrasted from the background, that is, if training is conducted in a novel apparatus. Learning about a new CS is enhanced only if training is done in a familiar environment; learning about an old stimulus is accomplished if training is done in a new unfamiliar environment. As the authors state "the most effective conditions for learning are those in which there is a contrast between the novelty of the environment and the novelty of the stimulus. Both a new stimulus in an old environment and an old stimulus in a new environment are relatively effective conditions for learning compared to an old stimulus in an old environment or a new stimulus in a new environment" (pp. 45–46).

Compound CS Conditioning

One of the most important areas of research in the last decade has been on the conditioning of compound CSs, conditioned stimuli that are comprised of *two* separate elements such as a light and a tone together. Traditional Pavlovian theory suggests that the strength that develops to either of the two elements depends upon the relationship between the compound stimulus and the US. That is, the fact that there are two elements involved in the CS should make no difference; each element should independently derive strength from the US and should not affect the other element. Under normal circumstances, it is true that conditioning takes place for both elements of a compound. For example, if a tone and light were presented together followed by shock, both elements would become conditioned exciters. Each gains approximately the same strength depending upon the initial intensity of the element. However, under other circumstances, one observes that the conditioning to one element depends upon the strength of the other element in the compound.

Let us consider the original work on this phenomenon. Kamin (1969) trained several groups of rats to press a lever for food. He then conducted a typical CER experiment using a compound CS (light and noise). The control group received the compound CS followed by shock; as noted above this group was expected to develop equivalent strength to both elements of the compound. The experimental group received the same compound conditioning procedure, but just prior to that phase subjects in this group received extra conditioning trials in which the noise alone was paired with shock. This group was used to assess the effect of prior conditioning to noise on later conditioning to the light. In summary, the principal groups differed only in one respect: the experimental subjects received noise-US pairings prior to their compound conditioning training, whereas the controls received only the compound conditioning trials.

In a third phase of the study all subjects received a test for the strength of the light element. As expected, for the control group the light derived strength during compound training. Suppression of the lever pressing, the typical CER measure, was rather intense. The results of the experimental group differed, however. These subjects, who had received extra noise-US trials *prior* to their compound CS trials, showed no evidence of conditioning to the light. Lever pressing remained stable during the light presentation. What had happened was that the conditioning of the noise element blocked or prevented later conditioning to the light element. This study illustrates that when a second CS element is added to another CS element that has already received conditioning trials, this added cue derives no strength. Even though the light was contiguous with the shock during compound training, the light failed to derive any strength because the noise element had already undergone conditioning. This failure for an added cue to become conditioned is known as blocking. It has been observed in a wide variety of species and for various Pavlovian reactions including eye blink CRs (Marchant & Moore, 1973) and taste aversion (Gillan & Domjan, 1977; Willner, 1978).

The blocking phenomenon is not limited to excitatory conditioning. Inhibitory CRs also can be blocked. This was demonstrated by Suiter and LoLordo (1971). All subjects were taught to press a lever for food as a preliminary to the CER procedure. Then Pavlovian conditioning was conducted in three phases. The experimental subjects received a light CS that consistently predicted no shock during Phase 1 (light was presented after shock offset). Therefore, the light was considered to be an inhibitory CS—. The control subjects, on the other hand, received only the shock presentations during Phase 1; no CS elements were involved for those subjects. During Phase 2, both groups of subjects received inhibitory conditioning using a compound CS (a light and tone together). Specifically, the compound CS predicted the absence of shock. Under normal conditions, as stated previously, the CS elements should become conditioned inhibitors. Finally, during Phase 3, the strength of the CS tone element was tested. This was accomplished by pairing the tone with shock and observing how quickly it acquired excitatory strength. Recall that excitatory conditioning is retarded to a stimulus if that stimulus is a conditioned inhibitor (see Chapter 2). If the tone had become a strong conditioned inhibitor during Phase 2, then excitatory conditioning should be retarded. On the other hand, if inhibitory conditioning to the tone had been blocked by the light, then the tone CS should readily acquire excitatory strength during Phase 3. Suiter and LoLordo discovered that excitatory conditioning to the tone took place readily for the experimental subjects. For this group, the inhibitory strength of the tone had been blocked during Phase 2, thus allowing it to become excitatory quite easily. The reason the tone was blocked was that the light was already a good conditioned inhibitor because it had received prior training.

Let us summarize this important effect called blocking. Any single stimulus that undergoes conditioning (stimulus A) will later block or prevent conditioning to a second added stimulus element (stimulus X). That is, an added cue will fail to derive any strength even though it is contiguous with the US if the other element in the compound already had undergone conditioning. It makes no difference whether the initial conditioning to CS_A involves excitatory or inhibitory procedures; either way, once the element possesses conditioned strength it can serve to block conditioning to an added element CS_X.

Several studies elaborated on the phenomenon of blocking (see Rescorla, 1971b; Rescorla & Wagner, 1972; Wagner, 1969b). Two important findings emerged in this early research. First, experiments revealed that the extra training trials afforded to the first stimulus in the blocking design do not have to precede the compound training trials. That is, the extra reinforcements for stimulus A can be mixed in with the normal training trials to the compound stimulus AX. Second, while extra A-US trials tend to block conditioning to the target element X, inhibitory conditioning, that is, trials pairing stimulus A with no US, do the opposite. Inhibitory conditioning to one stimulus element of a compound tends to *enhance* the conditioning of the other element.

Consider the study by Wagner (1969b). Three groups of rabbits were given eyelid conditioning. Animals in the normal control group received 112 trials during which a compound CS, designated as AX (light flashes plus a tone), was followed by a mild shock. Subjects in Group E received the same treatment, 112 compound conditioning trials, but in addition these subjects received extra trials interspersed among the compound trials. For Group E these extra trials involved the light element (designated as A) followed by the shock. In other words, the Group E animals received extra excitatory training to the light stimulus interspersed with their normal excitatory training to the compound stimulus. Finally, subjects in Group I received the same number of compound trials, but interspersed with those trials were *inhibitory* trials to the light element. In particular, Group I received additional light-only trials during which the light was presented without shock. In summary, all groups received the same number of compound conditioning trials. However, one group (Group E) received extra excitatory trials to the light stimulus, while the other experimental group (Group I) received extra inhibitory trials to this light element.

In the second phase of the study all groups were tested for conditioning strength to stimulus X, the tone. Recall that all three groups received exactly the same number of tone-US pairings (they all received the same number of compound training trials). As shown in Figure 3–10, the median percentage conditioned reactions given to the tone by the control group was about 30 percent. For Group E, however, the conditioned reaction to the tone was dramatically reduced: On only about 10 percent of the trials did

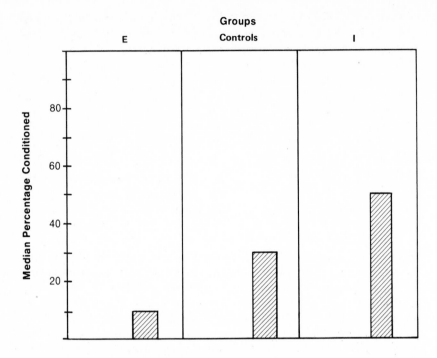

Figure 3-10.   Median percentage conditioned eye blink responses to the tone CS (element X) by the control group, Group E (extra excitatory trials to the light element), and Group I (extra inhibitory trials to the light). (From Wagner, 1969b)

the tone elicit the conditioned response. This finding, therefore, is very similar to the blocking phenomenon in that extra training trials to stimulus A (the light) reduced conditioning to the target element X (the tone). The opposite was true for Group I. For this group, conditioning to X was *enhanced* to about 50 percent because these subjects received extra inhibitory trials to the light stimulus, element A. In other words, extra trials in which the light stimulus was exposed without the US weakened that light stimulus (made it more inhibitory); this had the effect of enhancing the strength of the other element in the compound, the tone.

Let us summarize. The strength of the target stimulus X, the tone, depended not only upon the number of tone-US pairings, but also upon the strength of the other element in the compound, the light. If the light had become stronger because it received additional pairings with the US, then the tone, element X, suffered; its strength was weaker than normal. On the other hand, if the light had become weaker or inhibitory because it had been presented without the US, then the strength of the tone was enhanced. In short, conditioning appears to be due to two factors: the number of CS-US pairings *and* the strength of other stimulus in the compound.

Let us now consider the theory behind Kamin's (1969) blocking experiment and the general work showing that the strength of a CS depends upon the potency of other cues in the compound. The general concept that applies here is the notion of redundancy. According to Kamin, a US is effective in conditioning only if it surprises the subject, only if it is not redundant. Recall the original blocking experiment. During the initial phase the CS becomes a perfectly good predictor of the US. During Phase 2, when the compound CS is used, the added element, the one that is blocked, is redundant. The subject does not "need" this added element in order to predict the US presentation; the animal can rely upon the originally conditioned cue, element A.

But what if we surprise the subject by changing the US during these blocking trials? If a new US were used, or if some important condition were changed, then the added element would no longer be redundant; it would be a good predictor of these new conditions. If blocking failed to occur with such a change, one could be confident that surprise is necessary for conditioning. This is precisely what many researchers have shown. For example, Dickinson, Hall and MacKintosh (1976) performed a typical blocking experiment. During the compound trials, however, they "surprised" the subject by presenting an extra unexpected shock following the compound trial or, in another experiment, by omitting an expected shock. In all of these cases blocking was either not observed or was severely reduced. In other words, when the subjects already have a perfectly informative cue that signals the US, then added cues will not become conditioned; rather, these added cues will be blocked. However, if the added cue signals a *new surprising* condition, like the presentation of an extra shock or the omission of an expected shock, then blocking will not occur; an added cue *will* become conditioned. This idea is now believed to be a central concept in conditioning: conditioning only occurs when the US is surprising or unexpected (see Blanchard & Honig, 1976; Maki, 1979; but also Kohler & Ayres, 1979; and Maleske & Frey, 1979).

We should note parenthetically that the dependency of conditioning on surprise is perfectly consistent with an evolutionary perspective. Strong biological stimuli, USs, are those environmental events that the subject "needs" to predict for survival. If a good predictor is already established, then added cues, because they are redundant, need not be conditioned; they do not provide the animal with any added advantage. However, when the subject cannot predict important USs, then the animal must pay attention to the new arrangements and learn which cues predict the US.

How does surprise work to produce conditioning? The theory about surprise, developed mainly by Wagner and his associates (see Terry & Wagner, 1975; Wagner, Rudy, & Whitlow, 1973), is quite simple: surprise makes the animal rehearse the CS-US event just experienced. In other words, the surprising US compels the animal to think back over its recent experience and rehearse in its memory any salient CS-US episode. This seems to be the case because, as Wagner et al., (1973) showed, conditioning decreased if the rehearsal process was disrupted. Specifically, subjects were conditioned using a novel CS-US pairing. Some animals received a disrupting event immediately after each trial. The disrupting event was designed to eliminate or interfere with the rehearsal of the CS-US pairing. The results confirmed this idea: disruptive events reduced conditioning, whereas innocuous posttrial events did not disrupt learning (rehearsal was unaffected and the CS-US pairing was learned perfectly well). Numerous other experiments have also demonstrated that rehearsal is necessary for conditioning and that disruption of this rehearsal process interferes with conditioning (see Dickinson & MacKintosh, 1979; Kremer, 1979).

The studies cited above indicate that surprise is necessary for conditioning, but they do not provide any formal theory of the blocking phenomenon. The concept of surprise tells us how conditioning takes place; what we need is a concept or model that tells us why conditioning does not take place in the blocking experiment. Even more important, we need a theory that can tell us why the strength of a cue will be enhanced if the other cue in the compound undergoes extra inhibitory conditioning.

There are two general theories in this area. The first was proposed by Rescorla and Wagner (1972) who claimed that stimuli compete for strength. Each unconditioned stimulus has a maximum amount of conditioning strength that can be divided among various CSs. Stimuli compete for this strength. Therefore, if one stimulus in a compound is strong, then it "uses up" most of the power or strength that the US can offer; this leaves very little strength for the other cue in the compound and blocking is observed. Conversely, if one element of a compound is weakened, then the other element can benefit by deriving even more of the US strength for itself. In short, changes in the strength of one element based upon the power of the other element in the compound are due to the fact that the degree of strength available has changed. This model has received a great deal of support in recent years (see Kremer, 1978).

Another theory has been proposed by MacKintosh and his associates (see Baker & MacKintosh, 1979; MacKintosh, 1975a, 1975b, 1976; MacKintosh & Turner, 1971). MacKintosh's theory is quite simple: animals are constantly evaluating the predictive value of the stimuli in their environment and so they learn to ignore those stimuli which are irrelevant. In other words, conditioning is learning that a stimulus is relevant, that it predicts an important event like a US. Failures in conditioning (blocking) stem from the fact that the animal has learned that the added cue is irrelevant. The subject actively learns to ignore redundant and unpredictive cues.

This is quite a different theory from Wagner's model. Whereas Wagner and Rescorla claim that stimuli lose associability, that they become less salient and less able to become associated with the US, MacKintosh's theory stresses that blocking and related phenomena involve active learning: an animal is learning that a stimulus is irrelevant. Much of the evidence seems to favor the MacKintosh theory. For example, MacKintosh (1975a) conducted a typical blocking experiment, but he paid special attention to the amount of conditioning that developed to the added cue on the *first* compound trial. If Wagner and Rescorla were correct then even on that first trial the added element should show no conditioning because it will have lost its ability to become associated with the US. According to MacKintosh's theory, however, the animal has no way of knowing on the first compound trial that the added element is redundant. Therefore, conditioning should be observed. MacKintosh found that conditioning occurred on the first trial; thus it would appear that blocking is a matter of learned irrelevance.

Let us summarize this complex, but exceedingly important, area of research. The strength of a conditioned stimulus depends not only upon whether it has been paired with a US but also upon the strength of other stimuli that exist in the environment. If another stimulus in the environment, usually one deliberately compounded with the target CS, is strong, that is, if it is a good predictor of the US, then the strength of the target stimulus suffers. If the other stimulus in the compound is weak through extra inhibitory trials, then the target stimulus in the compound is unusually strong. Thus, stimuli derive strength from the status or strength of other stimuli in the environment. When an important event like a US occurs, the subjects rehearse this episode and conditioning takes place. Once the episode has been rehearsed and strengthened, then added cues are actively ignored. In other words, it is as if an animal is searching continuously for the best predictor of the US. If the animal discovers that certain stimuli are not good predictors, it will ignore those stimuli (they will be blocked). On the other hand, if the animal recognizes that one stimulus is a particularly *poor* predictor of the US, then the animal will place special emphasis upon other stimuli in the environment.

The idea that stimulus strength depends upon the strength of other stimuli in a compound has been used to explain a variety of Pavlovian phenomena. For example, consider the finding that preexposure of the US results in weaker conditioning later when the US is used in training (see Randich & LoLordo, 1979a, for a review). This phenomenon, called the US preexposure effect, has been demonstrated in a variety of Pavlovian conditioning situations, including eye blink response (Mis & Moore, 1973), CER conditioning (Randich & LoLordo, 1979b), and taste aversion learning (Batson & Best, 1979; Cannon, Berman, Baker, & Atkinson, 1975; Mikulka, Leard, & Klein, 1977). The work on compound conditioning explains the US preexposure effect in the following way. During Phase

1 when the US is being preexposed, conditioning is actually taking place although the experimenter does not observe it. During this phase, the US is getting attached to various stimuli in the environment. This could be called context conditioning because the context or apparatus is serving as a CS. Phase 2 involves the pairing of a specific CS and a US. But the context is still there, making this new CS an added cue. What is claimed is that during this conditioning phase, the context is actually blocking the conditioning to this added cue. Animals have already experienced the context and the US and so any additional cue, actually considered to be an element in a compound, would be blocked.

A number of other studies support the notion of context conditioning. For example, Batson and Best (1979) examined the effect of US preexposure on later taste aversion learning. They found that when the illness-inducing drug was given prior to training, taste aversion learning was reduced. However, if the US was paired with a distinctive environmental cue (a black box) during US preexposure, then taste aversion learning was retarded later, but only if it was conducted in that black box. In other words, if the US were *deliberately* paired with a distinctive cue during the preexposure phase and that context or distinctive stimulus was present during taste aversion learning, then the distinctive cue blocked conditioning to the new flavor stimulus. If the original cue, the black box, was *not* present during taste aversion learning, then blocking was not observed and the flavor aversion was learned quite easily. Thus, it appears that various Pavlovian phenomena can be interpreted or explained in terms of blocking. The US preexposure effect, of course, has other explanations, including the idea that the strength of the stimulus diminishes or habituates during preexposure, rendering it less effective in the future. Nevertheless, sufficient evidence does exist, in the form of context conditioning just mentioned, to suggest that the conditioning of a CS is not a static process; it depends upon the strength and status of other cues in the environment, including the context.

Summary

The strength of Pavlovian, or classical, conditioning must be assessed relative to appropriate control conditions. Otherwise pseudoconditioned reactions, that is, responses that are not due to proper forward conditioning procedures, can be erroneously interpreted as learned reactions. The appropriate forward technique for excitatory conditioning involves the presentation of the CS followed by the US, either prior to the CS offset (delay training) or after the CS offset has occurred (trace conditioning). Control procedures may involve simultaneous, backward, and random techniques. The last is considered most appropriate since it lacks the one essential ingredient for Pavlovian conditioning, namely, the predictive relationship between the CS and US.

Several training variables affect the strength of conditioning. As noted previously, conditioning is dependent on the information value of the CS. Performance is also positively related to the CS-US interval. That is, temporal contiguity is a crucial mechanism for conditioning. The optimal CS-US interval, however, depends upon the type of CR being studied. Skeletal CRs, like eye blink, are conditioned best with very short intervals; visceral reactions with longer intervals of about 5 to 15 seconds; and taste aversion learning with much longer US delays (as long as 12 hours). A third parameter affecting performance is CS intensity: conditioning increases as a function of CS strength because strong CSs tend to be more easily discriminated from background stimulation. The naturalistic quality of the CS and its temporal or spatial patterning are other important parameters.

Unconditioned stimulus variables include the intensity, duration, predictability, and quality of the US. All exert a strong effect on the strength as well as on the precise nature of the CR.

Finally, conditioning is related to the intertrial interval: the more the trials are spaced in time, the better the learning, although trial duration also has an important bearing on this relationship.

Research studies have revealed six basic Pavlovian phenomena that either extend the principles of classical conditioning or demonstrate exceptions to the rule. In second-order conditioning, a neutral CS gains strength not from a US but from an already powerful CS. Once the second-order conditioning has been accomplished, the strength is undiminished even when the originally strong CS is extinguished. Closely related to second-order conditioning is sensory preconditioning in which two CSs are paired prior to using one in a normal conditioning experiment. Later, the other CS, the one not used in normal conditioning, is found to elicit the CR. The importance of this effect, and of second-order conditioning, is that two CSs can get bonded even though neither one is a biologically powerful US.

A third phenomenon is inhibition of delay. This occurs when the US presentation is appreciably delayed. The subjects actually delay making the CR until just prior to the US; that is, they show temporary inhibition, then excitation on the same trial.

Latent inhibition is the retardation of excitatory learning following CS preexposure. It appears that the effect occurs because the animals are less attentive to the CS after preexposure and because they come to view the CS as irrelevant.

Research studies investigating the context effect show that conditioning occurs only when the primary CS stands out from the environment. If a CS has been preexposed, then conditioning will be retarded if it takes place in a familiar environment, but not if the background cues are novel. Similarly, novel cues are learned well only if they stand out from a familiar background.

The final phenomenon discussed concerns compound CS conditioning. Studies of this phenomenon show that if one CS element of a compound undergoes prior conditioning, then subsequent conditioning to an added CS element is blocked. Similarly, if one element of a compound CS undergoes inhibitory training, making it unreliable, then the other element of the compound gains even greater strength. This phenomenon indicates that conditioning to a CS depends, in part, on the strength of other cues in the environment that are exposed simultaneously. Furthermore, it suggests that in order for conditioning to take place, the US must be surprising or unexpected.

4

Response Learning

Introduction

Training Conditions That Influence Learning

Schedules of Reinforcement

Basic Instrumental Phenomena

Summary

Introduction

We will begin our discussion of response learning by reviewing some basic concepts from Chapter 2. Response learning, also called instrumental conditioning, is defined as the method of making the US presentation (or omission) contingent upon a particular CR. Unlike the case in classical conditioning, in instrumental conditioning, the organism's behavior is paramount. That is, the presentation of the US, or its omission, is dependent upon the execution of the response by the subject. Such an arrangement permits the organism to expect some future event, the US or no US, as a result of its own action. We know that the contingency between the animal's response and the outcome is paramount because we observe an increase in the rate or probability of the CR as a function of training. From this increase, we can infer that an expectancy has developed; there would be little reason for an organism to increase the frequency and probability of its behavior if it did not expect to receive the outcome that is correlated with its behavior.

Response learning represents a mechanism by which animals can change the world to their advantage. Since strong biologically active stimuli usually represent either valuable resources or threats to survival, then these stimuli must not only be predicted (stimulus learning provides one mechanism), but they also must be controlled. Organisms who have evolved the mechanisms that permit response learning can change the environment to their own advantage. They can acquire expectancies about future outcomes based upon their own behavior; and the responses they execute alter stimuli in ways that are important for survival.

Instrumental conditioning, like classical conditioning, has four major paradigms. Two of these paradigms involve appetitive or pleasurable USs. In particular, reward conditioning occurs when a response consistently produces an appetitive US, whereas omission training involves the withholding of reward following the criterion CR. There are also two additional paradigms based upon aversive USs. In the case of punishment, the response produces the aversive US (causing suppression of the behavior), whereas in the escape/avoidance situation, the CR terminates or avoids the presentation of the noxious stimulus.

Training Conditions That Influence Learning

In this section we will review the facts and theories concerning variables that influence instrumental conditioning. There are, of course, a great many such variables, several of which pertain to the characteristics of the reward.

Response-Outcome Correlation

Certainly one of the most important factors that affects learning is the correlation between the response and the reinforcing outcome. This relationship is, perhaps, more properly termed the reinforcement contingency. Almost any study in the instrumental conditioning literature illustrates this principle. Whenever the reinforcer is delayed or otherwise not contingent upon the organism's behavior, the appropriate increase in response rate or probability fails to occur. In short, if learning is to occur, the response must have a predictive value for the reinforcer in instrumental conditioning, just as in classical conditioning the CS must have a predictive value for the US.

CR Type

In addition to the correlation between the response and the outcome, the characteristics of the CR itself are important. First of all, it is clear that the principle of "least effort" applies in instrumental conditioning (Lewis, 1965). This means that organisms will seek to spend the least amount of energy in order to gain the greatest relative amount of reward. Since the CR is voluntarily executed during instrumental conditioning, the fact that animals spend as little effort as necessary is significant. The effort of the response, of course, has nothing to do with the expectancy that is learned;

the expectancy is based on the response-outcome correlation, regardless of how much effort is expended in the response.

A second point regarding the nature of the CR is that not all responses reach the same level of performance in instrumental conditioning (see Seligman, 1970; Seligman & Hager, 1972). This point will be discussed in much greater detail in Chapter 6. For the moment, however, let us simply note that some responses seem to be learned more easily than others, even though the reinforcement contingency is perfectly appropriate and the animal is quite able to execute both responses. Naturally, we would not expect to be able to teach a rat to fly, a pigeon to swim, or a mouse to play the trumpet; the range of behaviors for any organism is constrained by the physical makeup of the organism. But there are some behaviors which organisms apparently can perform physically but seem less able to learn. For example, rats have difficulty learning to press a lever to avoid shock although, in situations involving food rewards, lever pressing is a perfectly easy behavior for them to learn. There is no indication that these learning deficiencies reflect problems of acquiring an expectancy. Although the performance or execution of a particular response in a particular context may be difficult for an animal, nothing suggests that the animal is deficient in acquiring the expectancy that the particular response is correlated to a particular outcome.

In summary, responses are chosen by the investigator to serve as the criterion CR based largely upon convenience and upon the natural inclinations and talents of the subject. Some behaviors, however, are more difficult to perform than others. In addition, the subjects will naturally seek to expend the least effort that is permitted by the contingency.

US Intensity

The intensity of the US, that is, the magnitude of the reinforcer, is an important factor in response learning. Consider first the effect of appetitive reward magnitude. Many experiments have indicated that learning is positively related to the size of the reinforcement (Kintsch, 1962; Ratliff & Ratliff, 1971; Roberts, 1969). The rate of improvement as well as the terminal performance level both are positively correlated with reward magnitude. In addition, a number of studies have shown that the quality of the reward also plays a role; learning is a positive function of, for example, sucrose concentration (Kraeling, 1961).

The study by Roberts (1969) provides a good example of this relationship between the magnitude of the reward and response learning. Five groups of rats were trained to run down a straight alleyway for food. The magnitude of the food reward in the goal box varied between the groups. Specifically, subjects received either 1, 2, 5, 10, or 25 small reward pellets in the goal box. After about 30 trials, Roberts noted that the animals were running at their stable, maximum rate. Figure 4–1 shows how reward magnitude affected this rate of running. For all three response measures (the speed of leaving the start box, running down the alleyway, and running the last nine inches just outside the goal box), speed of response was significantly influenced by reward magnitude.

The fact that reward magnitude in appetitive instrumental conditioning is so potent a factor is perfectly consistent with an evolutionary perspective. An animal's investment of time and energy in order to gain the maximum amount of food is an adaptive strategy. The more food the animal receives, the healthier it is relative to its competitors.

This principle of response learning, that animals are powerfully influenced by reward magnitude, has been the focus of much research in the ethological literature (see Krebs, 1978; Krebs, Erichsen, Webber, & Charnov, 1977; Krebs, Kacelnik, & Taylor, 1978; Lea, 1979; Pyke, Pulliam, & Charnov, 1977). Generally, these authors claim that animals will perform in a way that is optimum in terms of energy received (reinforcement magnitude) and energy expended (response effort). For any given response, energy is lost; for any given reward, energy is gained. It would be maladaptive from an evolutionary perspective for an animal to spend more energy than it receives; it would soon die. Clearly, it is most appropriate for the animal to gain as much energy as possible, to work for the largest reward *relative*

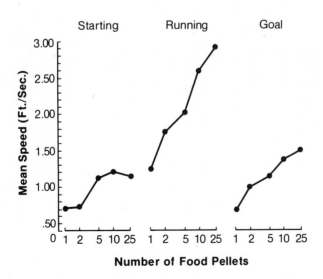

Figure 4-1. Mean running speed as a function of the reward magnitude measured by startbox speed, alleyway running speed, and speed of entering the goal area. (From Roberts, 1969)

to the amount of energy lost. Thus, the important factor is *not* simply magnitude of reward (energy gained), but rather the ratio of the energy gained to the energy lost. Research has supported this qualification: animals of all varieties, from birds to crustacea, will choose the most profitable food item (the reward yielding the largest energy gained to energy lost ratio), and they will forage in the most profitable feeding area (Krebs et al., 1978). In short, the general principle that response learning increases as a function of reward magnitude is true *only* if the response effort and time stay the same. If they do not remain the same, the relationship between performance and reward magnitude may change. For example, often in nature, larger food items require a greater time investment on the part of the animal (to extract the eatable portion) or greater risks in obtaining the food item. When this is the case, reward magnitude is only one determining factor; the time and energy investment must also be considered.

Consider now the effect of US intensity in punishment. Given the fact that the punishment procedure is identical to the reward procedure except that the US is aversive rather than appetitive, one would expect that responding should be an *inverse* function of US intensity. That is to say, the greater the punisher intensity, the more the CR is suppressed. This is, indeed, the case (see Camp, Raymond, & Church, 1967; Church, Raymond, & Beauchamp, 1967; Filby & Appel, 1966; Powell, 1970). In all of these studies, the aversive US was a shock delivered to rats following their lever press response. What was discovered was a graded relationship between suppression and shock intensity: the greater the intensity, the greater the suppression. Such an effect also occurs when aversive CRs, like avoidance reactions, are punished (Smith, Misanin, & Campbell, 1966). Moreover, punishers other than shock, for example, loud noises or temperature changes, have been employed and the same relationship is found: greater suppression results from stronger punishers.

Incidentally, it should be noted that duration of punishment is also an important parameter. Several investigators (Campbell, Smith, & Misanin, 1966; Church et al., 1967) have shown that response suppression is positively related to the duration of a shock.

The third instrumental conditioning method that has been considered in light of US intensity is the escape/avoidance procedure. First, consider escape conditioning, the situation in which the instrumental response leads to the termination of the aversive US. Many studies have shown that the stronger the US, the greater the improvement in escape performance. Although most studies have used shock USs (Franchina, 1969; Staveley, 1966; Trapold & Fowler, 1960), the relationship between US intensity and performance has also been observed in studies that trained animals to escape cold water (Woods, Davidson, & Peters, 1964), loud noise (Bolles & Seelbach, 1964; Masterson, 1969), and intense light (Kaplan, Jackson, & Sparer, 1965).

Figure 4–2 illustrates the results of the study by Franchina (1969). Rats were placed in a white start box and, after shock came on, had to jump into a safe compartment painted black in order to terminate the shock. Different groups received either 20, 50, or 80 volts. As shown in Figure 4–2, speed of responding was directly related to shock intensity; the higher the shock, the faster the escape response. Interestingly, animals that received all the shock intensities in a varied order showed even faster performance at each level of shock. As indicated in the figure, response speed to a 20-volt shock was much faster in those subjects that also experienced the other levels of shock than it was for subjects that experienced only the 20-volt level. The same pattern of results was observed for the other shock levels. In summary, it appears that escape performance is positively related to US intensity *and* that variability in intensity heightens response speed over all levels of the US.

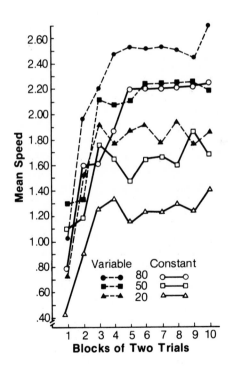

Figure 4-2. Mean speed of escape response as a function of training trials for animals that received either 20, 50, or 80 volts (open circles) and for subjects that received all three voltages (filled circles). (From Franchina, 1969)

The effect of US intensity in escape conditioning may not be strictly analogous to the effect found in reward or punishment training. The reason is that the US intensity level essentially defines the motivational level of the subject: the stronger the shock, the more motivated the animal is to terminate it. The superior performance, therefore, by subjects that receive higher intensity USs may have nothing to do with learning *per se* but rather it may reflect motivational differences. Indeed, the response *expectancy,* that the US will be terminated if the appropriate behavior is executed, is probably unaffected by US intensity, although *performance* certainly is affected.

Avoidance training is also influenced by US intensity but the effects are more complicated. Several studies (Boren, Sidman, & Herrnstein, 1959; Myers, 1977; Riess, 1970; and Riess & Farrar, 1972) have shown that lever-pressing avoidance behavior is positively related to shock intensity. For example, in the Riess and Farrar study, rats could postpone a .15-second shock for 20 seconds by pressing a lever; failure to press the lever resulted in shock being delivered every three seconds. Different groups of subjects received shock intensities of .25, .5, 1.0, 2.0, or 4.0 milliamps. The results showed that the improvement in lever-pressing behavior increased as a function of shock intensity. Although little lever pressing occurred at the lower intensities, substantial increases in performance resulted with higher shocks.

Pearce (1978) has shown that shock intensity and passive avoidance learning are positively related. This relationship applies, however, only when the instinctual, naturalistic reactions of the animal are measured. For example, Pearce shocked rats when they stepped off a safe platform onto the grid floor. Later, he tested the animals to determine if they would remain passively on the platform and thereby avoid shock. He found that when the safe platform was located at the center of the apparatus, performance was disrupted by intense shock; the rat's natural inclination to jump off and run toward the walls of the box competed with its avoidance learning. The rats tended to jump off the platform quickly, showing a low level of passive avoidance. However, when the safe platform was located near the periphery, this conflict was not present; under these conditions, passive avoidance learning, remaining on the safe platform, increased as a function of shock intensity.

The relation between US intensity and avoidance is quite different when the normal type of avoidance response, that is, shuttlebox avoidance, is tested. Recall that shuttle avoidance occurs when an animal runs from one side of the box to the other to avoid shock, and then, on the next trial, runs back to the original side to avoid shock once again. Several investigators have found that shuttlebox avoidance is *inversely* related to shock intensity (Bauer, 1972; Levine, 1966; Moyer & Korn, 1964; Theios, Lynch, & Lowe, 1966). That is, increases in the US intensity lead to a deterioration of avoidance performance.

The reason for this paradoxical finding (paradoxical because higher shock levels should normally increase the animal's motivation for avoidance and thus increase avoidance performance) has been the focus of a number of investigations. One of the most profitable theories was proposed by Theios et al. (1966). The authors argued that shuttlebox avoidance conditioning actually involves two conflicting tendencies. The first is the avoidance reaction, or the movement away from the locus of shock, and the second is the passive avoidance tendency, or the tendency to refrain from returning to the locus of shock on the previous trial. In other words, the rat is in a state of conflict. It is developing a tendency to flee from its present location, but at the same time it is developing a tendency to be wary of the other side of the box. After all, on the previous trial, that other side of the box was associated with shock. This conflict between response tendencies is claimed to cause the deterioration of avoidance learning with increasing shock intensity. The higher the shock intensity, the greater the passive avoidance for the previous shock side and therefore the greater the conflict.

Theios et al. supported their theory by showing that one-way avoidance learning is positively related to shock intensity even though, like shuttlebox avoidance learning, it involves running from one place to another. Recall that in one-way learning, subjects always begin in the same start box and run to a discriminably different goal box. The animals are then removed from the goal box and replaced in the starting area for the next trial. This procedure, while sufficient to develop the active avoidance reaction, would not allow the conflicting passive avoidance tendency to form. The goal box is never associated with shock, and therefore the animals never develop a reluctance to run to it.

This theory has been confirmed recently in a different way. Using a high, medium, or low intensity shock, Freedman, Hennessy, and Groner (1974) trained three groups of rats to make a shuttle avoidance response. Three additional groups were treated in the same way except that the shock intensity that was consistently associated with one side of the apparatus differed from the intensity on the other side of the apparatus. For example, some of these animals had low shock delivered to side "A" of the box, but high shock to side "B"; others had the reverse, or they had different combinations of low, medium, and high associated with one particular side or the other. The authors hypothesized that the passive avoidance component, the reluctance to return to the previous shock locus, would be weaker when the animals had to move to a place where shock was lower (relative to the shock it was currently receiving) than when the animal had to move to the box that was associated with the same or a higher shock. The hypothesis was confirmed. A subject performed a shuttle avoidance response much better when it was required to run to a chamber that had been associated with a lower shock intensity on the previous trial; the passive avoidance component was weaker in this case.

More recent work has clarified the issue even further. We know that animals develop a passive avoidance reaction during shuttle avoidance (in addition to the criterion CR), but we know very little about *how* the behavior actually deters avoidance learning. It is possible that the chamber to which the animal must run elicits some sort of freezing behavior that competes with active avoidance learning. Apparatus cues associated with shock certainly do elicit freezing (Blanchard & Blanchard, 1969a, 1969b; Bolles & Collier, 1976). But other explanations are certainly possible. Modaresi (1978), for example, claims that shuttlebox avoidance learning involves a lower overall magnitude of reinforcement than other forms of avoidance conditioning (see also McAllister, McAllister, & Dieter, 1976; McAllister, McAllister, & Douglass, 1971). To be specific, reinforcement is derived when the animals run from the fear cues they are experiencing at the moment to a place that is not associated with shock. The more they are able to rid themselves of fear cues and obtain "safety cues," the greater is their reinforcement. Since the side to which the animals run in a normal shuttlebox experiment is also fear provoking (because it was associated with shock on previous trials), the net reward level is rather low.

In summary, US intensity does appear to be positively related to avoidance learning in cases of lever press avoidance, passive avoidance, and one-way active avoidance, but not for shuttlebox avoidance; here learning is inversely related to shock intensity. There are various theories about this latter effect, one being that the subjects are not deriving sufficient reinforcement for the shuttle avoidance response. Running to a place that itself is associated with shock and is thus fear provoking is less rewarding than running to a place that is safe (but see Masterson, Crawford, & Bartter, 1978).

Delay of Reinforcement

If organisms learn to expect future events based upon their responses, then clearly the outcome of the response must follow rather quickly. If there is an appreciable delay of the US, then the response and outcome fail to become associated. Quite simply, it is difficult for animals to learn that a particular event is pending following the stimulus or response if that event is delayed too long in time (see Renner, 1964; Tarpy & Sawabini, 1974 for reviews). Delay of reinforcement is an important parameter, therefore, because it specifies the time over which the response memory must persist in order to become connected to the reward. In this sense, the reinforcement delay is analogous to the CS-US interval in classical conditioning.

In the case of positive reward, the important question historically was the extent to which a food reward could be delayed without precluding learning altogether. Little agreement was reached in the early research. For instance, Wolfe (1934) found that rats could learn the correct turn in a

T-maze even when the food reward was delayed 2 minutes. A decade or so later, other experiments found that learning was entirely absent if delays exceeded 30 seconds (Perin, 1943) or even 5 seconds (Grice, 1948).

Spence (1947) proposed a resolution of sorts. He argued that responses really had to be reinforced immediately. The reason for the discrepancy cited above was that apparatus cues, for example, smells and other stimuli in the environment, helped subjects to "bridge" the gap between the response and the delayed reward. Subjects used these stimuli to help maintain orientation and, more significantly, as sources of information about the pending reward. For this reason the subjects could learn, even with fairly long delays. In other words, learning was much easier if the subjects could establish some sort of information "chain." If a stimulus helped to bridge the gap between the response and the reward, then the connection was made more easily.

Recent research has tended to confirm the notion that cues are important in delayed-reward learning. For example, Tombaugh and Tombaugh (1971) trained animals to make a lever press response for a delayed reward. Some of the subjects were given a light stimulus during the delay interval. The researchers found that performance was greatly improved for these subjects relative to those that did not receive such a cue. If reward-related cues, such as the sight or smell of the food cup in the goal box of a maze, exist naturally in the apparatus, then learning could be expected to occur even with long delays such as found by Wolfe.

In instrumental escape conditioning, delay of shock offset has been shown to severely retard learning. Again, if the offset of shock is sufficiently delayed, subjects are unable to form an expectancy that their behavior leads to the offset (Tarpy, 1969). As in the case with positive reward, the precise length of the delay that eliminates learning varies somewhat. For example, Fowler and Trapold (1962) taught rats to run down an alleyway in order to terminate shock. The offset of the shock was delayed for various amounts of time for different groups of rats. As shown in Figure 4–3, learning was inversely related to delay: immediate offset led to the best acquisition scores, whereas delays of 8 or 16 seconds retarded escape learning.

These findings differ somewhat from those obtained by Tarpy and Koster (1970). Using a lever box, Tarpy and Koster showed that even a 3-second delay of shock offset virtually eliminated improvement in the lever press escape response. The cause for this discrepancy is similar to the one cited above. If animals are not provided with any sort of informative cue during the delay interval, their expectancy that their response did indeed produce the shock offset is severely weakened. However if a stimulus helps to bridge the gap between the response and the reward (shock offset), then learning is improved; learning can take place with appreciable delays. This was shown directly by Tarpy and Koster (1970). When a light was turned on during the delay interval, that is, between the lever response and the

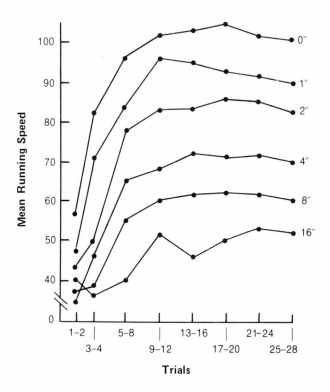

Figure 4-3. Mean running speed during escape learning in an alleyway as a function of trials for groups that differ in terms of the delay of shock offset. (From Fowler & Trapold, 1962)

shock offset, performance improved considerably. Speed of responding increased in *all* groups if the subjects were given the light, but only the 1.5-second group showed improvements in the no-cue condition. Returning to the study by Fowler and Trapold, we may assume that the goal area in the alleyway contained reward-related cues which served to sustain learning even with a 16-second delay.

Finally, consider the effect of a delayed punisher. Since punishment is similar to reward training except for the fact that an aversive US is employed, one would expect that a delayed US presentation would weaken the punishment and lessen the degree of suppression. Several studies demonstrated this finding (Baron, 1965; Camp et al., 1967; Randall & Riccio, 1969). In the Camp et al. experiment, the suppression of a lever press response, which was reinforced by food, decreased when punishment was delayed 30 seconds. That is, punishment was less effective when it was delayed. Therefore it would appear that delay of punishment operates like delay of positive reward: the effects are drastically reduced when the US is delayed. This happens even when aversive CRs are used. In particular, punishment of avoidance responding causes less response suppression if the punishment is delayed (Baron, Kaufman, & Fazzini, 1969; Kamin, 1959; Misanin, Campbell, & Smith, 1966). In summary, the effectiveness of punishment in suppressing instrumental behaviors, whether those behaviors are initially learned using reward or avoidance paradigms, is reduced if the punishment is delayed.

US Quality

Response learning is also affected by the type of US that is used. This principle is easily forgotten because so much of the research in animal learning employs strong USs such as food, water, or electric shock. But recall from Chapter 2 that we defined a reinforcer as *any* event that increased the probability or frequency of a contingent response. Such a definition suggests that numerous other conditions or stimuli could reinforce behavior.

One such condition is a change in the sensory input to an animal (see Berlyne, 1969; Eisenberger, 1972; Kish, 1966). For example, an early study by Kish (1955) illustrated that the presentation of a dim light contingent on a bar press was sufficient to support the acquisition of the bar press response. This phenomenon, termed sensory reinforcement, admittedly depended upon depriving the animals of all sensory inputs (achieved by placing them in a dark box for 25 minutes). Moreover, Barnes and Kish (1961) have shown that sensory reinforcement may be a transient phenomenon. Nevertheless, there is a sufficient amount of information to suggest that under many circumstances, particularly when animals have been deprived of sensory inputs (Fowler, 1967), a change in the sensory input can be a reinforcer. Thus, it is obvious from these experiments that response learning does not require the use of strong biological USs as reinforcers.

Sensory reinforcement is, in many ways, analogous to the sensory preconditioning phenomenon in Pavlovian literature. In both cases, the supposed "US" is an innocuous sensory event. That is, expectancies may develop based on a stimulus, in the case of sensory preconditioning, or based upon a response, in the case of sensory reinforcement, due to the high positive correlation between the first event, the stimulus or response, and the outcome, the sensory change.

During the past decade many studies have focused on the naturalistic learning patterns of animals (see Chapter 6). One fact that has emerged from this research is that animals are extraordinarily sensitive to various naturalistic stimuli. For example, the odor emanating from the urine of dominant mice is aversive to other mice (Jones & Nowell, 1974); the odor can serve as a punisher. Similarly, the odor of food itself can serve as a reinforcer (Allen, Stein, & Long, 1972; Long & Tapp, 1967). More specifically, rats will learn to press levers merely to receive the odor of food even though they never consume the food itself. Finally, rats generate specific odors during training; if they are not reinforced, they produce a "frustration odor." This odor contains a special chemical constituent which other animals find aversive. In particular, naive subjects will learn to jump a hurdle in order to avoid receiving this "frustration odor." All of this research should be no surprise. Rats, and other organisms, have a highly developed sense of smell which they use in procuring food and avoiding predators. Therefore, the finding that they use odors to identify dominant individuals (who could cause them harm) or individuals who have undergone frustrative nonreward is highly plausible.

A second principle that has emerged from the research on naturalistic learning patterns is that not all behaviors are learned equally well for a particular reinforcer. Reinforcers are often effective for some behaviors but not others. For example, Shettleworth (1975, 1978a) has shown that hamsters are easily taught to rear on their hind legs or scrabble (scrabbling is the activity of rapidly scraping the forepaws against the wall while standing erect), but they seem oddly inept at learning to face wash when the same reinforcer is used. The quality of the US may be said to influence the type of behavior that is learned. As discussed in Chapters 1 and 2, organisms have species-typical behavior potentials; these are energized or reinforced to various degrees by different USs.

Pattern of Unconditioned Stimuli

It is abundantly clear that the pattern of USs, both in terms of the temporal presentation as well as the magnitude, exerts a strong influence in response learning (see the following section on schedules of reward; see also Chapter 7). First let us note a simple but important fact. In response learning experiments, it is common—in fact, usual—to reinforce the CR only periodically; the intermittent reinforcement of behavior is called a partial reinforcement schedule. Up to this point we have concerned ourselves almost exclusively with the effects of continuous reinforcement (rewards are given on every trial or for every CR execution). But the use of partial reinforcement makes it possible to investigate patterns of US presentation.

It would appear that partial reinforcement reduces both the strength of the animal's response expectancy as well as performance. For example, the rate of learning a simple runway response deteriorates as the percentage of rewarded trials decreases (Ratliff & Ratliff, 1971). This deterioration in performance is probably due in part to the fact that the response expectancy is slow in forming. After all, the correlation between the response and the food outcome is weaker under a partial reinforcement schedule. Certainly the efficiency in learning is reduced with partial reinforcement; animals continue to make various extraneous responses, such as sniffing or sand-digging, long after the continuously reinforced animals have suppressed those behaviors (Wong, 1977). Overall, acquisition performance is slower under a partial reinforcement schedule, and this presumably reflects the slower formation of the response expectancy.

What about *explicit* patterns in the schedule, for instance, patterns of reward magnitude? Recent research has shown that these patterns make an important difference. Hulse and Dorsky (1977) trained rats to run down a straight alleyway in order to get food. For some animals, the reward magnitude in the goal box systematically decreased over a series of trials from 14 pellets to 0 pellets; specifically, the rats received 14, 7, 3, 1, then 0 pellets on successive runs. Control animals were given a random series of reward magnitudes; that is, the magnitude of reward did not consistently decrease over the series of trials. Hulse and Dorsky found that the reward magnitude pattern affected learning: the more prominent the pattern, the faster the animals learned to anticipate the magnitude and respond accordingly. The response expectancy developed more quickly when the reward pattern was consistent than when it was inconsistent; in fact, animals were able to anticipate the various quantities of food as shown by their faster running speed on higher magnitude trials and much slower running (compared to control animals) on the low magnitude trials. In general, then, the pattern of reinforcement, the schedule of partial reward, and patterns of magnitude influence expectancy and performance in instrumental conditioning studies (see also Hulse & Campbell, 1975).

Schedules of Reinforcement

In the preceding section we discussed the impact of reward patterns on response learning. We also noted that reward patterns invariably involve intermittent reinforcement (that is, reward follows only a portion of the CRs). In this section we will discuss reward schedules in greater detail (see also Chapter 7).

Schedules of reinforcement deserve separate and thorough discussion for several reasons. First, the schedule of reinforcement is one of the most powerful manipulations that has been investigated. It has important consequences for responding and thus has received considerable attention during the last few decades. Second, reinforcement schedules presumably reflect more accurately the natural environment of animals. Rarely in nature is a reward given following every response. Animals often must make several attempts, or wait a particular length of time, before reward is available. Consider, for example, the behavior of a small bird foraging for insects on pine cones. Each feeding site (cone) may or may not contain a desired prey item. Rewards are found only intermittently. The fact that behavior is controlled by intermittent reinforcement schedules is, therefore, an important concept that applies to virtually all species including human beings (see Schoenfeld, 1970, for a review).

There are four basic types of reinforcement schedules as depicted in Figure 4–4. For two of them, the two ratio schedules, reward is based on the number of responses that the subject makes. For the other two schedules, the interval schedules, a response is reinforced only after a certain period of time has elapsed. Each of these classifications, ratio and interval, are further subdivided according to whether the criterion is fixed or variable. That is, a fixed ratio schedule is one in which the reward is delivered following a particular number of responses; that number never varies from one reward to the next. In a fixed interval schedule, reward for a response is delivered following a given length of time; again the interval between rewards is constant. For a variable ratio schedule, reward is based upon a number of responses emitted, but the particular number varies from one time to the next. In variable interval schedules, the CR is reinforced after a certain period of time, but the precise length of time varies. Let us now consider each of these basic reinforcement schedules in greater detail.

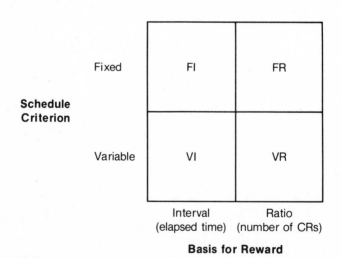

Figure 4-4.   Matrix showing the four basic types of intermittent reward schedules.

Fixed Interval Schedule

In the fixed interval (FI) schedule, where the subject is rewarded for executing a response after a fixed period of time has elapsed, the most efficient behavior for the subject is to wait until the interval has passed and then make a single response in order to earn the reinforcement. This strategy would allow the subject to expend the minimum amount of energy while it capitalized on all the rewards available after the fixed interval of time. Normally, however, subjects do not follow this strategy; they respond throughout the entire interval, although the rate of response varies markedly.

The basic pattern of responding on an FI schedule involves an increase in the rate following reinforcement. That is, subjects normally pause for some period of time following reward and then begin to execute the CR at a rather slow rate; as time passes, the rate of responding, the CRs/minute, increases to a maximum just prior to the next reward delivery (Schneider, 1969). The longer the interval, the longer the animal pauses before resuming its behavior. In addition, the more training the animal has, the longer the pause before it resumes responding (Caplan, Karpicke, & Rilling, 1973; Schneider, 1969).

Several investigators have tried to discover why animals exhibit this unique form of responding, why they pause immediately after receiving reward and then gradually and consistently increase their response rate to a maximum just prior to reward. As mentioned above, it would be more efficient for the animals to avoid responding altogether until the reward is due. Alternatively, animals could respond consistently throughout the fixed interval rather than demonstrate the increase in rate of response. It now seems clear that animals respond as they do because they develop a sense of the temporal interval between rewards. Quite simply, animals can learn to discriminate the passage of time, and they can do so fairly accurately (Church et al., 1976; Tarpy, 1969). While responding on an FI schedule, animals stop responding following reward because, in one sense, they do not expect to receive reward for some time; with greater experience, however, they become more accurate at timing their behavior. Their sense of the temporal interval between rewards improves with training (Caplan et al., 1973; Schneider, 1969). If the animals are given signals that help them to discriminate the passage of time, such as a light which increases in intensity throughout the fixed interval, then their temporal discrimination is even better; they show an even more pronounced change in rate of responding over the interval (Donahoe, 1970; Kendall, 1972). This result is evident in Figure 4–5. Caplan et al. had pigeons peck a key for food on an FI schedule. During the fixed interval a green light was gradually made brighter, thus making the temporal discrimination even easier. It is clear

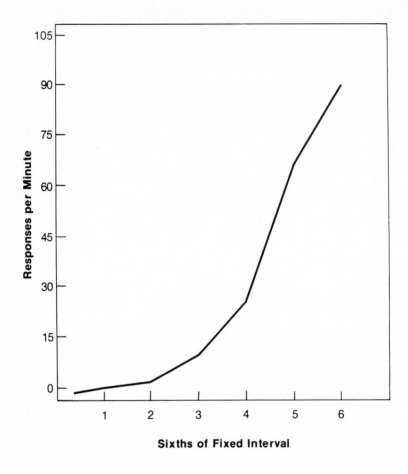

Figure 4-5. Mean response per minute as a function of sixths of the fixed interval. (From Caplan, Karpicke, & Rilling, 1973)

from the figure that response rate increased dramatically over the FI period. Finally Kello (1972) showed that merely exposing a subject to a fixed interval reward, without requiring a response, produces the typical FI behavior later when the subject is allowed to respond.

All these experiments show in one fashion or another that the pattern of behavior observed under an FI schedule is based upon a *stimulus* expectancy. The stimulus is the temporal interval which is acting as a discriminative stimulus indicating to the subject when it is appropriate to respond. As the subject learns more and more about the length of the interval, the response increasingly occurs in that period of time just prior to the reward presentation. In one sense, then, animals are approaching some kind of optimal behavior under FI schedules; they are learning to inhibit their behaviors (and thus to avoid wasting energy) until reward is available.

The pattern of behavior exhibited under an FI schedule really reflects the inhibition of delay phenomenon discussed in Chapter 3. First of all, as noted above, the behavior is largely controlled by the stimulus expectancy: the animals are using a stimulus, the temporal interval, to predict when food is available. The fact that they pause after each reinforcement and begin responding only slowly is evidence of some form of inhibition, in particular, inhibition of delay. This would seem to be true because a novel stimulus will disrupt the inhibition early in the FI interval and produce the CR (Singh & Wickens, 1968); such disinhibition is also found in more traditional Pavlovian conditioning studies.

Fixed Ratio Schedule

Under the fixed ratio (FR) schedule, the subject receives reward for executing a fixed number of responses. The overall rate of responding is much higher than that found in an FI schedule, although with very high FR values, responding often deteriorates. Deterioration of responding is in the form of long pauses following reinforcement; to this extent the response pattern for FR schedules is similar to that for FI schedules. One study by Felton and Lyon (1966) showed that the length of the pause was directly related to the size of the FR requirement. Pigeons were taught to peck a plastic disk for food (the schedule was FR–100). As shown in Figure 4–6, the animals paused for about 1.5 seconds after each reinforcement. When the response requirement was increased to FR–150, the pause lasted nearly a minute.

It is somewhat paradoxical that animals pause following the reward since the next reward depends upon their behavior, not the passage of time. The postreinforcement pause merely delays the next reward. Recent research has suggested that the postreinforcement pause following an FR reward is controlled by the temporal interval, just as it is following an FI reward (see Killeen, 1969; Neuringer & Schneider, 1968). In these studies, FI and FR schedules were directly compared by making reward available to the FI animal every time the FR subject earned reinforcement. The time between reinforcers and the frequency of reward, therefore, were identical for both subjects. The animals differed simply in terms of whether the reward was contingent on the execution of units of behavior or on units of time. Both animals showed the same postreinforcement pause, suggesting that the pause represented a temporal discrimination.

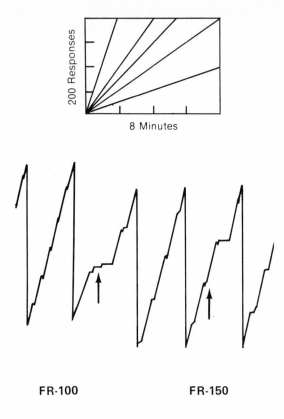

FR-100   FR-150

Figure 4–6. Cumulative record for a pigeon responding under an FR-100 and FR-150 schedule of reinforcement. Note that the steeper the line, the faster the response rate. Pauses result in horizontal pen tracings as indicated by the arrows. (From Felton & Lyon, 1966)

Variable Interval Schedule

The variable interval (VI) schedule is similar to the FI with one important difference: the interval of time between periods when reinforcement is available varies. When we refer to VI schedules, we note the average inter-reinforcement time. So, for example, a 30-second VI schedule, VI–.5, would permit the animal to earn rewards *on the average* every 30 seconds, although the actual intervals experienced by the subject could vary considerably from one reward to the next.

The characteristic style of responding under VI schedules is the maintenance of a stable but somewhat low rate of response. Recall that the frequency of reinforcement is not related to the rate of responding but rather to the passage of time. For this reason, responding is not as rapid as found for ratio schedules.

A number of experiments have indicated that rate of responding under a VI schedule is a function of the rate of reinforcement. Catania and Reynolds (1968), for instance, tested a variety of VI schedules, the average time ranging from 12 to 427 seconds. They found that responding rate increased as a function of the rate of reinforcements. Most of the subjects were making about 60 to 100 responses per minute under a VI–.2 schedule (this schedule yielded about 300 reinforcements per hour), but the rate dropped dramatically to about 20 to 70 responses per minute for a VI–7.1 schedule (which yielded approximately 8.4 reinforcements per hour). Although the rate of responding is fairly stable from one reinforcement to the next, it does increase slightly just before the next reward. The reason we do not observe a *larger* increase in rate prior to reward is that the subjects are unable to use the temporal interval to predict when reward is available. Quite clearly, the variability in the temporal interval prevents subjects from making this discrimination. Subjects are able to make only a crude estimate of the average temporal interval.

Variable Ratio Schedule

The final basic schedule is the variable ratio (VR) schedule. This is similar to the FR schedule except that the actual number of responses required to produce reinforcement varies from one reward to the next. We describe the VR schedule in terms of the average number of responses required. For example, if an animal were reinforced after an average of 10 responses, we would designate the schedule as VR–10, although the actual number of responses required may vary considerably.

Responding under a VR schedule is the highest of all four basic schedules, and it is normally very stable. The high rate of response is due to the fact that the subject's own behavior controls the frequency of reward: the faster the subject performs, the more frequently it is reinforced. The stability of the response rate is attributed to the fact that the actual number of responses required to produce reinforcement varies. Thus, the animal does not develop a very accurate expectancy about the number of CRs required for reward. If, however, the VR requirement is exceedingly high, subjects may stop responding for short periods of time following reward (Ferster & Skinner, 1957).

Comparison of Ratio and Interval Schedules

It is important to note that ratio schedules sustain a much higher level of responding than interval schedules. The reason for this, of course, is that the reward rate can be controlled under a ratio schedule (the faster the responding, the more frequent the reinforcers), whereas the reward rate cannot be affected by the response rate under interval schedules (no matter how fast the animal responds, reward is available only after the specified amount of time has elapsed).

Is the higher frequency of reinforcement under ratio schedules actually responsible for this disparity in rates? The answer to this question seems to be no. Killeen (1969) compared FR and VI reinforcement schedules. One subject, the VI animal, received reward only when the other subject, the FR animal, was fed. Both animals, therefore, received the *same* frequency of reinforcement, although they differed in terms of the response requirements. Killeen found that the FR animals performed at a much higher rate than the interval subjects. This discrepancy in response rate, therefore, could not be attributed to a difference in frequency of reinforcement.

A more suitable theory to explain the difference between response rates focuses on the interresponse time (the pause between each CR). Consider the likelihood of being reinforced following a very low response rate. With longer interresponse times, the probability of reward increases under an interval schedule. That is, the slower the animal responds, the more likely it is that the next response will be reinforced because the next response is always closer to the end of the time interval. The situation is different for ratio schedules, however. A very low response rate under a ratio schedule decreases the probability of reinforcement. Long interresponse times postpone the reinforcement and thus lower the probability. Let us state the argument in reverse: ratio subjects are reinforced for responding with short interresponse times (that is, with a high rate of response), whereas interval animals are not. This accounts for the discrepancy in rate.

This theory has been supported by several people, for example, Dews (1969). Using interval schedules, Dews demonstrated that the interresponse times just prior to reward for subjects rewarded on interval schedules were relatively long; correspondingly, short interresponse times were almost never followed by reward. For ratio schedules, however, short interresponse times were selectively reinforced. Quickness in responding was a virtue since it speeded up the point of reinforcement.

DRL Schedules

In the above section we suggested that animals can be selectively reinforced for responding with short interresponse times (see Shimp, 1967).

It is also the case that animals will learn to respond with unusually long interresponse times. This occurs under a DRL schedule, which stands for differential reinforcement of low rates of responding (see Kramer & Rilling, 1970). Under a DRL schedule, the subject must withhold its response until a certain period of time has elapsed. If the subject is successful in this task, then the *next* response it emits will be reinforced. An error, that is, a response prior to the criterion time, causes the timing apparatus to be reset and the interval begins again.

Initially, performance under a DRL schedule is inefficient. The animal is learning two conflicting habits: the response rate is *increased* because responding eventually leads to reward, but it is also *suppressed* because responding too soon postpones reward. With enough training, however, responding does stabilize and animals perform efficiently under the DRL schedule. Two factors must develop before efficient responding is possible. The first is inhibition; this tendency keeps the subject from responding. The second factor is temporal discrimination; the subjects must learn how long to wait before they can respond to receive food. This accounts for the high probability of responding once the appropriate time interval has elapsed (see Gage, Evans, & Olton, 1979).

One very interesting and unique feature about responding under DRL schedules is that subjects engage in collateral behavior while they are temporarily suppressing the CR. Some of the animals may assume idiosyncratic postures; others may nuzzle the food cup, sniff, or chase their tails (Laties, Weiss, & Clark, 1965). Psychologists have been somewhat puzzled by this phenomenon. There is some evidence suggesting that collateral behavior is a chain reaction; each response is an inherent part of this chain and serves as a signal for the next response. The collateral behavior is, in effect, the animal's way of timing the interval (see Laties, Weiss, & Weiss, 1969). Alternatively, other psychologists believe that these strange behaviors simply interfere with the performance of the CR; interference with the CR, of course, is precisely what is needed to increase the efficiency of responding under the DRL schedule. Animals can't perform the CR and the collateral behavior at the same time; a deterioration of the CR is, therefore, beneficial (see Schwartz & Williams, 1971). Although collateral behavior is still in many ways a mystery to scientists, recent evidence does suggest that it improves the efficiency of DRL responding by interfering with the CR rather than by playing a special mediating role (see Hemmes, Eckerman, & Rubinsky, 1979).

Complex Schedules

Psychologists are not limited to the four basic schedules discussed previously; more complex arrangements can be achieved by combining schedules in various ways. For example, when reward is contingent not just on the successful completion of one schedule but on completing the requirements of two distinct schedules at the same time, responding is said to be under the control of a compound schedule. One illustration of this might be an FI-FR schedule. Under this arrangement reinforcement is available only if the subject executes a specified minimum number of responses within a specified time. If either condition is not met, reinforcement is not delivered. Normally, behavior under such compound schedules is a combination of the response strategies found under each separate schedule. That is, for the FI-FR schedule, postreinforcement pauses are evident, an increase of response rate is shown following the pause, and the response rate just prior to the next reward is higher than would be found under an FI schedule only.

There are several ways of presenting schedules sequentially. Under a tandem schedule, the subject must complete the requirements of each individual schedule in succession before reinforcement will be given. An FI-FR tandem, for example, would require that the subject wait the fixed interval of time and then, subsequently, execute a fixed number of responses before reward would be presented. If each separate schedule is signaled by an external cue (for instance, a light is turned on during one of the schedules and a separate light during the other), then this schedule is called a chain schedule.

In a third arrangement, called a mixed schedule, the subject may obtain reinforcement during each individual component schedule, but various schedules are presented sequentially, in a random order. If each individual schedule is signaled by a discriminative cue in the apparatus, the arrangement is called a multiple schedule. Usually, the rate of responding is determined not only by the component schedule currently in effect, but also by the other component schedules which comprise the series. That is, the subject's response rate at any given time may be influenced by the schedule in effect at that time as well as by previous or subsequent schedules (see Williams, 1979; and the section on contrast later in this chapter).

Reinforcement Schedules in the Real World

It has long been claimed that the importance of studying reinforcement schedules lies in the fact that behavior in the real world is controlled by reward and that the reward is meted out under various complex contingencies or schedules. There is no doubt that the behavior of animals, including human beings, is controlled to a considerable degree by reinforcers. It is also clear that reinforcers are invariably given on an intermittent basis. We are not rewarded for every unit of behavior we execute.

Given this reasonable point of view, it is perhaps surprising to discover that good examples of reinforcement schedules in real life are difficult to cite. The clearest example is the behavior of gamblers who use slot machines. The rewards (payoffs) are delivered according to a VR schedule. The behavior of gamblers, that is, depositing the tokens, is maintained at a high and stable level by this schedule. Examples of other basic schedules are less clear, however. One's weekly or monthly paycheck is often cited as an example of an FI schedule. Similarly, payment by commission, called piecework, is, perhaps, an example of FR reinforcement: the person receives pay only after completing a fixed amount of labor. Finally, fishing could be cited as an example of a VI schedule. The reward is achieved after a certain period of time, although the exact time varies.

Part of the reason it is difficult to identify *unambiguous* examples of basic reinforcement schedules in the real world is that we often do not know what the CR is. Behavior, at times, is exceedingly complex, comprised of many components. Is the CR which is being reinforced the overall pattern or the individual components? Take fishing. What, precisely, is being reinforced? Casting the line? (If so, the rewards might better be described as under the control of a VR schedule.) Waiting patiently and watching for signs of a catch? (If so, our analysis breaks down because "waiting" is not the behavior that catches fish.) Take the example of one's paycheck. What precisely is being reinforced there? It certainly cannot be the work unit, since there doesn't appear to be a contingent relationship between the pay and those units. In summary, we often have difficulty identifying the behavior that is reinforced in the real world and, therefore, describing the schedule under which it is reinforced. While the study of reinforcement schedules continues to be an important focus in learning research, we must be cautious in applying our knowledge to the real world. The analogy between the reinforcement schedules studied in the laboratory and those found in the real world may not be as close as we have claimed in the past.

Basic Instrumental Phenomena

The discussion up to now has concerned many of the basic principles of response learning. In the following sections, we discuss a number of special phenomena.

Contrast

One of the most important and widely investigated phenomena in instrumental conditioning is contrast. Consider the following situation. Two groups of animals are trained to perform a CR for different levels of food reward. As noted earlier, the larger food reward invariably leads to a higher level of performance. After a sufficient training period, therefore, the animals in the large reward group will be found to run faster than those in the small reward group. Now imagine that we have two more groups.

Animals in one group receive the large reward, while those in the other receive the small reward. The difference is that after a sufficient degree of training, each of these groups is switched to the other reward condition. That is, the large reward animals are shifted so that now they receive small reward. Similarly, the small reward animals are shifted to the large magnitude. Finally, we compare the behavior of all four groups following the switch in reward level for two of them. One could hypothesize that the two groups that end up receiving large reward would perform at about the same level; likewise, the two groups that end up receiving the small reward should perform at the same level (but, of course, lower than the high reward animals).

This is not at all what is normally found. Usually the groups that are shifted *overshoot* the performance level of the constant groups. That is, the group shifted to high reward performs at a *faster* rate than the control group that received the large reward all along. Similarly, the group that is shifted from high to low reward performs even more poorly than the group that received low reward throughout training. The former is called positive contrast because the behavior is inflated beyond what normally occurs in the unshifted group. The second result is called negative contrast; when the reward conditions get worse, the animals overshoot and perform at an even lower level than normal (see Dunham, 1968). At the very least, contrast indicates that behavior is not due simply to the conditions that exist currently; performance is also influenced by previous reward conditions.

Contrast effects can be obtained in several different ways. The hypothetical experiment cited above exemplifies successive contrast. In this situation the animals experience one reward condition following their instrumental CR and then later are shifted to a different reward condition. Comparison usually is between two different groups (the terminal performance level for the shifted group relative to the performance level for an unshifted control group).

The second major technique for demonstrating contrast is called the simultaneous method. Here, experimental subjects receive a mixture of both high and low reward trials. Control subjects, of course, receive only one type of reward condition, either low or high. Throughout training, it is observed that the experimental subjects, who get both types of reward, perform at more extreme levels than the control subjects. That is, on the low reward trials the experimentals' performance is slow relative to that of the constant-low control subjects, but on the high reward trials, the experimentals perform at an even faster rate.

The most prominent factor that produces contrast is a shift in the amount of food reward (see, for example, Marx, 1969; Mellgren, 1972; Shanab, Sanders, & Premack, 1969). In a study by Mellgren, Wrather, and Dyck (1972) three groups of rats were trained to run down a straight alleyway to obtain food. Some of the animals received 8 food pellets in the goal box, while others got 1 pellet. In addition, a group of experimental subjects got a mixture of reward magnitudes; on some trials the reward was 8 pellets, on others it was 1. This is a study of simultaneous contrast since the experimental animals received both high and low reward throughout training. The authors measured the speed of running. The last two trials on each day's session were especially revealing. As shown in Figure 4–7, the Constant-8 group performed at a higher level than the Constant-1 group. This difference simply reflects the well-known fact that higher reward levels produce faster running speeds. The interesting outcome, though, was the performance of the experimental subjects. On trials on which they received 8 pellets, that is, the large reward, their running speed exceeded that of the controls. Similarly, when the experimentals got only the single pellet in the goal box, they ran even slower than the Constant-1 group. The data in Figure 4–7, therefore, illustrate both positive and negative contrast.

When the magnitude of reinforcement is manipulated in escape learning, contrast effects have also been observed. For instance, Nation, Wrather, and Mellgren (1974) conducted a study on successive contrast. They trained

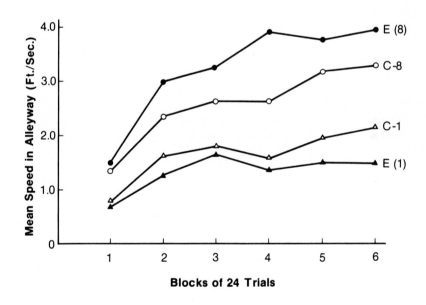

Figure 4-7.   Mean speed in the alleyway as a function of blocks of trials. Group Control-8 received 8 pellets in the goal box; group Control-1 received 1 pellet; group Experimental received both 1 and 8 pellets. (From Mellgren, Wrather, & Dyck, 1972)

rats to escape shock by running down an alleyway to a safe goal box. Shock intensity initially varied among groups (either .2, .4, or .8 ma) and, therefore, the magnitude of reward (amount of shock reduction) differed too. One group of subjects received the same shock intensity throughout (.4 ma), whereas the other two groups were shifted to the .4 magnitude of shock (reinforcement) after 20 trials. As shown in Figure 4–8, performance shifted following the change in shock level and overshooting occurred. The animals shifted downward performed even slower than the .4 ma controls, and the animals shifted upward did even better. Similar results have been observed by McAllister, McAllister, Brooks, and Goldman (1972) for subjects that were allowed to escape from fear-provoking stimuli.

Changes in the quality of the reinforcing US also can lead to contrast effects. Several studies (for example, Flaherty & Largen, 1975; Rabin, 1975) have shown that the rate of behavior varies as a function of sucrose concentration (higher levels of sucrose are more reinforcing for rats). When these levels are shifted, animals overshoot the control levels, demonstrating both positive and negative contrast effects.

A third factor that can lead to contrast effects is a shift in the delay of reward. Recall that a delay in reward decreases performance relative to immediate reward. If the delay level is suddenly shortened, thereby making the reward more favorable, performance shifts and overshoots the level of the control subjects that were maintained at that shorter delay all along. Conversely, if the reward is delayed even further, then the behavior deteriorates beyond the control level (Beery, 1968; MacKintosh & Lord, 1973; McHose & Tauber, 1972; Shanab & Biller, 1972).

Finally a great many studies have shown that changes in the reward schedule also lead to contrast (see MacKintosh, Little, & Lord, 1972; Wilkie, 1977). In the Wilkie experiment, for example, pigeons were trained to peck a plastic disk for food. After a certain period had elapsed, a second reward schedule came into operation. For some animals this second schedule produced the same frequency of reinforcement as the first schedule; for other subjects, however, the second schedule yielded a lower frequency of reward. When the first schedule came into effect again, Wilkie found that responding differed between these two groups. Although animals in both groups were now receiving the same frequency of reinforcement, the subjects that had just experienced a reduction in frequency actually responded at a faster rate. In other words, the pecking behavior increased in subjects that experienced both a high and low yield schedules relative to subjects that experienced only the favorable schedule.

How are we to explain positive and negative contrast? What sorts of processes might be occurring to make animals overshoot the performance level once the reward conditions have shifted? It has been claimed that incentive motivation underlies this phenomenon (see Chapter 10). Recall our emphasis on the survival role of unconditioned stimuli. Food represents

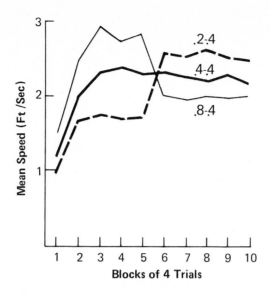

Figure 4-8. Mean running speed in the alleyway during forty shock escape trials. Groups received either .2, .4, or .8 ma shock followed by .4 ma. (From Nation, Wrather, & Mellgren, 1974)

a valuable resource for hungry organisms and so we might assume that the level of food reward affects the motivation of the animal. When the amount of reward is decreased, causing negative contrast, the new level is aversive; correspondingly, the enticement provided by the more favorable level of food reward, that is, the incentive motivation, is less. In fact, animals will avoid making a response that brings them into contact with this new low food level, presumably because it has taken on aversive qualities (see Eisenberger, Frank, & Park, 1975). Positive contrast, on the other hand, reflects "elation"; the incentive or enticement value of the new food level increases the animal's motivation and produces elevated performance. In other words, shifts in the magnitude or quality of reward change the overall motivational state of the animal, causing changes in performance level.

Although we have noted the role of incentive motivation, we have not really explained why overshooting takes place. It certainly would be plausible for animals to adjust their performance levels to accommodate new reinforcement levels without actually overshooting the performance level of control subjects. The fact that overshooting does occur is extremely important to our theory that instrumental conditioning is, fundamentally, the acquisition of a response expectancy. Quite simply, animals come to expect some level of reinforcement, and they are known to average their experience over time (McHose & Peters, 1975). When their expected level is not met, that is, when their response expectancy is violated, reward has an unusually large impact on performance. In short, animals adapt to a particular level of reinforcement and react with heightened emotion and motivation when these levels are changed. We believe that the reaction to a change in reward level is, indeed, an emotional one since investigators have failed to observe contrast effects in animals that were given tranquilizers which presumably blocked their emotional reaction (see Rabin, 1975).

How can we reconcile this phenomenon with evolutionary concepts? In what sense is it advantageous for animals to behave in this fashion, to perform at exaggerated speeds following shifts in reward? The answers to these questions are not easy because little work has been done on these issues. Nevertheless, it would appear that animals are attempting to maximize the reward value available to them. Frequent, immediate, and large rewards are clearly to their advantage; it would be adaptive to maximize those qualities while, at the same time, to minimize the energy expended for reward. But organisms also experience variability of reward in their environment. On some occasions food is abundant, while on other occasions it is scarce. It makes no difference whether we are referring to variations over a day, a month, or years. It is clearly beneficial for the animal to reflect upon this variability and to remember previous reward levels. If subjects do remember previous levels of reward, then they can appreciate the possibility that those levels will recur. For example, if an animal were adapted to a low level of reinforcement and suddenly were shifted to a higher level, then a wise strategy would be to harvest at an extra vigorous rate while the food was available; at any time the reward supply could revert back to the previous unfavorable level. Similarly, if an animal were used to an abundant food source and suddenly experienced low levels of reward, then it might be to the animal's advantage not to expend too much energy trying to earn food; it is perhaps likely (in nature at least) that the former high reward conditions will once again prevail. We do not know if this line of argument accurately reflects the underlying processes in contrast. Many other factors including the time and energy cost of responding, the degree of food variability, and so forth, are also important. But we do know that contrast can be an enduring phenomenon and that it reflects changes in the emotional and motivational state of the animal.

Biofeedback

Biofeedback is the use of instrumental conditioning techniques to train internal physiological responses such as heart rate, GSR, blood pressure, and others. This research is called "biofeedback" because it involves rewarding the subject, or otherwise providing some sort of feedback, contingent on a biological, usually autonomic, response. For human beings, the reward is often praise, money, avoidance of shock, or information concerning the correctness of the behavior. For lower species, various rewards have been employed including food, shock avoidance, and pleasurable brain stimulation.

Research on biofeedback is important for a number of reasons. First is the mere fact that biological responses can be instrumentally conditioned. For many years, the visceral reactions of the body were considered to be involuntary in nature. Such reactions can, of course, be classically conditioned because the strong USs that we employ invariably evoke them. In response learning, however, where the CR must occur prior to the reward, involuntary responses were not thought to be conditionable; after all, if these behaviors are involuntary, how can they be performed in order to achieve an instrumental reward?

After nearly two decades of research, it is abundantly clear that instrumental conditioning techniques are perfectly suited for conditioning visceral reactions (see Schwartz & Beatty, 1977). Consider a recent example by Kimmel, Brennan, McLeod, Raich, and Schonfeld (1979) who conditioned the skin conductance response in eight monkeys. The animals were placed in a harness and given a series of signals, either red or green lights. For some of them, the red light signaled an avoidance procedure: during the light any change in the skin conductance postponed shock for 40 seconds. During the green stimulus, however, these animals were punished with shock for emitting the skin conductance response. The authors found that the monkeys learned very readily, within one session, to respond accordingly. As shown in Figure 4–9, the average response rate was much higher during the avoidance segments than it was during the punishment segments of the session, indicating that those conditioning procedures had a powerful effect on the animal's behavior (see also Harris & Brady, 1974; Miller, 1978).

One important question that has received great attention concerns the mechanism for biofeedback. According to some early investigators (Crider, Schwartz, & Shnidman, 1969; Miller, 1969), the instrumental rewards were acting *directly* to change the autonomic responses; that is, visceral changes were not merely a by-product of some muscle behavior. More recently, the evidence seems to favor a different position (see, for example Black & Toledo, 1972; Goesling & Brener, 1972; Katkin & Murray, 1968; Miller & Dworkin, 1974; Obrist, 1976). These investigators claim that the somatic (muscle) and autonomic (visceral) systems are integrated; instrumental conditioning appears to act not on one or the other alone but on both in an integrated fashion. In short, visceral reactions can be conditioned, but muscle and skeletal movement seem also to be involved (but see Brener, Phillips, & Connally, 1977; D'Amato & Meinrath, 1979).

Instrumental techniques have been used to condition some rather specific and astonishing patterns of physiological reactions. Schwartz (1972, 1975) describes how various patterns were trained in humans. Some sub-

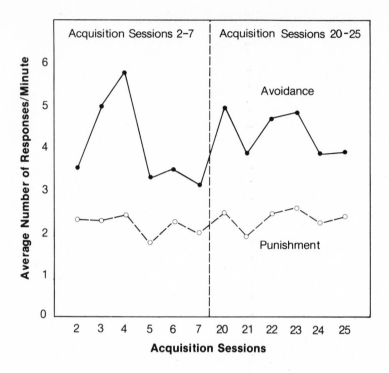

Figure 4-9. Average number of skin conductance responses per minute for eight monkeys as a function of the acquisition session under both punishment and avoidance contingencies. (From Kimmel, Brennan, McLeod, Raich, & Schonfeld, 1979)

jects were reinforced for changes in heart rate independent of changes in blood pressure. Some people learned to raise or lower both types of reactions while others raised one but lowered the other.

Biofeedback also has potential for use as a treatment of medical disorders. In laboratory investigations, some researchers have found marked improvement in patients who had serious problems in regulating their blood pressure (for example, Benson, Shapiro, Tursky, & Schwartz, 1971). Unfortunately, many of these patients were unable to maintain control over their visceral responding once they were living again in the hectic and rapidly changing world outside the laboratory (see Blanchard & Young, 1973; Fuller, 1978; G. E. Schwartz, 1973). To summarize, biofeedback has become an important frontier in learning research. Although instrumental techniques have been used to train visceral responses in animals, most of the focus in recent years has been on the applications of biofeedback techniques to human therapies.

Choice Behavior: Matching

In recent years there has been a considerable effort to generate *quantitative* laws of behavior. Psychology has had a number of theories about behavior, particularly the effect of reinforcement on responding, but precise mathematical formulations have not always been forthcoming. In the area of choice behavior, particularly regarding the choice between two schedules of reinforcement, a major breakthrough has occurred (see Baum, 1974; de Villiers, 1977; Herrnstein, 1970, 1974, for reviews). There is now a vast literature on a phenomenon called "matching." The matching law is a mathematical statement describing the relationship between rate of responding and rate of reinforcement. Specifically, animals are said to match their rate of behavior to the relative rate of reward for that behavior.

Let us first consider a typical experiment that shows response matching. One study on this topic was done by Herrnstein (1961). He trained pigeons to peck at two keys for food. Each of the response keys was associated with a different VI schedule. The bird was free to respond to either key, that is, free to choose between reinforcement schedules. Clearly, it would be in the subject's best interest to peck the key associated with the more favorable reinforcement schedule. What Herrnstein found, however, was that the subjects *matched* their responses to the frequency of reinforcement received on each schedule. That is to say, response rate on one key, relative to the other key, matched the frequency of reinforcement for that key, relative to the frequency of reinforcement for the other key.

In more formal terms the matching law is stated in the following equation:

$$\frac{R_a}{R_a + R_b} = \frac{r_a}{r_a + r_b}$$

In this equation the "R" terms stand for number of responses on key "a" or "b"; the "r" terms stand for the number of reinforcers received as a consequence of pecking key "a" or "b." Let us substitute some numbers and show how this equation operates. Assume that one schedule, schedule "a," is a VI-1-minute, whereas the other schedule, schedule "b," is a VI-2-minute. This means that the VI-1 schedule will yield twice the reinforcements of the VI-2 schedule. Given these facts, the frequency of reinforcement for these two schedules might conceivably produce, say, 50 reinforcers for schedule "a" and only 25 for schedule "b." This would yield a ratio of .67 according to the right-hand term of the above equation. What is found is that the responses match this ratio. That is, the number of responses to key "a" divided by the total number of responses (left-hand term in the above equation) would be .67 as well. In other words, when confronted with the choice between two schedules which differ in terms of the frequency of reinforcement, an animal will not simply respond to the better of the two;

it will divide its responses in such a way that the relative number of responses to one schedule will match the relative number of reinforcers for that schedule.

Since the formulation of this matching law, many studies have confirmed that it holds for a wide variety of conditions. The matching law appears to be a truly general and quantitative statement concerning the relationship between reinforcement and responding. For example, various studies have shown that subjects match their response rate not only according to the frequency of reinforcement, but also according to the amount of reinforcement received for each alternative (Catania, 1963), the duration of reinforcement (Keller & Gollub, 1977), the quality or type of reward such as different kinds of food items (Miller, 1976), and the immediacy of the reinforcement (Chung & Herrnstein, 1967). Anything that defines the quality or value of a reinforcer influences the response to the reinforcer (the choice of the animal); the choices or responding levels, in turn, tend to match the reinforcer qualities. The matching law does not even require that animals have contingent reinforcers. Even if the frequency of reinforcement is changed by the delivery of extra "free" reinforcers, the response rate will still match the frequency (Rachlin & Baum, 1972). In addition, matching does not depend upon merely the choice between two alternatives. Miller and Loveland (1974) showed, for instance, that pigeons would apportion their responses to five different response keys according to the relative reinforcement rate associated with each key.

Response matching is a quantitative law that applies to responses and aversive reinforcers as well as appetitive reinforcers. Several studies, for example, have shown that matching occurs in aversive situations (Baum, 1973; de Villiers, 1974). In the study by Baum (1973) subjects could stand on either side of a cage. Responding produced a time-out from shock, but the rate of reinforcers, the frequency of time-outs, differed for each side of the cage. Baum discovered that the subjects matched their relative time on each side to the relative frequency of time-outs. When one side was associated with twice as many time-outs from shock as the other side, the subject spent twice as much time on that side.

How do subjects actually match their responses to the relative reinforcement rate? One could imagine, for instance, that a subject could respond to two response keys an equal amount of time, but simply respond faster to one of them. Alternatively, the subject could respond at an equal rate to both keys, but simply spend longer responding to one alternative. The latter seems to be more consistent with the subject's behavior (see Baum & Rachlin, 1969; Brownstein, 1971). That is, subjects match their responses by spending a greater amount of time responding to the more desirable alternative. The greater the reinforcing value of one response, relative to other choices, the greater the time investment in that behavior.

The matching law is an important contribution to our understanding of behavior, specifically the quantitative relationship between reinforcement value and responding. It has given rise to a variety of formulations and laws that are designed to deal with other types of situations (see de Villiers & Herrnstein, 1976). Most important, however, the matching law can tell us a great deal about reward value. The study by Baum and Rachlin (1969), for example, showed that animals will allocate their time according to the payoff of the choices. In one sense, then, we have a way to assess the potency or value of a reinforcer. All one needs to do is to provide the animal with a choice between that reinforcer and some other reward and observe the relative response rate. Furthermore, we can manipulate the nature of the reward and observe how it affects the rate of responding. A study by Holland and Davison (1971) provided a good example of this. Their subjects could respond to one of two VI schedules. One schedule controlled the rate of food presentation; the other schedule of reinforcement programmed a different frequency of pleasurable brain stimulation. The authors found that matching occurred even when two completely different forms of reinforcement were provided. Using such a technique, one could determine how many food pellets were "equal" to a given number of brain stimulations.

Matching is a global measure; it assesses the total number of responses and the total number of reinforcers over an entire session. However, if one looks at the short-term rate of response, a moment-by-moment analysis *during the session,* then one sees a more complicated picture. For instance, Shimp (1966) found that subjects were choosing those responses that momentarily maximized the probability of reinforcement. That is, by switching from one response alternative to the other, subjects were actually increasing the probability of receiving reinforcement over the *short term.* What appears to the psychologist on a global level as a matching strategy, is actually a maximizing strategy when viewed in a more detailed fashion (see also Nevin, 1969, 1979; Silberberg, Hamilton, Ziriax, & Casey, 1978).

We should reemphasize the importance of the matching law as a general description of response strategies based on reinforcement frequencies. First of all, animals in their natural habitat invariably have choices to make. They are not constrained to a single source of food or a single type of response. Therefore, the study of choice is important because through it we can begin to appreciate how animals allocate their energies and time to various reinforcement choices. Second, the matching law helps us to understand how animals contend with competitors. Not only do animals have choices among resources, but they also must contend with other animals in the environment. It turns out that animals in groups tend to match the reinforcement frequencies. Graft, Lea, and Whitworth (1977), for instance, investigated the responding strategies of five rats housed together. The rats had access to four different levers, each of which was associated with a different VI schedule. The authors found that the rats, collectively, matched their responding to the relative frequency of reinforcement. That is, they responded to each lever in the same proportion that they received reinforcement from that lever. It is unlikely that momentary maximization could account for this phenomenon. The subjects were relatively unconstrained in this "seminatural" environment and therefore could afford to adopt any strategy they chose. The fact that matching occurred suggests that it reflects a fundamental characteristic of animal behavior.

Finally, the matching law provides the same sort of perspective that the Wagner-Rescorla model did in stimulus learning (see Chapter 3). Recall that the Wagner-Rescorla model demonstrated how it is possible for the strength of a stimulus to be affected by the strength of other stimuli in the environment. As one stimulus increases in strength, the other stimuli "suffer" and decrease in strength. The same is true for response learning: if one response increases in strength, that is, the correlation between the response and the desirable outcome increases, then responding is proportionately shifted toward that choice and away from the others. Different responses, in a sense, "compete" for strength based upon the correlation between the response and the reinforcing outcome. The number of responses to choice "a" is influenced not only by the frequency of reinforcement associated with "a" but *also* by factors that are associated with choice "b." Choice "a" will suffer or benefit depending upon the reward parameters associated with schedule "b." In summary, then, response learning and stimulus learning appear to be quite similar in this regard. In both cases, when alternatives are available, the strength of one alternative, either a stimulus or a response, depends upon the strength of the other alternative.

Summary

Response learning, or instrumental conditioning, involves the acquisition of an expectancy based upon the organism's own behavior. The strength of the expectancy, in turn, depends upon the response-outcome correlation. When the correlation is high, the response has predictive value and the behavior reflects strong conditioning. There are other important parameters of instrumental conditioning as well. The type of behavior designated as correct by the experimenter has an effect on the ease of conditioning. In particular, animals will spend as little effort as necessary to gain the greatest relative amount of reward; moreover, not all behaviors will be learned with equal facility.

The US intensity affects instrumental conditioning as it does classical conditioning. Positive relationships have been observed between the amount learned and the US intensity in reward, punishment, escape, and several forms of avoidance learning. Shuttlebox avoidance, however, is inversely related to shock intensity because in that situation the tendency to avoid shock is in conflict with an opposing tendency to avoid the other side of the apparatus (the other side having been associated with shock on the previous trial).

Delay of reinforcement is another potent variable in instrumental conditioning. The longer the US presentation is delayed, the poorer is reward conditioning. In punishment training, delay of punishment results in less suppression. Short delays make an important difference to performance although cues that intervene between the response and the US may help to bridge the gap and, thus, attenuate the effect of the delay.

The fifth variable is US quality. Most reinforcers are potent, like food or shock, but sensory changes and various naturalistic stimuli may also serve as reinforcers in instrumental learning.

Finally, the pattern of the US magnitude or the schedule is another factor that influences learning.

One of the most important determinants of behavior is the schedule on which the reward is given. Intermittent reward may be given after the execution of a fixed or variable number of responses or after a particular length of time has elapsed. These basic reward schedules have characteristic response patterns associated with them. The feature that affects rate of response appears to be the selective reinforcement of either fast or slow interresponse times. This factor accounts for the differences observed between ratio and interval schedules. In addition to the basic schedules, various "special" schedules, such as the DRL schedule, or combinations of schedules may be used. Although schedules are exceedingly important in learning research, it is difficult to cite unambiguous examples that exist in the real world.

There are three basic instrumental conditioning phenomena of note. One of the most important is contrast, or the finding that animals, for whom the reward magnitude is changed, overshoot the performance level of control subjects that received that magnitude throughout training. Specifically, an increase in reward produces faster levels of responding than found among the constant-high control subjects; conversely, a decrease causes the experimental subjects to perform even slower than animals maintained at that constant-low level. Similar changes are observed when quality, delay, or schedule are manipulated. The reason for such an effect appears to be a change in motivational state brought about by shifts in reward levels.

Biofeedback, the second important phenomenon, refers to the use of instrumental training techniques to condition visceral reactions, even subtle physiological patterns. This phenomenon is of interest because such reactions were thought to be involuntary and therefore not amenable to instrumental procedures.

The third noteworthy phenomenon concerns response matching. Animals that have a choice between two reward schedules will match their response rate for one schedule, relative to the rate for the other, to the ratio of rewards obtained on the two schedules. Such an outcome provides an accurate quantitative description of choice behavior.

5

Stimulus-Response Interactions

Introduction

Simultaneous Stimulus-Response Learning

Two-Factor Theory

Conditioned Emotion

Interaction Experiments

Summary

Introduction

In the last three chapters we have considered both stimulus and response learning, but we have only briefly discussed the possibility that these two learning processes occur together and interact (see Chapter 2). In actuality, response and stimulus learning take place together under all but the most constrained and specialized laboratory conditions. This is a most important lesson. Stimulus learning, or Pavlovian conditioning, and response learning, or instrumental conditioning, *always* coexist in any learning situation. There is virtually no case where the type of learning is purely response learning, nor is there any example of pure stimulus learning. Both processes are simultaneously present *even though* the experimenter may be interested in only one of these processes. In other words, an experiment that ostensibly focuses on stimulus learning also entails response learning even though the experimenter may not be interested in that process at that particular time. Similarly, all response learning experiments also involve simultaneous Pavlovian conditioning even though the investigator may not measure that process directly.

Simultaneous Stimulus-Response Learning

Consider the prototypical Pavlovian experiment. A dog is trained to salivate to the sound of a buzzer through the repeated presentation of the buzzer and food powder. An instrumental response component is certainly present. The dog, for example, may adjust its body position in order to avoid spilling any of the precious food; it may position its mouth in a particular way, or it may remain motionless so that no food is lost. In this way the animal's behavior may influence the quantity of the US it receives. It may optimize the immediacy or amount of the food. *Any* response on the part of the animal that influences the US presentation is, by definition, an instrumental or response contingency.

The argument becomes even clearer when we consider aversive conditioning. Imagine an experiment in which a light is paired with shock. The Pavlovian response would be paw flex, heart rate acceleration, or any of a number of CRs. Covertly, however, there exists an instrumental contingency. In particular, the dog may adjust its body position in order to minimize the impact or aversiveness of the shock; it may freeze or even try to struggle in order to lessen the pain when the shock is delivered. Such reactions would be learned instrumentally.

We can make the identical claim for instrumental conditioning studies. All experiments on response learning also involve Pavlovian conditioning. Take the simplest example of instrumental conditioning—learning to press a lever in order to receive food. Superimposed on this otherwise instrumental study is a stimulus learning paradigm. The sight of the lever (CS) is paired with food (US); other stimuli in the apparatus, or stimuli that are internal to the animal such as those that arise from food deprivation, are also correlated with the reward presentation. The presentation of two stimuli independent of the animal's behavior is, by definition, an example of Pavlovian conditioning. In short, it is impossible to conduct a "pure" response learning study without, at the same time, involving Pavlovian contingencies.

The fact that the two types of conditioning procedures occur together makes it virtually impossible to separate the processes. Consider a common example: a child approaches a stove which has a red-hot burner. For one reason or another, perhaps out of curiosity, the child touches the hot burner. The hand is immediately withdrawn; the child cries and runs away. On future occasions, when the child sees the red-hot burner, the child may show signs of anxiety, may maintain a very healthy distance from the burner, or may even run from the room to seek comfort. Quite clearly, the child has learned a variety of things, but what elements are due to response learning and what elements reflect stimulus learning?

In any complex learning situation, including naturalistic situations, the only way to identify the response and stimulus components is first to identify stimuli that are presented independent of the organism's behavior (the Pavlovian component) and then to identify responses that clearly modify or influence the US presentation (the instrumental component). In other words, to analyze complex situations one must go back to the original operational definitions given in Chapter 2: the Pavlovian components occur when a stimulus and an outcome are correlated independent of the organism's behavior, whereas the instrumental components involve a contingency between the US presentation and the response, that is, the response influences the US in some fashion.

Using these operational definitions, we may find it somewhat easier to analyze the example given above. When the child sees the burner, we can assume that a "CS" is presented. Touching the burner is one of many responses that the child could perform. The feeling of pain is the US arising from touching the burner, and it is paired with the sight of the burner (stimulus learning). The US is also a consequence of touching the burner and so it leads to response learning. To be more specific, the response of touching is punished by the US; the response expectancy is that if the response is made, the outcome is the punishing US. The child has, in effect, learned two things: red stove burners are associated with aversive USs (stimulus learning) *and* touching burners is correlated with aversive USs (response learning).

When the child encounters the bright red stove burner in the future (CS), we may again observe the operation of both stimulus and response learning. The sight of the burner, the CS, may elicit various Pavlovian reactions such as anxiety, accelerated heart rate, and the like, as well as behavior reactions that would be analogous to limb withdrawal. In terms of instrumental conditioning, there is the expectancy that if the burner is touched, pain is felt. This expectancy gives rise to an avoidance response: the child remains at a distance or even runs away in order to avoid getting burned. It may be more accurate to say that the child runs away in order to terminate the fearsome CS that developed. That is, the CS, because it induces anxiety, is itself aversive, and the child will respond instrumentally to get rid of it. In any case, there is an avoidance response (instrumental), and physiological or emotional reactions are elicited (Pavlovian) when the child encounters the CS.

Let us consider this instrumental/Pavlovian interaction in somewhat more abstract terms. Observe the diagram shown in Figure 5–1, which is similar to the diagrams in Chapter 1. Figure 5–1 represents the joint occurrence of stimulus and response learning. An animal is confronted with a stimulus complex comprised of cues from the environment, cues such as flashing lights or tones deliberately presented by the experimenter, and even internal cues arising from the animal's physiological state. In addition, the organism is able to perform a variety of behaviors; unless the animal is compelled to perform a reflex, it can choose among a number of responses. When the animal performs the "appropriate" reaction, it is presented with a US. One can see from Figure 5–1 that both the stimulus complex *and* the response are correlated to the US presentation. The stimulus and US are paired independent of the animal's behavior *and* the response is paired with the US. Both those elements should, later, elicit expectancies. The final element in this simple model is the feedback (the dotted line in Figure 5–1). Whenever the response has produced the US or modified it in some significant fashion, the organism and the environment are significantly affected. That is to say, the US presentation affects the subject and, at the same time, changes the environment. Depending upon the nature of the US, internal stimuli are also altered. The ingestion of food is a good example; the food would eliminate the "hunger stimuli," such as stomach contractions, and cause changes like an elevation of blood sugar. In summary, Figure 5–1 indicates that *both* stimulus and response learning elements are present in any learning situation. Furthermore, the entire sequence is not merely the pairing of a stimulus, or a response, with a US presentation; it also involves the feedback and subsequent alteration of the stimulus complex by the US.

Incidentally, we should note that although Figure 5–1 is intended to illustrate the joint occurrence of stimulus and response elements and the feedback loop, nevertheless the diagram represents only one of a variety of

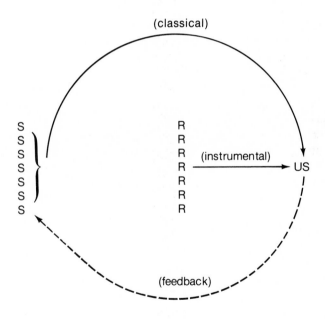

Figure 5-1.   Schematic diagram showing that a stimulus complex is associated with the US, while at the same time the response (or response complex) is correlated to the US. The feedback line shows that responses may affect the environment. See text for further explanation.

ways to look at this problem. It is not important at this point to consider all the various possible mechanisms for response-stimulus interactions. Some of these will be discussed later in this chapter. For the moment, it is merely appropriate to note that both elements invariably exist in any given learning experiment and to acknowledge that a thorough and accurate understanding of learning entails the analysis of how they interact.

Two-Factor Theory

We can elaborate upon the argument that Pavlovian and instrumental procedures are always present together by discussing in some detail an important theory of avoidance learning. The two-factor theory of avoidance (Mowrer, 1947) was designed to explain instrumental avoidance responding, but the theory has implications that extend far beyond that single learning paradigm. Let us begin by describing the precise sequence of events in an avoidance experiment. On the first few trials, an animal experiences the presentation of a discriminative stimulus, the tone or light, and, because it has not yet learned to avoid the shock, the US presentation 5 or 10 seconds later. The shock, of course, is painful and therefore the animal runs to the other side of the apparatus to terminate it. According to Mowrer, two things occur during these early trials. First, the discriminative stimulus is being paired with the shock US; this is a stimulus or Pavlovian learning component. The second process that is occurring is instrumental escape learning; the running response terminates the shock US and therefore is instrumentally reinforced.

These two forms of learning, Pavlovian and instrumental, continue to be present for a number of trials. The result is the strengthening of both the stimulus and the response expectancies. That is, the CS becomes more powerful; it elicits greater fear and aversion and probably elicits a variety of internal physiological reactions like accelerated heart rate because it's paired with shock. Secondly, the instrumental escape response is gradually becoming stronger because it produces reward (shock offset). Other potential instrumental responses, in contrast, are not reinforced and therefore do not increase in strength.

The important aspect of the theory, however, deals with the avoidance response that begins to occur after the initial escape trials. According to Mowrer, the Pavlovian fear that develops during escape learning increases to a point where it energizes the animal. Many investigators believe that fear and other motivational states energize behavior (see Chapter 10). In this case the predominant response, relative to all other responses, is the escape reaction. Therefore, when fear builds up and energizes the animal's behavior, it is the escape response that is triggered. After all, why should fear trigger other behaviors that are irrelevant to this situation? When fear increases sufficiently to energize the escape response *prior to* shock presentation, the response, of course, is called an avoidance reaction.

But how is the avoidance response sustained? What reinforces the animal for responding prior to shock onset? The answer, according to Mowrer's two-factor theory, is fear reduction. If the stimulus provokes fear in the animal and an avoidance reaction terminates the fearful CS, thereby enabling the animal to avoid shock, then the animal should feel some relief; its fear should be reduced by the offset of the fear-provoking stimulus.

Certainly we can explain avoidance responding in terms of Mowrer's theory of emotion, but we could also translate the situation into the language of expectancy. Specifically, we could say that the stimulus expectancy is one of shock presentation; the presentation of the CS induces the expectancy that shock will soon follow. Second, the execution of the avoidance reaction, for whatever reason initially, results in the presentation of "no shock." That is, the response expectancy is one of shock avoidance. Thus, on any given trial, the animal expects to receive shock when the CS is presented but then counters that expectancy by executing the appropriate response. Either way, the stimulus and response expectancies interact in a complex fashion. The animal can be said to control the situation insofar as it has knowledge of the contingencies, that is, sufficient information to "solve" the problem.

The main point here is that Pavlovian/instrumental interactions can be analyzed or viewed from both perspectives. The emotional as well as the cognitive or informational aspects of stimulus and response learning are both involved in avoidance learning. Recall that expectancies are inherent in stimulus and response learning situations because both forms of learning involve correlations between the element and an outcome (expectancy learning). Emotions are appropriate too since powerful biologically relevant USs are normally used in both stimulus and response learning experiments.

Support for Mowrer's Theory

Mowrer's two-factor theory was modified somewhat by Schoenfeld (1950). He agreed that the CS was becoming an aversive cue because of its association with shock presentation, but he assumed that the avoidance response, rather than producing a reduction in fear, was simply an escape response; it terminated an aversive secondary cue, the CS. In other words, what Mowrer considered a fear-reducing response, or an avoidance CR, Schoenfeld described as an aversion-eliminating response, or an escape CR.

This change in emphasis from avoidance to escape was a useful device because it allowed psychologists to focus on the actual point of reinforcement. For example, if the animal really were escaping the aversive CS rather than avoiding it, then the CS offset should be the locus of reinforcement. And if that were the case, then delaying the CS offset should result in a retardation in learning (performance is always adversely affected by delay of reward).

This is exactly what was found by Kamin (1957a, 1957b). He delayed the CS offset for a few seconds following each successful avoidance response; that is, the animals could make the avoidance response, but they would continue to experience the CS for several more seconds. Kamin found that the rate of avoidance deteriorated as a function of the delay. The actual data are shown in Figure 5–2. The figure clearly indicates that the no-delay animals learned the avoidance response quite readily; after all, those animals did not experience a delay of reward (that is, a delay of the CS offset). However, animals for whom the CS offset was delayed 10 seconds showed almost no increase in avoidance performance over the training session. This experiment, as well as others by Kamin (1954, 1956) provided powerful evidence for the two-factor theory of avoidance, and, more specifically, for Schoenfeld's version of that theory. The general belief that avoidance learning could be explained by an interaction between Pavlovian emotional states and instrumental conditioning was widely held by psychologists in the 1940s and 1950s. In addition, the more specific notion that the CS offset was the actual source of reward for the avoidance reaction was generally accepted.

Challenges to Mowrer's Theory

Within a decade of Kamin's work, however, several important aspects of Mowrer's two-factor theory were severely challenged (see Bolles, 1970a; Herrnstein, 1969 for reviews). Some evidence dealt with the function of the CS and the CS offset. Sidman (1953a, 1953b) showed that an animal could learn to avoid a shock even when a CS was not provided. (As described in Chapter 2, in Sidman avoidance the animal presses a lever to postpone shock; if it responds at a sufficiently high rate, then all shocks are avoided. The important point, however, is that deliberate CSs, such as lights and tones, are not usually used in these experiments.) On the surface, Sidman avoidance raises a problem for the two-factor theory: If CSs aren't used, then there can be no CS offset; and if there is no CS offset, there can be no reward for the avoidance response. This problem, of course, is not really a serious one because, as pointed out by Anger (1963), the animals develop a sense of the time interval between shocks. In other words, subjects use the temporal interval itself as a CS. This should be no surprise. Animals are known to use temporal intervals in a variety of situations, including fixed interval schedules (see Chapter 4). In the end, therefore, Sidman's challenge was not a significant one; two-factor theorists merely pointed out that there really was a "CS" although it was not an external cue.

The problem of specifying what constituted the CS offset in Sidman avoidance was more difficult. If the temporal interval served as the CS, as

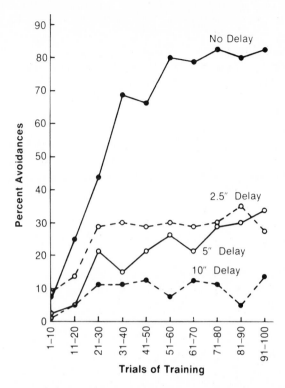

Figure 5-2. Mean percent avoidance as a function of training for groups that received different durations of CS offset. (From Kamin, 1957a)

Anger claimed, then what exactly was the CS offset? Anger extended his argument and hypothesized that internal aversive stimuli, the subject's own sense of discomfort, built up during the temporal interval; these stimuli, in turn, were suddenly eliminated following a lever press. The sudden decrease in these stimuli, therefore, served as the CS offset. In other words, the subject experienced more and more internal aversive stimuli as the time for the next scheduled shock approached. When the animal executed the response, these internal stimuli were dispelled, and the avoidance behavior was reinforced.

Anger's hypothesis seems to have salvaged the two-factor theory from the challenge posed by Sidman avoidance. At least it provided a plausible account of how the temporal interval functioned as the CS and how the sudden dissipation of the internal aversive feelings could act like the CS offset. But there were still problems for Mowrer's theory. For example, Sidman (1955) and Keehn (1959) permitted animals to respond to avoid shock (lever press in the former case and wheel run in the latter study). In both experiments the shock was preceded by a CS which *also* could be avoided. In other words, the subjects could postpone not only the aversive shock but also the aversive CS that preceded shock. In both studies the animals failed to avoid the CS. Rather, they tended to wait until the CS came on, and then they responded to avoid the shock. If the CS really were aversive, as Mowrer and Schoenfeld had suggested, then the subjects in these experiments should have avoided that too.

The conclusion from these studies was that the subjects were using the CS as a source of information, that is, as a discriminative cue that simply signaled the appropriate time for a response. This conclusion is in sharp contrast to the two-factor position which claimed that the function of the CS was actually to create fear which then would later be reduced by the avoidance reaction.

If the CS onset serves as a source of information to the animal, then it is reasonable to suggest that the CS offset is also a source of information. In this case, the offset would signal to the subject that the correct response had been made. Such a signal would be important because, on avoidance trials, there is no other source of information. If the shock is never delivered, as on avoidance trials, then how else is the animal to know that it has made the correct response? The only change in the environment that results from the animal's behavior on an avoidance trial is the CS offset; hence, it is the only event that could provide such information.

The notion that the CS offset is a source of information, rather than a source of reinforcement, has been confirmed by Bower, Starr, and Lazarovitz (1965). In their experiment, the investigators noted two important results. First, avoidance performance improved as a function of the degree of change in the CS following the response. A partial reduction in the noise CS following an avoidance response proved to be less effective than complete termination of the noise. The more salient the CS offset, the better the information source. Second, avoidance performance could be improved if a special information cue were deliberately presented after the response (see also Bolles & Grossen, 1969; D'Amato, Fazzaro, & Etkin, 1968). In the Bower et al. study three groups of rats were trained to avoid a shock. Animals in Group R could terminate the tone CS on avoidance trials in the normal fashion; Group R, therefore, was the "regular" trained group. For Group D (delayed), a response enabled the rats to avoid the US, but the CS offset was delayed for 8 seconds. Learning in these subjects should be quite poor just as it was in Kamin's study earlier (see Figure 5–2). Animals in the third group, Group DL (delay plus light), also received the 8-second delay of CS offset, but they were given a second stimulus during that delay. More specifically, after animals in Group DL made an avoidance response, the tone CS continued for an additional 8 seconds but another cue, a light, was turned on.

As shown in Figure 5–3, Group D performed the worst. The result confirmed Kamin's earlier study: delay of CS offset seriously debilitates avoidance performance. Group DL and Group R, however, performed normally. Although Group DL received the 8-second delay just as Group D did, the behavior of the subjects in Group DL was not affected. The reason, of course, is that the Group DL subjects were given a light cue during the delay interval. This cue functioned as a source of information for the animals in the absence of the CS offset. In other words, the Group DL animals did not have the immediate tone offset as a source of information, but they *could* use the light stimulus as a source; that cue came on just after the avoidance response (a change in the environment), thus confirming their expectancy that the avoidance reaction was, indeed, the response that produced the nonoccurrence of shock.

In summary, these studies indicated that the CS functions as a dis-

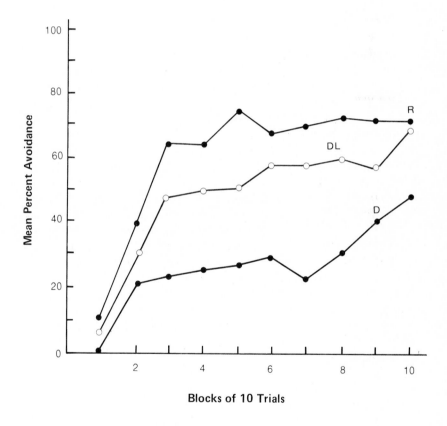

Figure 5-3. Mean percent avoidance as a function of blocks of ten training trials for the "regular" trained group (Group R), the CS offset delayed group (Group D), and the delayed group that received the extra light stimulus (Group DL). (From Bower, Starr, & Lazarovitz, 1965)

criminative stimulus. Its presentation elicits the avoidance response because it indicates when that response should be made. In addition, the CS offset acts to inform the subject that the correct response has been made; thus, it is a source of information, but it is not the locus of reinforcement.

Additional behavioral studies, which also tend to refute the two-factor theory, have been conducted. For example, Bolles, Stokes, and Younger (1966) ran five studies in which the various reinforcement contingencies were examined. Four different groups of rats were used. Animals in one group could terminate the CS, avoid the US, and escape shock on those trials on which avoidance did not occur. Subjects in a second group could avoid the shock and terminate the CS but, if they failed to avoid, they could not escape the shock, which was presented for a fixed period of time. Animals in a third group could escape shock if it came on and they could terminate the CS, but they could not actually avoid the shock, which came on regardless of their behavior. Finally, animals in a fourth group could only terminate the CS by making the response; shock came on and lasted a fixed period of time even though they performed the proper behavior.

The results showed that the most critical factor for learning was the ability to avoid shock, not the ability to terminate the CS. Immediate CS termination, of course, did help performance, but it was not as important as shock avoidance itself. In summary, the study by Bolles et al. indicates that US avoidance, not the CS offset, is the "true" reinforcer for avoidance learning. Although the CS offset does allow the subject to perform more efficiently by giving it immediate feedback concerning the response, it is not the source of reinforcement.

Finally, Herrnstein and Hineline (1966) found evidence against the two-factor theory. A brief shock was administered to their rats according to two separate schedules. If the subject simply sat passively in the testing apparatus, it would receive shocks periodically according to Schedule 1. If the subject, however, responded by pressing the lever, the next shock would be delivered according to Schedule 2. The important feature of this design was that Schedule 2 involved a lower average rate of shocks. That is, by responding, the subject would receive, on the average, fewer shocks because the average time between shocks was greater on Schedule 2. It is important to keep in mind that no external CS was used in this experiment, nor was there a fixed temporal CS (the intershock intervals were all randomized). No escape or avoidance of shock was possible either. The subjects simply had a choice between receiving a relatively high frequency of shock pulses from Schedule 1, or responding and thereby receiving a somewhat lower frequency of shock pulses from Schedule 2.

Herrnstein and Hineline found that the rats did, indeed, learn. They pressed the lever not to escape or even avoid shock, nor to terminate a CS, but merely to reduce the frequency of shocks. According to these researchers, it is quite unnecessary to postulate sources of fear and reinforcement as the two-factor theorists had done. Rather, they suggested that avoidance could be explained more simply as a behavior strategy that reduced the frequency of aversive stimuli.

The studies cited above, and others, provide evidence that avoidance learning does *not* depend on fear onset and offset as Mowrer suggested. While it is true that the CS is capable of eliciting fear reactions (for example, as in typical CER studies), it does not seem to function simply as a fear-arousing stimulus in avoidance learning tasks. Nor does the CS offset function as the reinforcer, the locus of fear dissipation. Rather, the CS onset and offset both appear primarily to be information cues signaling, respectively, when the response should be made and that the response was successful.

Conditioned Emotion

Let us now reconsider the two-factor theory in light of the discussion of Pavlovian conditioning in Chapters 1 through 3. Classical conditioning, of course, involves the development or learning of overt reactions: animals learn to salivate, flex the paw, change the heart rate, or perform one of a great many other such conditioned responses. But, as discussed in Chapters 1 through 3, a subject also learns an expectancy, a cognitive reaction. Based upon their correlation with events, conditioned stimuli become signals for future events. The cognitive or mental state of the animal is altered in a fundamental fashion during stimulus learning.

But consider another argument presented in Chapter 1, that all stimuli can be classified along a pleasure-pain continuum. The stronger the stimulus, the more pain or pleasure it inflicts on the organism (see Figure 1–3). Alternatively, the stronger the stimulus, the more likely it can be classified as a US. After all, USs are defined as strong, biologically relevant cues that elicit immediate reflexes.

The psychological counterpart to pain is fear. Stimuli that produce physical pain also lead to the development of fear. Similarly, the psychological counterpart to physical benefit is pleasure: when a strong stimulus induces physical well-being, such as food does for a hungry dog, then there is a corresponding emotional state, a state of pleasure.

Let us summarize these arguments. All USs are biologically powerful and therefore they induce either painful or beneficial reactions in the organism. The psychological conunterpart to these reactions, that is, the corresponding mental state, is one of emotion; fear in the case of aversive USs and pleasure in the case of positive or appetitive USs.

Now consider once again the significance a CS acquires during conditioning. As discussed above (and in previous chapters), a stimulus that is paired with a US comes to elicit an expectation of that outcome; it elicits a cognitive reaction. But a CS *also* comes to elicit an emotional state. It is paired with a strong emotion-producing stimulus, the US, and thus is able to produce the emotion by itself. In short, the CS is a signal; it informs the subject about a future outcome (usually the US). At the *same* time, the CS evokes emotions based on the quality (aversive or pleasurable) of the US.

Only recently have psychologists studied the joint occurrence of the cognitive and affective components of the conditioned reaction. But the evidence confirms that CSs acquire *both* a signaling and emotional function (see, for example, Fowler, Fago, & Domber, 1973; Ghiselli & Fowler, 1976; Overmier & Bull, 1970). Ghiselli and Fowler (1976), for instance, conducted an aversive Pavlovian experiment in which they first paired a CS+ with shock, and a CS− with no shock. The CS+, according to our theory, would signal the future presentation of shock *as well as* induce fear. The CS−, in contrast, would signal the nonoccurrence of shock (the expectancy component) *as well as* induce an emotional reaction, which we might call relief.

In a second phase of their study, the CS+ was paired with food for some subjects and paired with no food for others. Similarly, the CS− was paired with food for some animals but with no food for others. Ghiselli and Fowler discovered that the CS+ signaling shock was rather easily converted into a CS+ for food. The emotional characteristics of the reaction, of course, changed radically; the CS+ no longer induced fear but rather some kind of hope or pleasure following the change from a shock US to a food US. But in both phases of the study, the CS+ signaled the presence of the US. Therefore, the signaling function of the cue remained the same. In other words, the fact that it was easy to convert an aversive CS+ into an appetitive CS+ indicates that the signaling and emotional functions were both present, but independent, *and* that the emotional function dissipates or extinguishes more readily than the signaling function.

The same result was found for the CS− when it was converted from a CS− signaling no shock to a CS− signaling no food. The emotional quality of the response changed, but its signaling function remained the same and therefore the transition was rather easy. This research is important because it demonstrates directly that a Pavlovian conditioned reaction has two separate aspects: cognitive and emotional. Both of these aspects are conditioned simultaneously but can vary independently.

We can expand our analysis here in two important ways. First, note that *all* CSs can elicit emotions; cues that signal aversive USs (and thereby induce fear reactions), of course, are dramatic examples. But in theory, all four major types of CSs, as depicted in Figure 2–3 in Chapter 2, also elicit conditioned emotional reactions.

Second, the arguments concerning conditioned emotions can apply to responses. Recall that in Chapters 1 and 2 we discussed the concept that responses, as well as stimuli, that were associated with strong USs came to elicit expectations for those events. Again, because the USs are biologically strong, an emotion is also a result of the conditioning process. All sources of information, therefore, both stimuli and responses, involve emotions if for no other reason than the fact that strong USs are involved in conditioning.

Mowrer (1960) was the first to develop the claim that all informative CSs elicit an emotion. Here, we are extending this claim to responses as well. Consider the matrix in Figure 5–4. Represented are the four types of Pavlovian and four types of instrumental learning situations. In the case of excitatory/reward conditioning, CSs that signal food or other appetitive USs are said to provoke an emotion of hope; similarly, responses that lead to food reward may also be considered to involve the emotion of hope; after all, a response also may be used to predict an outcome, in this case, food. Both the CS and the response in this example elicit expectancies of food presentation and, at the same time, an emotion termed hope.

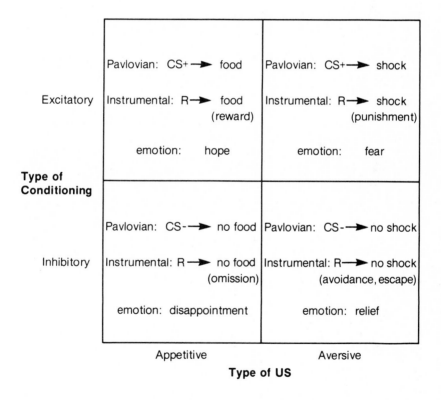

Figure 5-4. Matrix giving examples of Pavlovian and instrumental conditioning and the accompanying emotion that is presumably generated.

Now consult the upper right-hand portion of Figure 5–4. In this case a CS is paired with shock or a response is followed by shock (punishment). For both the stimulus and the response, the underlying emotion associated with the expectancy of shock is fear. Such a fear CS, of course, was claimed to be the basis for avoidance responding according to the two-factor theory.

The remaining two examples of conditioned emotion are depicted in the bottom cells of Figure 5–4. On the left is the example of Pavlovian inhibitory conditioning (CS followed by no food) and omission training (response leads to the omission of an appetitive US). In each of these situations an emotion is generated, specifically the emotion of disappointment. That is, the presentation of a CS that signals no food or the execution of a response that signals no food induces an emotional state in the animal, one of disappointment.

Finally, the remaining examples in Figure 5–4 are inhibitory conditioning with aversive USs. In the case of stimulus learning, it is a CS followed by no shock; for instrumental conditioning, it is a response followed by no shock (avoidance or escape learning). In each of these cases, the emotion is one of relief. That is, the stimulus and the response, because they signal the absence of an aversive US, induce the state of relief in the subject.

In summary, we started with the specific example of avoidance learning and discussed how avoidance could be conceived of as an interaction between emotions and behavior. Other forms of instrumental conditioning, and the other forms of stimulus learning, also can be conceived of as involving emotion. During Pavlovian conditioning, the US produces not only a specific motor reaction or visceral response but also an emotional state; the particular kind of emotion depends on the nature of the US (aversive or appetitive) as well as the type of conditioning (excitatory or inhibitory). Similarly, instrumental conditioning involves the acquisition of an overt motor reaction, an expectancy, and an emotion. Again, the nature of the emotion depends upon the type of US and the type of conditioning paradigm.

Interaction Experiments

What does all this mean in terms of the two-factor theory? Does the fear motivate the avoidance response as Mowrer claimed? Let us consider once again the problem of avoidance learning. We have claimed that *some* emotional state underlies avoidance responding; indeed, emotions underlie all response learning situations (see Figure 5–4). We might guess that the actual emotion is fear but how can we demonstrate that fear underlies the avoidance reaction? This is the question to which some of the interaction experiments are addressed.

The basic idea of the interaction experiments is as follows. If the emotional state is, indeed, fear, then we ought to be able to increase or decrease the fear level by adding extra fear or by detracting from that fear in some fashion. Consider this analogy. If we placed two wires into a solution of liquid and applied an electric current between the two wires we might "observe" that electricity is conducted from one wire to the other. Furthermore, we might guess that the reason for this transmission of electric current is the presence of salt ions, although we may not be entirely sure. In order to confirm that salt ions are responsible for the transmission, we might increase them with a known quantity of salt. If the substance in the solution and the known salt additive are compatible, then the behavior of the system, that is, the flow of current, should increase. On the other hand, if we subtract from the substance in the solution by diluting the salt ions, then we should observe a decrease in the current flow.

This is essentially the idea of interaction studies. It is easiest to conceive of them as involving conditioned emotions, although, as we shall discuss later, this is not necessarily the case. For now, let us assume that the interactions are emotional in nature. Return to the example posed by the two-factor theory: does fear underlie avoidance learning? One way to address this question is to add extra fear and observe an increase in the avoidance behavior. Such an increase would mean that the known Pavlovian emotion, fear, and the emotion on which the avoidance response is based are compatible. Adding extra fear increases the emotional state already present, thus producing an increase in the avoidance behavior.

This procedure was used in a number of experiments (see Grossen & Bolles, 1968; Martin & Riess, 1969; Rescorla & LoLordo, 1965; Scobie, 1972; Weisman & Litner, 1969a, 1969b). A good example was the Martin and Riess (1969) experiment. In Phase 1, they trained rats to press a lever in order to postpone shock (Sidman avoidance). After a stable response rate was achieved, the researchers gave their subjects classical fear training. Specifically, in this second phase, a light CS and a shock US were paired; this is the standard technique for producing a fear cue (aversive CS+). Finally, in the third phase of the experiment, while the animals were once again performing their Sidman avoidance reaction, the aversive fear signal alone was presented. The authors discovered that the avoidance rate increased and the amount of increase was a function of the intensity of the shock during Phase 2 of the study. In other words, when a fear stimulus is superimposed upon an avoidance reaction (the motivation for which is presumed to be fear as well), the avoidance reaction increases. The conclusion is that the underlying emotional state and the known emotion, fear, are compatible; they are, in effect comparable emotions. Moreover, the greater the extra fear (due to more intense shock levels during Pavlovian conditioning), the greater the change in avoidance rate.

The opposite result has also been found (see Grossen & Bolles, 1968; Rescorla & LoLordo, 1965; Weisman & Litner, 1969a, 1969b). While animals were engaged in an avoidance reaction, an aversive CS— was superimposed upon the behavior; such an aversive CS— evoked the emotion of relief (see Figure 5–4). And since relief and fear are counteracting emotions, they tended to cancel each other out. In other words, the known Pavlovian emotion, relief, counteracts or detracts from the fear state that underlies the avoidance reaction. Since we know the CS— induces relief and we know that it detracts from the state that maintains the avoidance reaction (behavior declines), then the underlying state must indeed be fear.

Both these effects were shown in the study by Weisman and Litner (1969a). In Phase 1 rats were trained to turn a small wheel to postpone shock. Then in Phase 2 they were given Pavlovian training. For one group of subjects, a tone CS was always followed by a shock US. Another group received the reverse training procedure: the CS was never followed by the US, thus establishing the tone as an aversive CS—. A third group received the CS and US presentations in a random fashion; conditioning was not expected to take place for these subjects and, accordingly, the CS was not expected to elicit any particular emotion. During Phase 3, the test phase, all animals were given the 5-second tone CS while they were performing the wheel-turn avoidance task. As shown in Figure 5–5, the CS+ (fear cue) caused the avoidance behavior to increase; this is the same result observed in the Martin and Riess study discussed above. And the conclusion is the same too: the CS+ augmented the fear that already was present during the avoidance task; that's why the behavior increased.

The CS— (the relief cue) had the opposite effect on behavior; it caused a reduction in avoidance responding because, as noted above, the relief induced by the CS— counteracted the fear state that supported the avoidance response. Finally, note in Figure 5–5 that the CS presentation caused no change in behavior for the random group. Since the CS did not bear any predictive relationship to the US for subjects in this group, it could not evoke an expectancy for the US nor could it elicit an emotion. As a result, therefore, it had no effect on the ongoing avoidance reaction.

A Matrix of Interaction Studies

The general technique described here can be used to investigate the underlying emotions for any instrumental learning task (see Rescorla & Solomon, 1967). The basic question asked above in connection with avoidance can be generalized to all forms of instrumental conditioning: what is the nature of the motivational state underlying the response?

We have already encountered a technique which draws upon this methodology—the CER study. Recall that the CER technique involves training an appetitive response, such as lever pressing for food, and then superimposing a Pavlovian fear cue while the animal is pressing the lever. As shown many times (for example, by Annau & Kamin, 1961), the fear

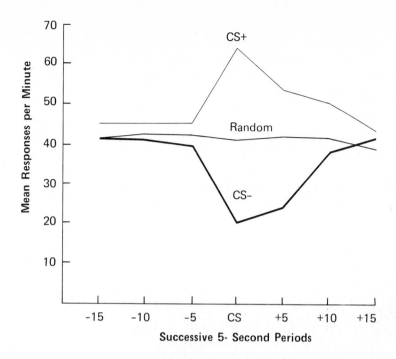

Figure 5-5. Mean avoidance responses per minute as a function of successive 5-second periods (CS presented at point marked "CS") for subjects that had received excitatory (CS+), inhibitory (CS−), or random Pavlovian training. (From Weisman & Litner, 1969a)

cue causes disruption of lever pressing. Furthermore, the degree of disruption reflects the intensity of the superimposed fear: the greater the shock intensity during the Pavlovian phase, the greater the disruption. The general CER technique, therefore, is analogous to the research cited above. The only difference is that in the former studies the ongoing behavior was an aversive reaction, whereas in the CER studies the ongoing behavior is based upon appetitive rewards. These appetitive rewards presumably sustain the behavior by inducing the emotion of hope. And since hope and fear are incompatible reactions, one (fear) ought to conflict with the other (hope), thus causing a reduction in the behavior.

Consult the diagram shown in Figure 5–6. This scheme, adapted from Rescorla and Solomon (1967), shows *all* the possible interaction experiments. The first example we discussed, the addition of extra fear to an ongoing avoidance task, is depicted in cell 6 of Figure 5–6. As indicated, the Pavlovian state that intrudes upon the instrumental task is induced by an aversive CS+ (fear cue); the effect of that superimposed Pavlovian stimulus is shown by the arrow. In this case the stimulus causes an increase in the behavior. Similarly, an aversive CS— (relief) produces a decrement in avoidance responding as shown in cell 8 of Figure 5–6.

The example of the CER task is shown in cell 5. The presentation of a Pavlovian fear cue, while the animal is engaged in an appetitive instrumental task, induces a decrement in that behavior. Correspondingly, cell 7 indicates that an aversive CS—, one that induces relief, should enhance appetitive responding. Such results were found in a study by Hammond (1966). In summary, the emotions of fear and relief interact in predictable ways with the emotional states that underlie the appetitive and aversive tasks. When the emotions are compatible, the behavior is augmented (cells 6 and 7), but when the intruding state is incompatible, it detracts from the ongoing emotions and causes a decrement in performance (cells 5 and 8).

The types of interactions depicted in the top four cells in Figure 5–6 have also been studied in recent years. In each case, Stage 2 of the interaction experiment involves appetitive Pavlovian conditioning. That is, a CS is paired either with food (CS+, causing hope) or with no food (CS—, causing disappointment). When those stimuli are later superimposed on ongoing instrumental tasks, the emotions, the known emotional state resulting from the Pavlovian stimulus and the emotion underlying the instrumental behavior, interact and a change in performance is observed.

As indicated in cell 1 of Figure 5–6, an appetitive CS+ (hope) increases the rate of responding on an instrumental appetitive task (Bolles, Grossen, Hargrave, & Duncan, 1970; LoLordo, McMillan, & Riley, 1974), while an appetitive CS— cue (inducing disappointment) causes the ongoing instrumental reaction to decline in rate (Bolles et al., 1970). The study by Bolles et al. (1970) illustrates both these findings. Rats were first trained to run down an alleyway to obtain food in the goal box. After 30 trials, they were placed in another apparatus and given Phase 2, that is, Pavlovian appetitive conditioning. For one group of subjects, the CS+ was followed by food delivery; for a second group, the CS— signaled no food for at least 30 seconds. A third group got the tone CS and food US in a random fashion; no conditioning, excitatory or inhibitory, was expected to occur in these subjects. In the third phase of the study, the tone CS was sounded while the animals were in the start box of the alleyway. Although the speed of running in the alleyway was not affected by the CSs, extinction responding was. Specifically, the CS+ caused the subjects to run faster during extinction, whereas the CS— produced a suppression in start box speeds. Both these effects were relative to the behavior of the random control subjects; for them, the CS caused no appreciable change in behavior.

Finally, as indicated in cells 2 and 4 of Figure 5–6, appetitive CSs can influence aversive responding. If, as depicted in cell 4, the interacting emotions are essentially compatible, such as disappointment and fear (Bull, 1970; Grossen, Kostansek, & Bolles, 1969), then avoidance behavior increases. This is not an obvious result, but the interaction studies have

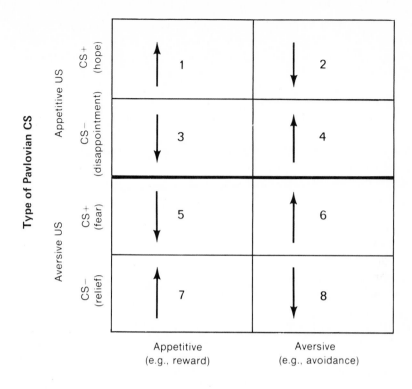

Figure 5-6. Matrix illustrating the various types of interactions between Pavlovian emotional states and instrumental behavior. Arrows indicate whether the superimposed Pavlovian CS facilitates (↑) or inhibits (↓) the instrumental performance. Numbers identify the cell for discussion in text.

confirmed that an appetitive CS− causes an improvement in avoidance. In contrast, an appetitive CS+ that induces hope causes an avoidance reaction to decrease, depicted in cell 2 of Figure 5–6 (see also Bull, 1970; Davis & Kreuter, 1972; Grossen, Kostansek, & Bolles, 1969). Here again, the intruding Pavlovian emotion (hope) is incompatible with the emotional state underlying the avoidance reaction (fear); the net effect is a lowered level of fear and, consequently, a reduced rate of avoidance responding.

In summary, these interaction experiments are extremely important because they show how Pavlovian emotional states intrude upon and influence instrumental behavior. Presumably, the emotion that already exists during instrumental training influences the rate of performance. When that state is altered, by adding a compatible emotion or by detracting from the original state, then performance changes. Thus, it can be claimed that instrumental contingencies *do* involve underlying emotional states; in this sense, the most general principles of the two-factor theory appear to be true.

Alternative Explanations

We have explained the above results in terms of interacting emotions. This would seem to be justified since Pavlovian conditioning invariably involves strong emotion-producing USs. Other explanations have been offered though (see Dickinson & Pearce, 1977; Overmier & Lawry, 1979). According to one explanation, the interaction experiments merely reflect the interaction of two instrumental responses, rather than central emotional states. The argument goes as follows. In Phase 1 of the study, the animal learns the instrumental reaction. Then, in Phase 2, the Pavlovian conditioning phase, the theory claims that the animal is merely learning to make another skeletal muscle reaction. (We have already noted that Pavlovian experiments always involve instrumental contingencies; see the first section of this chapter.) For instance, the animal may be learning to adjust its body posture to minimize the impact of the US. Regardless, the theory claims that in the third phase of the study, when the CS is given while the animal is again performing the original instrumental response, the resulting change in behavior is caused merely by the interaction of conflicting muscle movements. For example, if an animal were responding to avoid a shock and a fear cue were presented, then the conditioned muscle spasms in the leg (which were covertly conditioned during Phase 2 of the study) could enhance the running reaction already being exhibited by the animal. Conversely, the CS— could cause the animal to freeze or tense the muscles, thus competing with the ongoing instrumental avoidance response.

Several experiments have shown quite clearly that the interaction experiments cannot be explained in this fashion (Overmier, Bull, & Pack, 1971; Scavio, 1974). For example, in Scavio's study, rabbits were first given Pavlovian eyelid conditioning using a tone CS+ and shock US. Then they were trained to make a jaw movement CR by pairing the CS with water presentations. Scavio discovered that the aversive eyelid conditioning interfered with learning the jaw movement reaction; having undergone the aversive training task (the eyelid reaction portion of the study), the rabbits found it difficult to learn appetitive conditioning in Stage 2. Scavio showed that the interference was not due to conflicting muscle movements (eyelid responses do *not* conflict with jaw movement), but rather due to conflicting emotional states: Phase 1 involved fear, while Phase 2 involved hope.

There are other kinds of experiments that also demonstrate this point: that the interaction studies reflect alterations in the central emotional state of the organism. These experiments show that various Pavlovian phenomena can cause complex and *precise* changes in behavior (Rescorla, 1967a; Zamble, 1969) not just gross changes which, conceivably, could be explained by conflicting motor reactions. In Rescorla's study, for instance, the US presentation was delayed for a considerable length of time during the Pavlovian phase of the experiment. As we noted in Chapter 3, such an arrangement leads to inhibition of delay: the withholding of the Pavlovian reaction (inhibition) until just prior to the US presentation at which point the CR is given (excitation). As shown in Figure 5–7, when Rescorla later superimposed the fear cue on the avoidance response task, he observed inhibition of the avoidance behavior early in the CS interval (cell 8 of Figure 5–6) but an improvement in the avoidance rate during the latter portion of the CS interval (cell 6 of Figure 5–6). The presence of the Pavlovian state, inhibition of delay, was evident from the animal's instrumental response rate. This result suggests, therefore, that the precise Pavlovian-induced state does, indeed, influence instrumental responding. In other words, classically conditioned states seem to underlie instrumental behavior since even *specific* Pavlovian phenomena, such as inhibition of delay, may be detected using the interaction approach.

Although we have interpreted the interaction experiments almost exclusively in terms of conflicting or compatible emotions, recall that both stimulus and response learning involve the formation of expectancies as well (Ghiselli & Fowler, 1976). Therefore, we should note that it is possible to explain the interaction study results in terms of a compatibility or incompatibility between expectancies, rather than emotions. Consider the case of avoidance learning, in which a fear cue enhances the avoidance rate and an aversive CS–, a relief cue, produces a decrement in the rate. One could claim that the animal, upon being presented with the fear CS+, expects shock soon (after all this is what the CS came to signal during the Pavlovian phase) and therefore increases its rate of responding to compensate for this pending shock. Similarly, the presentation of a CS– induces the expectation of no shock; if the animal is avoiding and suddenly receives such a stimulus, then it is plausible to assume that the avoidance reaction would decline in rate. Continued avoidance responding is unnecessary if the animal no longer expects to receive shock.

There is some work suggesting that such an interpretation is appropriate, at least in some cases (see Fowler, Fago, Domber, & Mochhauser, 1973; Overmier & Schwarzkopf, 1974; Wielkiewicz, 1979). In the Overmier and Schwarzkopf study, for example, two groups of animals were used. Animals in one group were trained to jump over a barrier in order to avoid shock whenever a discriminative stimulus was given. Animals in a second group were also taught to jump over the barrier when the same cue was presented, but these subjects were reinforced by food, not shock avoidance. Note that the underlying emotions should be quite different: fear in the former case

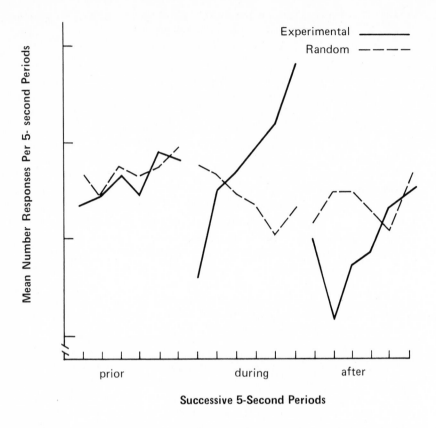

Figure 5-7. Mean number of avoidance responses per 5-second period as a function of successive 5-second periods prior to the CS, during, and after the CS. (From Rescorla, 1967a)

and hope in the latter case. During Phase 2 of the study, all the animals were trained to make a standard shuttlebox avoidance reaction. Finally, in Phase 3, while the animals were avoiding shock, the discriminative stimulus from Phase 1 was presented. Overmier and Schwarzkopf found that the avoidance reaction was facilitated for *both* groups of animals. This clearly is not the result of interacting emotions. If it were, then the avoidance rate for the food-rewarded animals should have gone down (not up) since the cue presumably elicited hope in those subjects. Hope and fear should have been incompatible, causing a reduction in avoidance behavior (cell 2 of Figure 5–6). The facilitation, therefore, must have been due to the signaling function of the cue. The appetitive CS+ signaled the availability of reward, the response to which was compatible with the shuttlebox avoidance behavior. In summary, this study and other similar experiments indicate that the results of interaction experiments could be due to conflicting or compatible expectancies, not necessarily interacting emotions.

Let us summarize this complicated but important research area. Interaction studies usually have three phases: (1) the instrumental response is established which is based on an unknown motivational state; (2) Pavlovian conditioning is conducted during which a *known* emotional state is conditioned (for example, fear, hope, etc.); (3) the Pavlovian CS is presented while the animal is performing the original instrumental task. If the emotional state underlying the instrumental behavior and the known Pavlovian state are compatible, then behavior improves. If they are not compatible, then behavior is disrupted. We can describe the compatibility as a compatibility between two emotions (see Figure 5–6). For example, fear disrupts behavior based on hope (the typical CER study) but augments aversive behavior such as avoidance responding. Alternatively, the compatibility between the emotional state and the known Pavlovian state can be viewed as an interaction between expectancies. If the animal is avoiding shock and receives a signal that predicts shock (a Pavlovian aversive CS+), then the avoidance rate increases. We do not really know whether interaction studies reflect interacting emotions or interacting expectancies. Most surely, they reflect the interaction of *both* emotion and expectancies. Regardless, these studies indicate quite clearly that Pavlovian and instrumental processes coexist and interact, that control of behavior is a result of some complex central process, not simply peripheral motor reactions.

Summary

It is important to note that stimulus and response learning occur jointly and interact. In virtually all experiments both Pavlovian CSs and instrumental CRs are conditioned. Stimuli are associated with outcomes and responses modify those outcomes. One area of research that historically has been concerned with the joint occurrence of stimulus and response learning factors is avoidance research. According to the two-factor theory of avoidance, Pavlovian fear (stimulus learning) motivates an organism to make an instrumental avoidance response which then terminates the classical fear (response learning). Subsequent modifications of this theory identified the CS offset in avoidance learning as the actual locus of reinforcement (that is, the point of fear reduction). Such a modification made the theory somewhat more testable, but research in this area tended to discount this notion. While animals surely experienced Pavlovian fear in these situations, the CS onset and offset appeared to act more as information cues than points in the training sequence where fear suddenly came on and then went off again. Other work also tended to discount some of the specific points raised by two-factor theory.

The two-factor theory, however, was reconsidered and supported in light of studies on emotion and Pavlovian/instrumental interactions. Fear is but one emotion stemming from Pavlovian conditioning. Hope, disappointment, and relief are others that correspond to stimulus learning procedures. When cues that arouse these emotions are superimposed on ongoing instrumental behaviors, predictable changes in performance are observed. This procedure is termed an interaction approach. From these performance changes we can infer that the instrumental behavior is, indeed, supported by an underlying emotion that is compatible (or incompatible) with the superimposed Pavlovian state. For example, fear and disappointment disrupt appetitive instrumental tasks like lever pressing for food but facilitate aversively motivated behaviors like avoidance. Similarly, hope and relief cause appetitive tasks to increase in rate but cause disruption in aversive behaviors. The interaction experiments, therefore, are important because they illustrate the way in which Pavlovian emotional states mediate instrumental behaviors. It should be noted, however, that the interaction results also could be due to changes in expectancies. That is, the superimposed states could cause congruent or conflicting expectancies that produce the appropriate changes in performance. Most likely both emotions and expectancies are involved.

6

Biologically
Specialized Learning

Introduction

Naturalistic Response Patterns

Biologically Specialized Learning Phenomena

Social-Oriented Learning Phenomena

Summary

Introduction

We have stressed the idea that the learning abilities of most creatures have undergone changes throughout evolutionary history. When the structures that permitted learning to take place were present and the resulting behavior was adaptive, then those structures were selected for and passed on to subsequent generations. In other words, learning reflects the genetic endowment of the organism to the extent that the *capacity* to become changed, to learn, is an inherent property of the species.

We must not forget this lesson. Much of the work concerning stimulus and response learning that we have discussed in previous chapters seems to have little bearing on the animal's "natural" behavior. After all, rats never experience levers or shock grids in the real world. Although "artificial" or arbitrary responses like lever pressing can serve as useful *analogies* to real behavior, nevertheless we must try to relate the principles of learning discovered in the laboratory to more naturalistic conditions. The principles of the acquisition of stimulus and response expectancies must, ultimately, be consistent with the species-typical behavior potentials of the species. Only then will we be able to see the true nature of learning as a cognitive process that permits organisms to know, order, and manipulate their environment in ways that increase their chances for survival.

Why must we insist on such an attitude in learning research? What is wrong with the idea of studying lever pressing in order to generate universal laws of learning? Quite simply, many of the general principles that we have discovered do not apply uniformly to all behaviors. That is, when we examine more naturalistic situations, we often find that our laws of learning are inadequate; they don't predict the behaviors that actually get performed. The general claim, therefore, that learning principles must be viewed in the context of evolutionary development, that principles of stimulus and response learning depend upon the biological and environmental contexts, represents a cautious but wise policy (see Hinde & Stevenson-Hinde, 1973; Seligman, 1970; Seligman & Hager, 1972; Shettleworth, 1972).

Let us examine a famous example of the failure to show learning even though appropriate reward contingencies were available. Breland and Breland (1961) reported their efforts to condition a number of different species to perform circus tricks. In one instance, they tried to teach a raccoon to deposit tokens in a metal box to obtain food reward. The animal appeared to have no problem with this task when it was given only one token. However, when the raccoon was required to deposit two tokens, this "simple" and arbitrary behavior was learned very poorly. The raccoon had trouble learning this response because, instead of releasing the tokens into the box, it rubbed them together, often dipping them into the opening only to pull them back out again. It would appear that the reinforcement contingency was perfectly appropriate for this task; furthermore, depositing tokens was a response that the raccoon otherwise found easy. But for some reason the behavior of depositing two tokens was sufficiently difficult that the animal resorted to its "instinctual" behavior patterns. That is, the raccoon "rubbed and washed" the tokens as if they were food. This behavior, because it was resistant to conditioning, was termed a misbehavior. These results challenged conventional learning theory because they showed that our normal procedures don't always work even for such simple responses. Animals are not molded simply by the reinforcement contingencies; rather, their behavior is influenced enormously by their naturalistic response tendencies.

The misbehavior noted above exemplifies the failure of response learning techniques. Similar results have been found for stimulus learning. For example, several studies have shown that animals anticipate the presentation of food when it is given at the same time each day (Bolles & Moot, 1973; Bolles & Stokes, 1965) by becoming active just prior to its presentation. The stimuli that the animals use to predict food are internal cues stemming from the daily hunger cycle. These cues build up over a 24-hour period and are easily associated with feeding. Such a result makes sense: becoming active when food is available increases the chance of coming into contact with the food. However, when a comparable study using shock USs was done, rats failed to learn. For instance, Bolles, Riley, Cantor, and Duncan (1974) gave a mild shock to rats at the same time each day for 30 days. The animals could avoid the shock by jumping onto a safe platform. The authors found that the animals did not learn to anticipate the shock. The internal stimuli associated with the time of day, even though they were perfectly adequate signals for food presentation, were ineffective cues for shock. In the real environment, predators and other sources of danger do not appear at fixed times. It makes no sense for an animal to be able to anticipate dangers by the time of day or by its sense of hunger.

In conclusion, all these experiments suggest that there may be specialized sensory abilities and unique organizational patterns that have developed in many species about which we are still very ignorant. To discover that reinforcement is ineffective in a particular situation, as Breland and Breland did, is not to deny the utility of all learning principles, but it certainly does offer a serious challenge to current thinking. It suggests, at the very least, that our current list of principles is inadequate. The solution to this problem, therefore, would seem to be to broaden our research programs to include new areas of study, notably naturalistic behavior patterns. In doing so, we should avoid the temptation to view these biologically specialized behaviors in isolation from other typical learned responses. Rather, we should try to address the problem of how various principles of learning and species-typical behavior potentials interact.

Naturalistic Response Patterns

Many organisms, particularly the "lower" species, respond to special stimuli in their environment by performing fixed response sequences. A list of these fixed patterns would be extensive; indeed, much of the field of ethology is devoted to the study of just such behaviors (see Eibl-Eibesfeldt, 1975). More relevant to the study of learning are those response patterns that are pliable, that is, action patterns that are not rigidly fixed reactions to specialized stimuli.

Much of our knowledge about these action patterns stems from the work of Shettleworth (1975, 1978a, 1978b, 1978c). First let us describe some of the naturalistic behaviors of the golden hamster that occur both in the home cage as well as in various testing or observation environments. Hamsters may walk, sniff various objects, rear on the hind legs in the open area of the cage, rear on the hind legs while touching the wall of the cage, scrabble (scrape the forepaws against the wall while standing erect, sometimes hopping up and down), dig in the sawdust, push sawdust backward with the hind feet, gnaw, stand motionless, mark with their scent gland, face wash, groom the fur, stretch, and so forth. This is not an exhaustive list, but it does provide a rather good account of some of the many simple action patterns performed by these animals. Again, note that these are neither fixed reflexes nor arbitrary behaviors specified by the psychologist; rather they are naturalistic action patterns that occur in the animal's own environment.

Shettleworth was able to specify how these action patterns change as a result of a variety of conditions including learning and motivational manipulations. In one study, Shettleworth (1975) observed the frequency of these behaviors in both the home environment and a much larger "open field"; in addition, hunger was manipulated. The results were enormously complex as Shettleworth found that the two major factors, hunger and testing location, didn't affect the behavior patterns uniformly. For example, hungry subjects walked around more, reared on their hind legs both in the open and near the wall, and scrabbled more than nondeprived subjects, especially in the home cage. Scent marking was low in the food-deprived animals but somewhat higher for the nondeprived subjects when tested in the open field. Exploration increased as a function of hunger and in the open testing environment. Scrabbling decreased in the open environment, although it increased when the animals were hungry. In summary, "maintenance" behaviors decreased when the animals were hungry, whereas "exploratory" behaviors increased with food deprivation.

Shettleworth also investigated whether the frequency of these action patterns would change as a function of reinforcement. The behaviors that were explicitly reinforced were scrabbling, digging, rearing in the open, marking, washing, and scratching. As shown in Figure 6–1, the first three of these behaviors showed large and immediate increases in rate as a function of contingent food reinforcement. The latter three action patterns showed weak and temporary effects. It is clear that those action patterns that were influenced most by hunger were also the ones most influenced by food reinforcement. That is to say, the action patterns that were readily conditioned were those that increased in hungry animals, while the hard-to-condition behaviors were depressed in hungry animals. The implication is that certain behaviors, such as scrabbling and digging, are especially "relevant" to or influenced by hunger and food reward: when hunger increases, it is adaptive for the animal to engage in behaviors (scrabbling and digging) that increase its probability of coming into contact with food. At the same time, these behaviors have an affinity for food reinforcement; patterns that are performed in anticipation of food are more susceptible to reward. In summary, not all behaviors, certainly not all naturalistic action patterns, are susceptible to reward or to changes in the level of hunger (see Figure 6–1). But there is an important similarity between those action patterns that are responsive to changes in hunger and those that are responsive to changes in food availability.

Shettleworth (1978b) also studied the effect of punishment on these action patterns. The general finding was that punishment reduced scrabbling to a substantial degree and little recovery was observed. Face washing, on the other hand, was suppressed by punishment but not permanently so.

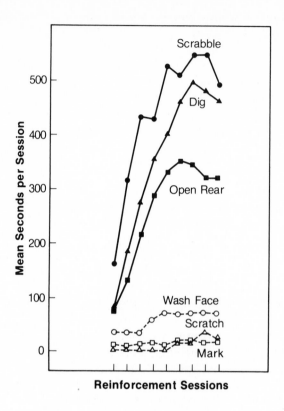

Figure 6-1. Mean number of seconds per 1200-second session as a function of reinforcement session for groups receiving food reward for different action patterns. (From Shettleworth, 1975)

Shock punishment was least effective for suppression of rearing in the open part of the cage. The results, however, were somewhat different when punishment was given on an intermittent basis or on a noncontingent basis. Free shock didn't affect many action patterns such as digging or gnawing; in fact, it increased some, for example, scent marking, standing motionless, and urinating. One conclusion was that various action patterns differed in their responsiveness to shock. The more important conclusion, however, was that the pattern of suppression was different than the pattern of change found for food reinforcement. In other words, punishment suppresses behavior, but different types of actions are affected by punishment than by food reward.

Finally, Shettleworth (1978c) looked at the effect of Pavlovian cues on these action patterns. Conditioned stimuli that were followed by food influenced various behaviors. For example, rearing was increased, but grooming and scrabbling declined in frequency. Pavlovian CSs that predicted shock also had complex effects on certain action patterns, including a dramatic increase in freezing. Again, the effects of Pavlovian fear cues on the performance of action patterns differed from the effects of Pavlovian appetitive cues. Even more interesting was the finding that Pavlovian CSs changed the behaviors differently from reinforcers. That is, inducing fear in the hamsters by means of a Pavlovian aversive CS+ affected behavior in a different way from punishing various patterns directly. Similarly, the presentation of Pavlovian appetitive CSs, which according to our discussion in Chapter 5 induce hope, changed action patterns differently from direct reinforcement of those patterns.

These studies are immensely complicated because the behavior is multidimensional and the treatments are complex. Nevertheless, they signify an increase in the interest in studying naturalistic behavior patterns. In terms of learning, Shettleworth's research indicates that punishment, positive reinforcement, and conditioned Pavlovian cues all have dramatic effects on action patterns, but the type of effects is not the same nor could the outcomes be predicted from simplistic principles derived from the study of other "irrelevant" behaviors like lever pressing. In addition, this research confirms one important theme that we have stressed: there is a relationship between learning factors (such as contingent reinforcement or the presentation of Pavlovian signals) and adaptive behavior. In particular, reinforcement for behaviors that are unrelated to the survival needs of the animal in its natural environment are least likely to be responsive to punishment or reinforcement manipulations, whereas those behaviors that are naturally related to environmental pressures are most likely to change as a function of such manipulations.

Biologically Specialized Learning Phenomena

The notion that learning mechanisms are like other biological characteristics because they are subject to natural selection suggests that specialized forms of learning should have evolved. This is, indeed, the case. In this section we will review several major topics of learning research that deal with specialized learning abilities. In each case, learning is involved although the underlying mechanisms which determine that learning appear to have evolved in specialized ways.

Acquired Food Preferences

One of the most interesting and important areas of research in the last decade has focused on how animals regulate their diet. This, obviously, is a crucial problem for animals if they are to remain healthy and competitive. The ones that do not receive sufficient nutrients, or those that for one reason or another ingest harmful substances, are clearly at a selective disadvantage.

There are numerous strategies for regulating diet. Some animals exist perfectly well by specializing in a single food item; for them, diet selection seems to be innate. But other animals have a number of substances in their diet and learning is definitely involved in their food selection. Learning which substances are most suitable provides a flexibility for these species. They can change their eating habits if the environment no longer contains the foods of their former diets, they can seek out new sources of food that are more nutritious than their present food, and they can learn to avoid harmful foods. Those animals that do not specialize in one type of food or another are, therefore, behaving adaptively when they learn about the nutritional value of the available food items.

Let us consider an example of diet regulation. Imagine that, for some reason, the available food supply for an animal became deficient of some nutrient that the animal required. In the long run, the animal might even die from lack of this nutrient. It would therefore be in the best interest of the animal to recognize new food sources that contained the necessary nutrient. Generally, animals are quite capable of correcting such imbalances by selecting new and appropriate diets. This learning ability has been examined in the literature on acquired flavor preferences (see Rozin & Kalat, 1971).

One important study was performed by Rozin (1969). Rats were placed on a diet that was deficient in thiamine. When such a procedure is followed, rats (and most other organisms) become ill and lose weight because they fail to eat a sufficient quantity of the deficient food. The rats were then given a choice of four distinctively flavored diets, only one of which contained thiamine. Some of the animals developed an immediate preference for the enriched diet. Others took somewhat longer, but eventually they too began to prefer the enriched food. The important point was that the feeding strategy tended to maximize the animals' chances for learning about the new foods: specifically, they would eat one food at a time as if they were attempting to assess the consequences of that meal. Once they had isolated which was the enriched food, they maintained their preference for it. This is an astonishing talent. To be able to learn which food will cure the thiamine deficiency from a selection of different foods when, in fact, the additive has no detectable taste, is an example of a remarkable, biologically specialized, learning capacity.

How are food preferences acquired? How do animals learn which food item is appropriate? The answer appears to be that rats, and other species that are not food specialists, are able to associate the qualities of food, for example, its taste, with illness. In the case of a thiamine deficient diet, the illness results from the deficiency. Stated somewhat differently, flavor preferences result from slow-onset taste aversions (see Chapter 3). The flavor of the food is associated with malaise. There are several factors supporting this belief. First of all, the familiar (deficient) diet is treated as if it were aversive. For example, Rozin (1967) found that rats tend to spill the old diet just as they do when they encounter a diet that they find distasteful. Second, the animals will gnaw on the wire cage, indicating aversion, or they will show fear of new objects or foods if they have been poisoned in the recent past (Galef, 1970; Rozin, 1968).

If the animals are developing an aversion to a deficient diet, then they ought to show preferences for novel diets. This is precisely what Rozin, Wells, and Mayer found in their (1964) study. Moreover, rats will prefer novel foods even if those foods are deficient too. Rodgers and Rozin (1966), for instance, placed the vitamin supplement in the old food and then gave the animals a choice between the old and a novel food. The rats preferred the novel food, even though it was now the deficient one.

Preference for a novel food substance over a familiar deficient substance is one way of solving the problem of a vitamin deficiency. Another way, however, would be showing a preference for a known safe substance. This was demonstrated in a study by Rozin (1968). He used three phases in his experiment. In Phase 1, a familiar safe diet, containing all of the essential nutrients needed by the subject, was presented. In Phase 2, a second diet which was deficient was given. Finally, in the test phase of the experiment, subjects were given a choice between the familiar "safe" diet, the training deficient diet, and a novel diet. Rozin found that the subjects preferred the familiar "safe" diet over the other two choices. In other words, after they have been ill, animals do *not* choose the novel food unless their choice is only between a novel and a deficient diet; rather, their strongest choice is for a diet that they know to be safe.

In summary, subjects are unable to identify the appropriate diet merely by taste (Rodgers, 1967), but with the appropriate feeding strategy (discrete and isolated meals of separate food items), they learn which food has the beneficial consequences. They are motivated to learn this because they have developed an aversion to the old deficient food substance. Quite clearly, this example of adaptive specialization of learning is fundamentally similar to the taste aversion reaction discussed in Chapters 2 and 3. One difference is that flavor preferences are slow to develop, whereas taste aversions are faster acting, that is, based upon the injection of a poison. At any rate, the notion of flavor preferences suggests that many animals can regulate their diet by learning about the consequences of their food items (see also Chapter 10).

The research on acquired food preferences indicates that food items become aversive when a slow-acting illness is induced. But the opposite phenomenon has also been demonstrated: certain flavors or food items will become preferred when they are associated with a beneficial consequence (Garcia, Ervin, Yorke, & Koelling, 1967; Zahorik & Bean, 1975; Zahorik

& Maier, 1969; Zahorik, Maier, & Pies, 1974). In these studies, a flavor was paired with an illness-curing substance, that is, the antidote to an illness, and a preference was established for the flavor. Specifically, in the study by Zahorik et al. (1974), several groups of rats were maintained on a thiamine deficient diet during which time they received a flavored water. It was anticipated that the flavor would become associated with the vitamin deficiency and therefore become aversive. During a second phase of the experiment, animals were given a different flavor and then injected with thiamine. Thiamine was, in a sense, the medicine that cured the deficiency. Since this new flavor was associated with the cure, it was expected that the new flavor would become preferred. During the test phase the authors measured preference for this recovery flavor and found that it was, indeed, preferred to all other flavors, including familiar flavors that presumably were judged safe. In short, these experiments demonstrate that flavors which are associated with recovery from illness will be preferred even more than flavors which the animals know are safe.

Acquired Taste Aversions

It should be obvious that learned taste aversion is a topic closely related to flavor preferences (see Barker, Best, & Domjan, 1977; Logue, 1979; Riley & Clarke, 1977). Taste aversions reflect an aversion to a flavor or odor which has been paired with some fast-acting illness-inducing drug rather than a slow-developing deficiency. As such, taste aversion learning provides an important mechanism for the regulation of diet. There are two historical results that have made conditioned taste aversion an interesting learning phenomenon; they are the belongingness and long delay results. We shall discuss belongingness first.

In conventional Pavlovian conditioning, we often use arbitrary, but convenient, stimuli such as lights, tones, and shocks. This practice can have the unfortunate consequence of suggesting that *all* stimuli should work in the same fashion. This is not the case, however. Evidence indicates that some stimuli are more easily associated than others. For example, early research on taste aversion learning found that flavors and illness "belonged" together (see Domjan & Wilson, 1972b; Garcia & Koelling, 1966; Garcia, McGowan, Ervin, & Koelling, 1968; K. F. Green, Holmstrom, & Wollman, 1974; L. Green, Bouzas, & Rachlin, 1972; Larsen & Hyde, 1977). In the Garcia and Koelling (1966) experiment, thirsty rats were allowed to drink a saccharin-flavored solution while a light and noise were being presented. The authors thought of this procedure as providing a "bright, noisy, and tasty" CS for the subjects. One group of animals was then given a mild poison US; a different group received a shock US. In a second phase of the study, Garcia and Koelling tested the animals for their acquired aversion. Half the animals that had received the poison US were given the "tasty" solution, the saccharin-flavored solution, while the other half were given the "bright, noisy" solution, plain water accompanied by the light and noise stimuli. Drinking during this time was recorded so that if an aversion existed, drinking would be suppressed either by the "tasty" CS or by the "bright, noisy" CS. The other subjects, those that received the shock US, were also divided into two groups; half were tested with the flavor CS and the other half with the light-noise CS.

As shown in Figure 6–2, Garcia and Koelling found that the conditioning to the flavor was strongest when that CS had been followed by poison; similarly, conditioning to the light-noise CS was strongest when that stimulus had been followed by shock. The other combinations, paradoxically, did not produce very strong conditioning: flavors did not become especially aversive when they were followed by shock, and the light-noise CS did not become strongly aversive when it was followed by poison. In other words, the rats easily learned to associate the flavor and the poison, and they were equally good at associating the light-noise and the shock. However, they were poor at learning the association between flavor and shock as well as the light-noise and poison association.

This dramatic result led a number of theorists to speculate that flavors and malaise naturally "go together"; they belong to the same feeding system. The two types of stimuli are uniquely associable because stomach malaise almost surely would be due to having eaten a harmful substance, that is, one with a taste. In contrast, exteroceptive stimuli such as lights and noises do not "belong" to internal reactions such as illness; they are irrelevant to one another. Conditioning would not be expected to occur.

Garcia and Koelling's finding that all cues are not equally associable with certain USs provoked the notion that taste aversion is a unique form of learning. And it is an appealing idea. After all, illness usually results from the ingestion of foods, but never from exposure to external stimuli such as lights and noises. Indeed, although ingestion itself is not absolutely necessary, ingestion of the substance does enhance taste aversion learning (Domjan & Wilson, 1972a). This argument bolstered the concept of belongingness. Evolutionary principles seemed to suggest a sort of belongingness too: animals that have a specialized ability to identify ingested flavors that caused illness were selected for; their chances of eating the food again and thus of being fatally poisoned were reduced. In short, flavors and gastrointestinal events belong together to the extent that flavors (for rats and many species) are the primary cues associated with food. If rats use flavor to identify foods, then it is plausible that they have developed a unique capacity to associate the flavor of foods with their consequence (illness).

This position, that taste aversion is a unique form of learning because not all CSs are equally associable with all USs, has been modified by recent research. At least the strong form of the argument has been softened a good deal. For one thing, numerous experiments *have* demonstrated that ex-

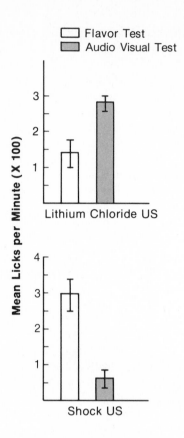

Figure 6-2. Mean licks per minute ($\times$ 100) of the "tasty" solution (open bars) or the "bright, noisy" solution (filled bars) for the groups that were injected with poison or given the shock US during training. (From Garcia & Koelling, 1966)

teroceptive cues, such as lights and tones or other environmental stimuli, can, indeed, be conditioned using a poison US (Best, Best, & Mickley, 1973; Mitchell, Kirschbaum, & Perry, 1975; Revusky & Parker, 1976). The more correct generalization, therefore, appears to be that flavor and odor cues overshadow exteroceptive cues (flavors *are* conditioned more easily than lights and tones), but exteroceptive cues are, by no means, impossible to use in these studies. They are conditioned quite readily if the experiments are conducted in the appropriate manner, that is, if precautions are taken to maintain the novelty and salience of the exteroceptive cues (Mitchell et al., 1975).

All this suggests that stimuli have various qualities: they have a taste, a texture, a visual appearance, an odor, or, perhaps even a noise. Depending on the context, animals are able to utilize these various qualities to different extents. In feeding situations in which gustation is the predominant sense, taste and odor are most salient; visual and auditory cues are not normally part of the food acquisition process and therefore they are not conditioned as well.

But let us reexamine this whole issue in greater detail. Is taste really the predominant sense in poison-aversion studies? What if we tested an animal for whom exteroceptive cues, specifically colors, were more important than taste for recognizing food items? Wouldn't these animals show the opposite result: color-illness associations would be learned more easily than flavor-illness associations? The answer to this question would appear to be yes as shown in a study by Wilcoxon, Dragoin, and Kral (1971). Both rats and quail were given a sour-flavored solution that also had been colored blue with food dye. Thus, the animals experienced two types of CS dimensions, flavor and color. Then, all the animals were poisoned. In the test phase of the study one group of rats and quail received blue water, while another group of rats and quail received a colorless, but sour-flavored, fluid. The rats and quail that were given the blue water were being tested for their aversion to the color of the solution, while the rats and quail that were given the sour solution were being tested for their aversion to the flavor. The results were striking. As shown in Figure 6–3, the rats primarily avoided drinking the flavored solution, whereas the quail avoided drinking the colored solution. In short, rats attended to the flavor more than quail; quail attended more to the color of the solution. These results support the notion that specialized sensory and attentional systems have evolved. For rats, smell and taste are important for food procurement; for quail, visual stimuli are important in foraging. Both species, therefore, are able to identify potentially poisonous foods by focusing upon different stimulus dimensions. In summary, the Wilcoxon et al. study seems to indicate that belongingness is based more on specialized sensory or attentional processes than on unique learning capacities.

This point was elaborated in an important study by Gillette, Martin, and Bellingham (1980). The purpose of the experiment was to evaluate the notion that animals associate eating-related cues with illness. Young chickens were given aversion training using a variety of CSs. In one experiment, a red dye was added to the food (the CS) and it was followed by an injection of lithium chloride (the illness-inducing US). The authors found that the chickens had no trouble forming an aversion to the colored food. Other studies, however, demonstrated that the animals were very poor at learning an aversion to flavored food or colored water, that is, when those CSs were paired with illness, a conditioned aversion failed to develop. Finally, their research showed that the flavor of the water was the most salient cue in aversion learning. That is, chickens rapidly learned to associate a sour flavor

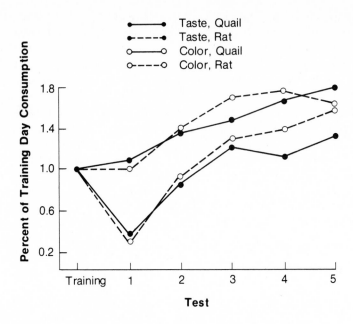

Figure 6-3. Mean percent of training day consumption as a function of training and test sessions for quail (solid lines) and rats (dotted lines) that were tested with the sour water (closed circles) or blue water (open circles). (From Wilcoxon, Dragoin, & Kral, 1971)

in the water with illness. In summary, Gillette et al. demonstrated a rather remarkable finding: chickens easily learn to associate the color of food and the flavor of water with illness, but they do not easily learn to associate the flavor of food or the color of water with a poison US.

In order to explain their findings, Gillette et al. had to examine very closely the actual behavior of the chickens as they ate and drank. When eating, the bird would position its head inside the feeding trough and maintain visual contact with the food throughout; it had to "look" at the food in order to peck at it accurately. On the other hand, when the chicken drank, it did not really orient visually with the water; rather it dipped its beak into the water cup and immediately retracted its head to allow the fluid to run down its throat. In other words, the bird did not utilize the visual aspects of the water in consuming it. These results indicate, therefore, that *whatever* cue is used by the subject in securing and consuming the food or water is the one most likely to be conditioned. In the case of food, the chickens used vision; the colored CS therefore was most strongly conditioned. In the case of drinking, however, they used vision less but were sensitive to the taste; the flavor therefore was the most salient cue.

On the surface, these results appear to conflict with those found by Wilcoxon et al. (1971) who used quail. Recall that those authors discovered that quail learned an aversion to the color of the water more strongly than to the flavor. The resolution of this discrepancy by Gillette et al. is important to our understanding of biologically specialized learning abilities. Gillette et al. speculated that the most potent CS in an aversion study would be the one that the animals actually used to procure the CS itself. They noted that Wilcoxon et al., had used clear tubes to hold the colored water, whereas they had used a covered container with a water cup inside. According to Gillette et al., then, if the birds were "forced" to locate the water using visual cues, such cues might become conditioned more strongly. In a final study, Gillette et al. used drinking spouts similar to the ones used by Wilcoxon. They confirmed that the color of the water was indeed conditioned more strongly than the taste.

The study by Gillette et al. indicates that taste or color cues do not necessarily "belong" to illness independent of the context in which they occur. What appears to be important is the method by which subjects procure food. The cues that are used in that process are the ones likely to become strongly conditioned. For many animals, including rats, taste is a very important stimulus dimension in feeding situations. Birds, on the other hand, use visual cues more often. But it is the actual context that ultimately determines which stimulus dimension is most important. If the situation is structured so that the animals are exposed to the flavor of the water, then taste will become conditioned more strongly than other attributes of the CS. However, if the context requires the use of visual cues for locating food, then color will be the predominant characteristic attended to. It is inappropriate, therefore, to state that color will always be most salient in birds simply because they are visually oriented animals that often use visual stimuli in procuring food. Color *may* be most important, but flavor, too, can be the salient dimension if it is critical to the animal's feeding behavior.

This general conclusion fits well with an evolutionary perspective. Rats normally use taste to identify food; birds normally use visual cues. This is why such species differences are observed in poison-avoidance studies like the one by Wilcoxon et al. Our reconsideration of the belongingness notion has, therefore, led us to this resolution. Unfortunately, however, this resolution may be too simplistic an answer. In fact, recent evidence suggests that there may indeed be some *special* relationship between taste and illness, quite independent of the suggestion by Gillette et al. that the CS modality most susceptible to conditioning is the one that is used at the time of food ingestion.

Consider for a moment which sense modality is most likely to be stimulated just prior to illness. It is more likely to be taste than vision. That is, taste surely would be the "last" sensation an animal would have after it has eaten. Even for a bird, the visual characteristics of a food item would have long since disappeared from view when the taste of that item is still present. From an evolutionary perspective, therefore, perhaps there should be something "unique" about taste after all. Perhaps taste, because it is the last sense to be stimulated before ingestion, ought to become conditioned to illness especially easy.

We know that taste will potentiate the conditioning of exteroceptive cues; it seems to be special in that regard (see, for example, Galef & Osborne, 1978; Morrison & Collyer, 1974). In addition, we should note that both Wilcoxon et al. and Gillette et al. *did* show that birds could learn to avoid flavors. More important, however, are the findings of a recent study by Lett (1980). Pigeons and quail were poisoned after drinking flavored water, colored water, or both flavored and colored water. Then they were tested for their aversion to either the flavor component or the color component. Lett found that the birds easily learned to avoid the flavor. Color was much less important. In fact, the addition of a flavor to the colored water enhanced conditioning to the color. The birds avoided drinking the colored water *more* if the water also had contained a flavor component during training. Conditioning was much weaker when the subjects were trained and tested with color only (see also Clarke, Westbrook, & Irwin, 1979).

Let us summarize this complicated but important area of research. First, it appears that rats and birds can learn an aversion to flavor *and* exteroceptive CSs. Flavors admittedly are predominant, especially for rats, but visual cues are conditioned with relative ease too. Second, as shown in the study by Gillette et al., the sense modality used by the animal to procure food at any given time is important. Birds especially may use visual cues when confronting foods or fluids (and thus become strongly conditioned to them) even though later, when consumption has occurred, taste may predominate. Finally, as shown by Lett, flavor ultimately may be the most salient stimulus for poison avoidance learning and, thus, the one most readily identified by the subject as the cause of illness. In short, we see no reason to dispute either of these factors: taste may indeed play a unique role in poison avoidance learning, but animals also may be especially susceptible to acquiring an aversion to a nontaste cue that played a role in their choosing the food item originally.

Let us now consider a second important characteristic of taste aversion learning, namely, the fact that learning can take place even when the CS and US are separated by a considerable period of time (Andrews & Braveman, 1975; Garcia, Ervin, & Koelling, 1966; Kalat & Rozin, 1971). In the study by Kalat and Rozin, already discussed in Chapter 3, different groups of rats were given sucrose water and then injected with a mild poison (lithium chloride) after either .5, 1, 1.5, 3, 6, or 24 hours. Two days later the subjects were tested for their aversion to the sucrose by being presented with a choice between sucrose and water. As shown in Figure 3–2 in Chapter 3, the authors found that the percentage preference for sucrose, the substance the rats normally prefer, increased as a function of the CS-US interval. That is, a strong aversion was evident for the groups that received lithium chloride within 1 hour following consumption, but after 6 hours, no aversion was evident, that is, those subjects preferred sucrose as strongly as the 24-hour subjects.

The fact that taste aversion occurs even when the US is delayed for several hours differs from the findings of most Pavlovian studies that have studied "conventional" CRs like conditioned salivation or heart rate. In most Pavlovian studies, except for the CER procedure, conditioning fails to occur with CS-US intervals longer than 10 or 20 seconds. To this extent, then, taste aversion is a "unique" form of learning. More importantly, however, this characteristic demands that we adopt an evolutionary perspective: animals in the distant past that could bridge the long interval between tasting a food item and its poisonous consequence were at a selective advantage; if they survived, they were more able to remember the poisonous flavor and avoid consuming that food again. Animals that could not tolerate the long delay were those that were unable to profit from experience; they did not develop the aversion to the flavor and were likely to consume the flavor a second time. Perhaps, such a second experience was fatal. In short, learning to associate a poison with its qualities, either flavor or visual characteristics, reflects a specialized learning ability that has been selected for. Poisoning, of course, takes time to occur; the effects do not occur immediately upon ingestion of the food.

The fact that taste aversions can be learned when the US is delayed for several hours has intrigued and puzzled psychologists. The problem is that associative learning is thought to require close temporal contiguity. How is it that an animal can associate a taste with a feeling of illness when the two are separated by several hours? Although the typical CS-US interval gradient is observed, that is, the strength of the aversion decreases with increasing CS-US intervals, nevertheless the problem remains: the interval can be many hours and still learning will take place.

For a number of years, investigators tried to explain long delay learning by making adjustments to contiguity theory. The adjustments or assumptions were designed to salvage the contiguity theory by attempting to show how the contiguous occurrence of the CS and US really might occur (for example, the "aftertaste" lingers on the tongue and is present later when the US is given). These efforts, however, were never very successful. More recently, Kalat suggested a different sort of explanation, termed the learned safety theory (see Kalat, 1977; Kalat & Rozin, 1971). This theory basically claims that during a long CS-US interval animals are learning that the flavor is safe. The animals do not later show an aversion because they have learned that the CS flavor is safe, not because they have failed to associate the two stimuli.

Let us consider this theory in more detail. It is well known that many animals, such as rats, display neophobia. They fear new objects, environments, and even novel flavors (Mitchell et al., 1975; Nachman & Jones, 1974). Neophobia, of course, is an adaptive reaction to novel objects. If organisms didn't show caution when they encountered some new food, they could commit a disastrous mistake by consuming too much of the food. After all, unknown foods could be poisonous. Therefore, animals react pessimistically to new food and eat only a small portion. After repeated experience with the food, however, consumption increases (Nachman & Jones, 1974) presumably because the animals have judged the food to be safe. If the animals experience illness during this time, however, neophobia is reinstated (Mitchell et al., 1975; Rozin, 1968). According to Kalat's theory, then, the reduction in neophobia following exposure to a food is the same as the acquisition of a sense of safety; extinction of fear of the substance reflects the knowledge of the safety of the substance. The typical CS-US interval gradient, therefore, does not mean that the animals failed to associate the CS and US after longer intervals. Rather, it reflects an increasing reduction in neophobia and correspondingly an increase in the animal's sense of confidence in the safety of the CS. Consistent with such a notion is the fact that taste aversion conditioning is strongest when using novel flavors (Ahlers & Best, 1971; Revusky & Bedarf, 1967; Siegel, 1974). Familiar flavors are considered safe and therefore are not judged to be the cause of illness when they are later used in a taste aversion experiment.

Perhaps the clearest support for the learned safety theory was provided by Kalat and Rozin (1973). Three groups of rats were used. One group was given a novel flavor 4 hours prior to receiving the poison US. A second group received the CS only a half hour before the poison administration. Based merely on the contiguity of the two stimuli, one would expect that the second group would condition more strongly than the first; this, indeed, is the normal finding. The important group, however, was the third group: subjects in this group were given the CS flavor twice, once 4 hours prior to the poison US and again a half hour before the US. The contiguity theory would predict that this group would be most similar to Group 2; after all, in both cases the CS-US interval was only a half hour. The learned safety theory would predict the opposite result: the group that received the CS twice would be equivalent to the 4-hour group because both those groups had the same amount of time in which to discover that the flavor was safe. The poison was not administered for 4 hours and so the degree of learned safety should be equivalent. The results favored this latter theory. Groups 1 and 3 showed approximately the same degree of aversion, indicating that aversion learning decreased as a function of the opportunity to learn that the CS is safe.

This theory has been criticized and altered in recent years (see Kalat, 1977). First of all, it is known that aversions can be conditioned to thoroughly familiar flavors (Nachman & Jones, 1974). The learned safety theory, of course, would predict that such familiar flavors could never become aversive, or at least could become so only with very great difficulty. Second, Best (1975) found that preexposing a flavor CS reduced the ability for that CS to become either a conditioned exciter or a safety signal. If the animals truly had learned to perceive the flavor as safe, then they should have shown an enhanced preference for that flavor later on. This was not the case, however. A preexposed flavor was not easily converted into a preferred safe flavor.

The resolution has been to change the focus from learned safety to learned irrelevance (see, for example, Baker & Mackintosh, 1979). By doing this, one avoids the implication that "safety" is conditioned during the CS-US interval. Quite simply, the learned irrelevance theory maintains that animals show a natural pessimism or neophobia for new flavors. Their genetic programming has made them wary of the possibility of injury or sickness. Gradually, if the animals maintain their health even after consuming some of the CS, they come to believe that the flavor is meaningless. They apparently do not endow the flavor with any special properties such as safety, but they do learn that the flavor does not predict any illness which they originally suspected could arise. In this regard then, the learned irrelevance theory of taste aversion is nearly identical to the associative theory of latent inhibition (see Chapter 3). In fact, the processes seem identical: preexposure of a CS, whether it is an exteroceptive CS like a tone or an internal cue like a flavor, leads to a reduction in future excitatory learning because the preexposure results in learned irrelevance.

It should be clear that taste aversions and flavor preferences are important mechanisms for the regulation of diet. A diet that slowly poisons an animal or a food substance that quickly induces malaise (taste aversion) or even a diet that brings rapid relief or recovery to a sick animal (taste preference) all promote conditioning which, in turn, determines which foods the organism will choose in the future. The avoidance of aversive foods, either slow or fast acting, and the preference for beneficial foods, especially those that are rich in calories (Bolles, Hayward, & Crandall, 1981), then, is a highly adaptive behavioral potential.

Let us consider one final example of taste aversion as it operates in the natural environment. We have claimed that these learned reactions, aversions or preferences, guide the subject in choosing an appropriate diet. Furthermore, we have emphasized the notion that the stimuli which become predominant are those which are naturally used by the organism to locate food. These arguments are confirmed and highlighted in many studies on naturalistic feeding patterns. For example, in the natural environment there are some insects that are toxic to birds; eating them causes malaise. In particular, blue jays learn to avoid ingesting monarch butterflies by associating the coloration and/or taste of the butterfly with malaise

(Brower, 1969; Brower, Ryerson, Coppinger, & Glazier, 1968). But often experiments have shown that birds don't even have to experience illness in order to learn to avoid a visually distinctive object. The prey item merely has to have a bitter taste. Gillan (1979), for instance, showed that birds will suppress drinking during a visual stimulus when that visual stimulus signals the presence of quinine (quinine is exceedingly bitter). The animals do not suppress their drinking during other visual stimuli that are not associated with quinine. In conclusion, taste aversion and food preferences operate in the real world although an animal need not become ill to learn them.

Avoidance Learning

In Chapters 2, 4, and 5, we discussed avoidance learning, claiming that it was an example of instrumental conditioning, and furthermore, that animals learn to avoid noxious stimuli, in part, because they were conditioned to fear such stimuli. Here, we will extend that discussion by providing a biological perspective on avoidance learning. Just as many appetitive behaviors reflect learning specializations, such as food selection and taste aversions, avoidance learning also involves specialized mechanisms. This makes good sense. Defense from predators or other harmful stimuli is an important activity for organisms in their effort to behave adaptively. From a biological point of view, it is clear that animals have evolved morphologically to defend against predators: camouflage, special coloration, special sensory and motor systems, all provide a measure of protection for many species. Behavior, therefore, or, more exactly, learning potential, ought to have evolved in specialized ways too. Defense against harm is simply too important not to have evolved specialized mechanisms.

There is a growing literature indicating that this, indeed, is the case. For example, an experiment by B. Schwartz (1973) illustrates this point. A response key, or plastic disk, was illuminated with one of two colors. One color signaled the presentation of food and the other color signaled the presentation of shock. A response, a peck, would prevent the US presentation. So, for example, if the pigeon pecked during the food light, food was omitted; similarly, if the pigeon pecked during the shock light, shock was omitted. Schwartz found an astonishing result: subjects couldn't seem to refrain from pecking during the food light, but they couldn't seem to manage efficient pecking during the shock light. This outcome illustrates two points: that some responses are not learned very well when they avoid shock (they are poor defensive or avoidance reactions) and that the same responses are biologically linked to appetitive behaviors (see the material on autoshaping in Chapter 3). In other words, Schwartz's experiment shows that the pecking behaviors were not governed by the reinforcement contingencies. If they had been, the animals would not have pecked during the food light (pecking caused food to be omitted), but they would have pecked consistently during the shock light (because a peck omitted shock). The opposite was found, however. Food presentation seemed to compel pecking, because it is a natural reflex or unconditioned reaction to food, whereas the pigeons seemed quite unable to learn to peck in order to avoid shock. When the pigeon is allowed to make a more *naturalistic* response to avoid shock, such as wing flapping, learning is much better (MacPhail, 1968).

Another illustration showing that animals cannot utilize simply any arbitrary response as an avoidance reaction is the finding that, under many conditions, rats have considerable difficulty learning to press a lever in order to avoid shock (D'Amato & Schiff, 1964). Lever press learning is routine for food reinforcement but not for shock avoidance. If special procedures are used (for example, D'Amato & Fazzaro, 1966; Delprato & Holmes, 1977; Giulian & Schmaltz, 1973; Kulkarni & Job, 1970; Scheuer & Sutton, 1973), then bar press avoidance improves. But the very fact that special procedures must be used suggests that lever pressing, ordinarily an easy response for the rat, is not an effective defensive reaction.

Avoidance learning and evolutionary specializations were considered in a remarkable paper by Bolles (1970a). He argued that because certain responses, such as lever pressing for rats, are very poor avoidance reactions, then avoidance learning in general must reflect species-specific tendencies. Bolles went even further. He claimed that avoidance behaviors actually are innate or unlearned defensive reactions that are particular to a given species. Survival from predators is far too important to be dependent upon some gradual learning process. If an animal had to learn gradually to avoid its predators, it very likely would not survive long. Effective avoidance, therefore, must be learned very rapidly (or be innate) and must reflect the specialized talents and learning abilities of that particular species. Specifically, Bolles states: "The parameters of the natural environment make it impossible for there to be any learning of defensive behaviors. Thus, no real-life predator is going to present cues just before it attacks. No owl hoots or whistles five seconds before pouncing on a mouse. And no owl terminates his hoots or whistles just as the mouse gets away so as to reinforce the avoidance response. Nor will the owl give the mouse enough trials for the necessary learning to occur. What keeps our little friends alive in the forest has nothing to do with avoidance learning as we normally conceive of it or investigate it in the laboratory" (pp. 32–33).

Bolles thus claims that animals have species-specific defense reactions (SSDR) which are innate responses for dealing with predators and harmful stimuli. For rats, the normal SSDRs are fleeing or running, but freezing, threat displays, or even burying of aversive objects (Terlecki, Pinel, & Treit, 1979) may be observed. According to the SSDR theory, each species possesses a hierarchy of species-specific defense reactions. When confronted with aversive stimulation, the animal naturally performs its most preferred SSDR. If that particular response is not permitted by the reinforcement contingencies, then the subject begins performing its next most probable SSDR. For example, in a typical avoidance situation where the response is running from the start box to the goal box, rats learn perfectly well because the experimental procedure permits them to perform their most preferred SSDR. In other situations, however, where the preferred SSDR is not permitted, such as in lever press avoidance studies, then learning is very poor; the rat must suppress its natural inclination in favor of unnatural defense reactions. Bar press avoidance would be learned only after all the animal's natural SSDRs are suppressed. It is certainly true that bar press avoidance is learned either not at all or only after very protracted training experience. It is also true that lever pressing is not in the least like the rat's natural reaction to aversive cues such as shock.

A number of investigators have tested several facets of the SSDR theory in recent years. Although rats will run when shocked, their natural inclination to fear stimuli is to freeze or crouch (Blanchard & Blanchard, 1969a; Bolles & Collier, 1976). In fact, some investigators have suggested that freezing is innate (Bolles & Riley, 1973). Nevertheless, freezing certainly is an SSDR when an animal is fearful; avoidance learning will take place rapidly if the freezing response is compatible with requirements of the experiment. This point was demonstrated by Brener and Goesling (1970). Every time the subjects in the immobile group moved, they received a 2-second tone followed by a shock. They could avoid the shock by simply remaining motionless. However, rats in another group could avoid the shock by being active. That is, if body movement was not detected by the special electronic sensing device, a 2-second warning signal was sounded followed by shock. As shown in Figure 6–4, the freezing group learned the task exceedingly well, much better than the active avoidance subjects who didn't even show an increase in activity despite the fact that activity itself was reinforced. In other words, since freezing is the natural reaction of a fearful rat (the rat has been shocked in this particular environment and it is therefore fearful), then the animal rapidly learns freezing as an avoidance reaction. Rats normally do not become active when they are frightened and so the active avoidance response was learned poorly. This study, therefore, indicates that the efficacy of any given reinforcement contingency in an avoidance task depends upon whether the criterion response is compatible with the animal's SSDR. Stated somewhat differently, avoidance learning reflects the learned specializations of a given species, the behavioral patterns that have evolved as specialized mechanisms for defensive reactions.

Another important confirmation of the SSDR theory was provided by Grossen and Kelley (1972). They first showed that rats will become thigmotactic when frightened; that is, they will tend to stay in close contact with objects or walls, never venturing into open spaces. From an evolutionary perspective this thigmotactic, or wall-hugging, behavior is highly adaptive: ancestors that were too bold and ventured into the open

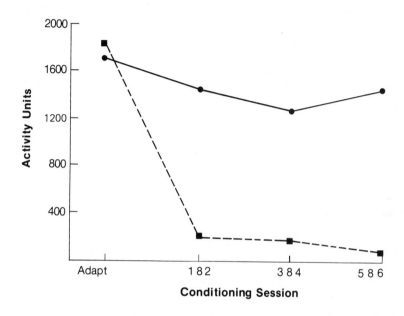

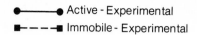

Figure 6-4. Mean activity score during the adaptation and conditioning sessions for groups that were reinforced for making an active avoidance reaction or an immobility reaction. (From Brener & Goesling, 1970)

spaces were more likely to be spied and captured by, say, birds of prey. However, animals that were timid and avoided venturing into the open spaces were more likely to survive and pass this timidity on to future generations. So aversion or fear should heighten this thigmotactic behavior, which is exactly what Grossen and Kelley found. Furthermore, they discovered that rats could easily learn to avoid shock if they were allowed to run to a safe platform located at the periphery of the environment; this was compatible with their thigmotactic inclination. However, rats were poor at learning to avoid shock when forced to run to a platform located in the center of the apparatus; running to the center was inconsistent with their thigmotactic tendency and therefore not learned easily. In summary, this experiment indicates that natural defensive reactions play an important role in an animal's avoidance learning. Not only does active avoidance learning depend upon the compatibility between the avoidance response permitted by the experimenter and the animal's own SSDR, but passive avoidance learning does as well (Pearce, 1978).

What about the environmental conditions at the time of learning? Do these physical constraints play any role in avoidance learning or does avoidance depend exclusively upon the animal's innate predispositions? The answer appears to be that the environmental factors as well as the reinforcement contingency do play an important role in avoidance learning. Blanchard and Blanchard (1971) found, for example, that rats would learn to run away from an approaching cat. The rats would even learn to run across an electrified grid in order to avoid coming into contact with the cat. If, however, the rat were confined in a small wire cage when the cat was placed nearby, the rat would freeze; it would not attempt to run away. In one case the SSDR was an active response, whereas in the other case it was a freezing response. Both were learned quite effectively; both are naturalistic SSDRs. However the constraints of the environment influenced which SSDR was performed. One could imagine an animal in the natural environment running away from a predator unless it was cornered, in which case it might freeze to try to avoid being detected.

Social-Oriented Learning Phenomena

In the preceding section we discussed certain forms of survival behaviors that involved learning specializations: diet regulation and defensive reaction learning. The adaptive behavior of animals, however, is not exclusively concerned with the procurement of food and the avoidance of danger. Other things are important as well, principally social activities related to mate selection and communication. In this section we will review several phenomena that show specialized learning mechanisms in a social context.

Imprinting

Consider for the moment the problem that an animal faces in learning to recognize its own species. One could imagine, of course, that species recognition might be innate; after all, it would seem strangely maladaptive to be unable to recognize a member of your own species immediately after birth. But research has clearly shown that species identification, at least in terms of preference, is learned. The learning is called imprinting and it has provided an interesting and complex challenge for experimental psychologists and ethologists.

Imprinting is most noticeable in precocial birds. These are birds who are well developed at birth, who can utilize their motor abilities to nearly their full extent right after birth, and whose parents leave the nest soon after the birth of the chicks. The normal pattern is as follows. After a short period of caring at the nest, the bird, say a duck hen, vocalizes recurrently and then slowly walks away from the brood. The chicks begin to follow within a short period of time. This is imprinting: the development of filial attachments as shown by the following behavior of the young birds. Quite clearly, imprinting is an important process from an evolutionary point of view. The young chicks, by staying near the hen, are able to maintain their security and are exposed to important stimuli in their environment. In fact, it is believed that imprinting is, in some way, involved in later sexual selection. Young birds learn to identify and associate with members of their own species, and such learning leads to proper mate selection later on.

As noted above, it would seem that animals would naturally know who their mother is and innately follow her when she leaves the nest. This is not the case, however; in fact, research on imprinting demonstrates that young chicks will follow virtually *any* moving object soon after birth. A typical example is the following. Upon hatching, chicks are isolated in a housing area prior to the experiment. In Phase 1 of the experiment the chick is exposed to a particular object. For example, it might be placed in a large arena and permitted to observe various visual targets. The targets can vary enormously: geometric shapes in two or three dimensions are common, as are more bizarre objects such as toy trains, tennis balls, watering cans, foam-rubber dolls, or stuffed animals. Lorenz (1937), the noted ethologist who first discovered imprinting, even showed that precocial chicks, greylag geese to be specific, could be imprinted on human beings; they would follow Lorenz just as readily as the normally reared bird would follow its own mother.

In the second phase of the typical imprinting study, the bird is placed into a choice apparatus and allowed to approach either the preexposed object or some novel object. Inevitably, the bird demonstrates the effects of imprinting by approaching the preexposed (imprinted) stimulus. Indeed, the animal will show a strong and lasting preference for the imprinted object over the animal's own natural mother.

Early research suggested that imprinting was a unique form of learning, quite unlike the typical processes that underlie classical and instrumental conditioning (Hess, 1959a). Let us highlight some of the interesting findings which suggested that imprinting might be a unique process. First, imprinting was thought to be irreversible. If a bird were imprinted to an inanimate object, such as a revolving cardboard disk, then its preference would be permanent; it would be impossible to substitute a different object for that imprinted stimulus. Such a belief would make intuitive sense; once an animal followed its mother and became imprinted on her visual appearance, then it would be highly maladaptive for the young animal to adopt a different object.

Second, and perhaps more importantly, there was thought to be a critical period, or a sensitive time, for imprinting. This was shown in the work of Hess (1959b). Ducklings were imprinted at various ages after hatching. The task was to follow a duck decoy around a circular alleyway for a period of 10 minutes. The strength of the following response was measured. As shown in Figure 6–5, Hess found that some animals would imprint immediately after birth. However, the degree of imprinting increased to a maximum at 13 to 16 hours of age and then, more importantly, declined with age. In other words, Hess claimed that animals could only be imprinted at a particular time in their life. If the animals failed to be imprinted during this sensitive period, then the opportunity was lost and imprinting would be difficult to achieve. The fact that animals possessed only a short period of time during which this sort of learning was possible suggests that the learning was quite distinct from normal associative conditioning.

A third characteristic of imprinting, which also suggested that it was a unique form of learning, was the finding that the more effort required on the part of the young animal, the stronger the imprinting. Subjects that had to run greater distances to affiliate with the object, climb over hurdles, or walk up inclined surfaces (thereby expending greater energy) seemed to imprint more strongly than subjects that were not required to expend effort. Normal conditioning, of course, shows the opposite effect. Animals appear to learn less well when greater effort is required.

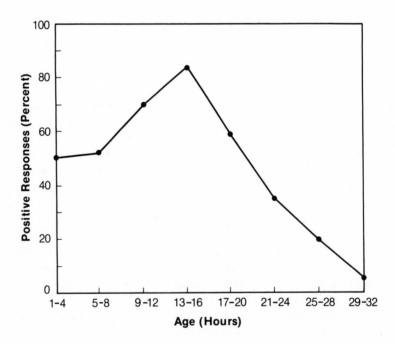

Figure 6-5.   Mean percentage following response as a function of age after hatching. (From Hess, 1959b)

Finally, Hess (1959b) showed that imprinting was stronger if the animal were given electric shock. This result, of course, provided another reason for believing that imprinting was unique. Shock normally suppresses behavior (see Chapters 4 and 11), but in the case of imprinting, it enhanced responding.

In summary, Hess and others claimed that imprinting was essentially a fixed pattern of behavior that was "released" by the appropriate stimulus. The releaser for the imprinting response could vary widely in its stimulus characteristics, but once the animal had been exposed to the object during the initial period, its affiliation was confined to that specific stimulus.

During the past two decades, there has been a great deal of research on the imprinting phenomenon (see Bateson, 1966; Hoffman & Ratner, 1973; Sluckin, 1964). Imprinting is no longer viewed as a highly discrepant process, but rather as a biologically specialized form of learning, not entirely unlike other forms of conditioning. First of all, it is true that a wide diversity of objects can be used as the imprinting stimulus; the objects can vary in shape, color, texture, etc. But not all objects are suitable. In fact, several investigators have discovered that the subject must see the object in motion if it is to be imprinted upon it; merely viewing the static, motionless object is not enough (Hoffman, Eiserer, & Singer, 1972). Moreover, the sound made by a stationary stimulus appears to have little effect (Hoffman, Stratton, Newby, & Barrett, 1970; but see Gottlieb, 1973), although the sound made by a moving object does gain control of the subject (Eiserer & Hoffman, 1974b). Finally, tactile contact between the subject and the imprinting stimulus can enhance the imprinting process (Clements & Lien, 1975).

Perhaps the greatest change in our understanding of imprinting has been in terms of the critical period and the irreversibility notions. Modern research has indicated that there is no critical period *per se.* Several investigators (Gaioni, Hoffman, DePaulo, & Stratton, 1978; Ratner & Hoffman, 1974) failed to show that imprinting is restricted to a particular time in the animal's life. Much older subjects, for example, five- to ten-day-old ducklings, can be imprinted without difficulty. Granted the older birds must be given a longer exposure period to the novel imprinting stimulus, but nevertheless imprinting does occur. The earlier belief that imprinting was restricted to a few hours after birth was based on experiments that used very short exposure periods.

It has also been found that the imprinting process is reversible; with long enough exposure to a second imprinting object, subjects are able to shift their attachment to the new stimulus (Hoffman, Ratner, & Eiserer, 1972). The shift represents a true change in preference since the occasional presentation of the first imprinting object retards the switch (Eiserer & Hoffman, 1974a).

Finally, consider the curious finding that shocking the animal enhances the imprinting process (Hess, 1959a). We now know more about this phenomenon. It appears that shock has two effects: it can cause an enhancement of the following behavior (Ratner, 1976) *and* it can cause the animal to shift its preference away from the imprinting object to a novel stimulus (Barrett, 1972; Ratner, 1976). In other words, shock modified the ducklings' behavior by enhancing the motivation for following the imprinting object, but, at the same time, shock endowed the imprinting object with aversive qualities. It is interesting to note in this regard that if the shock is contingent on a particular *response,* such as pecking, as opposed to the imprinting *stimulus,* then the subject does not develop a preference for a novel object (DePaulo, Hoffman, Klein, & Gaioni, 1978). It is as if the subject attributes the shock to its pecking behavior rather than to the imprinting stimulus. Regardless, it is believed that the effect of aversive stimulation upon the following behavior is due to the fact that such stimulation enhances the arousal of the animal; in an aroused state, the animal is more attentive to any conspicuous moving object.

Scientists are not entirely clear about the mechanisms underlying the imprinting phenomenon nor are they in agreement about the theoretical explanations (see Rajecki, 1973). Some have argued (Bateson, 1966) that imprinting is an example of perceptual learning. Mere exposure to the environment and to the imprinting stimulus produces familiarity and preference. As a particular object becomes more familiar, novel objects tend to elicit more neophobia; the subject follows and prefers the familiar object but rejects the novel stimulus.

A more current and integrated theory of imprinting was proposed by Hoffman and Ratner (1973). They claimed that imprinting is based on a conditioning process. First, Hoffman and Ratner assume that precocial birds are innately disposed to respond to certain kinds of stimulation; the animals find this stimulation reinforcing and therefore they show filial behavior toward the object. Second, these authors claim that the stimulus features of the imprinting object, its visual appearance and perhaps sound, come to acquire the capacity to elicit the filial behavior too. Through a process of conditioning and association, the mere sight of the object is sufficient to elicit the imprinting reaction. Third, precocial birds develop fear of novel objects as they grow older. Initially, they do not fear these objects (they find them reinforcing), but a day or so after hatching, the neophobic reaction becomes quite strong. As birds age, they become more and more frightened by novel objects. Indeed, if the birds are deprived of viewing their environment by the placement over their eyes of special "goggles" which do not permit the viewing of patterned light, then the age at which the birds begin to show this fear reaction is increased (Moltz & Stettner, 1961). In other words, if an object is given to a very young subject, then the suppression of distress is immediate; the precocial bird does indeed find the object innately reinforcing during the early part of its life (Hoffman et al., 1970). In contrast, an older subject reacts to novel stimuli with fear rather than filial behavior (Gaioni et al., 1978); it takes a much longer period of exposure for the stimulus to become reinforcing. In summary, the reinforcement model of imprinting claims that two processes are evident: early in life, precocial birds find certain forms of stimulation innately reinforcing; at the same time there is a growing tendency to fear stimuli that are unfamiliar. Since the reinforcing (imprinting) stimulus can alleviate the distress, the visual characteristics of the stimulus become reinforcing. In fact, the presentation of an imprinting object can be used as a reinforcer for routine responses like pecking (Bateson & Reese, 1969; Hoffman, Searle, Toffey, & Kozman, 1966) and its withdrawal can be used to suppress these behaviors (Hoffman, Stratton, & Newby, 1969).

The final claim of Hoffman and Ratner's (1973) theory is that the behavior of the bird is a resolution of the two competing tendencies aroused by the imprinting stimulus: the filial or approach tendency and the fearful or neophobic reaction. The animal will follow the object if it has learned about the visual characteristics, but it will avoid the object and show distress if the object has not been exposed on previous occasions. The degree to which the animal will follow or avoid, then, is a combination of the two tendencies.

Hoffman and Ratner's reinforcement model can explain a good deal of the imprinting literature. Filial behavior is restricted to the exposed stimulus because that stimulus is familiar and rewarding, whereas other stimuli, presented later in training, are novel and aversive. This theory also deals with the critical period idea. Older animals simply require a longer period of time to become familiar with a particular object. Once familiarity is achieved, then the novelty-induced fear reactions subside. Finally, recent literature indicates that imprinting is not permanent. For the same reason used to explain the critical period, subjects can switch their attentions to new imprinting stimuli. In conclusion, the notion that the visual characteristics of the imprinting object become associated with the reinforcing effects of that object is the sense in which imprinting is described as based on a conditioning model. Otherwise, the theory claims that imprinting is a function of two processes working jointly: the filial reinforcing process coupled with the increasing neophobic reaction.

There are other theorists who do not agree entirely with the reinforcement model of imprinting. Some claim that we must investigate imprinting more as it occurs naturally than has been true of past research (Gottlieb, 1973). In fact, Gottlieb emphasized that when imprinting is viewed in a more natural context, one discovers certain important discrepancies. For example, Gottlieb (1971) showed that precocial birds identify their species on the basis of auditory perception. Subjects seem to be "tuned" to the peculiar qualities of the mother's call; without these qualities filial behavior does not occur. It is true that most imprinting studies do not control for the various noises that occur in the environment; the hatchlings are normally maintained in visual but not auditory isolation prior to the experiment. It may be, therefore, that important factors have not been adequately considered.

An important point to note is that imprinting is not restricted to precocial birds. It is claimed to occur in many species including human beings (Bowlby, 1969) and infant monkeys (Mason, 1970). To this extent, then, imprinting is a pervasive and important form of learning, one that influences the social development of the animal and one that reflects biological specializations.

Animal Language

There is, perhaps, no other behavior as important as communication among animals, at least among highly developed and complex animals like mammals. This is especially true of animals that live together socially. They require some form of communication so that efforts to secure food and defend resources are not duplicated or lost. Communication among animals is extremely varied (see Sebeok, 1977). Types of communication include odors or scent markings by animals as diverse as wolves, hamsters, and worms; threat displays; special coloration and appearance; gestures of various kinds; and, most important for our present discussion, noises. We are not concerned here with all of the aspects of animal communication; that topic is much too vast for adequate coverage. Rather, we will focus on specialized forms of communication that depend upon the learning process.

While scent trails and specialized coloration are not learned characteristics, some primitive communication systems do depend upon experience. Most notable is song learning in many bird species (see Nottebohm, 1972). Research has demonstrated that some birds learn their songs, whereas other birds appear to need no experience; they sing the appropriate song without ever having heard it before. The white-crowned sparrows are among those that require experience to learn their song (Marler, 1970). If the young birds are isolated when they are between 10 and 50 days of age so they cannot hear the singing of an adult white-crowned sparrow, the isolated birds later sing a very poor version of the song. If, however, the young birds are exposed to their species-specific song during this critical period of their life, they are later able to sing the song appropriately. This type of learning is obviously different from the stimulus and response expectancy models that we have previously discussed. Nevertheless, it illustrates that some form of observational learning is required for many birds to develop adequately their vocal communication system.

The very best example of communication learning in animals comes from the literature dealing with language in apes. The answer to the question "can apes learn language?" is complex and has generated a good deal of controversy. Traditionally, it was believed that human beings were the only species on earth capable of language (Chomsky, 1959). It is certainly true that humans have a unique ability to use language in a flexible way relative to any other species. But many scientists now believe that apes are also capable of learning and using a true language. By language we mean the use of arbitrary symbols to communicate messages about things that are not present; to arrange these symbols in such a way that various rules, such as those governing syntax, are obeyed; and to derive meaning from the message in a flexible way.

The first efforts to teach apes language met with failure (Hayes, 1951). Despite extensive training, the chimpanzee seemed unable to utter more than "mamma" and "cup." One of the reasons for this poor performance was the fact that chimpanzees do not have the vocal capabilities that humans have. Humans have a species-specific mechanism that allows for speech, but this mechanism is not duplicated in the nonhuman primates (Lieberman, Klatt, & Wilson, 1969). If apes were to learn language, therefore, it would be necessary to use a system different from vocal speech.

The breakthrough in this area came when Gardner and Gardner speculated that chimpanzees could learn sign language. After all, chimps were very facile with their hands and they were known to gesture to one another in the wild in order to communicate primitive messages. Furthermore, humans had developed an elaborate system of sign language for use by deaf people, called the American Sign Language (Ameslan). Thus, there existed a common set of hand symbols that various people could use to help train the animals.

The first animal chosen for language training was Washoe, a young female chimpanzee (see Fouts & Rigby, 1977; Gardner & Gardner, 1971). For four years Washoe spent all of her waking hours with a team of trainers, all of whom could sign in Ameslan. They taught her various symbols by reinforcing her with bits of food for making the correct gestures and by molding her hands into the appropriate sign. Some of the symbols, however, were learned strictly through observation. For example, Washoe learned to sign the symbol for "sweet" (which is simply touching the tip of the tongue and lower lip with the index and second finger extended from the hand) merely by observing the trainer perform it. In four years, Washoe learned a total of 130 symbols, not a great deal in terms of the average adult human vocabulary but still quite sufficient to demonstrate language acquisition.

One of the most important results that confirmed language acquisition was the fact that Washoe could combine words using appropriate syntax. For example, she would say "gimme sweet" and "come open" when she wanted to convey those particular messages. Note that while the symbols themselves were arbitrary, their combination and, most importantly, their order or syntax, were appropriate.

Perhaps a more impressive demonstration of chimpanzee language was the finding that new combinations of words were created by the animal itself. Lucy, another chimpanzee, created several novel combinations of words. Although her trainer consistently used two symbols "water" and "melon" to refer to a "watermelon," Lucy devised her own set of signs, specifically the signs meaning "drink fruit" and later, "candy fruit." This suggested that Lucy thoroughly understood the meaning of those separate terms. On a different occasion, after eating a radish for the first time, Lucy signaled "cry hurt food." Again, this novel combination suggested that Lucy, indeed, possessed the underlying cognitive components typically associated with human language. In summary, the extensive research on Washoe and her friends suggests that apes can acquire language. Through their communication with human beings, although it was rudimentary and dealt only with very simple ideas, the chimpanzees displayed the critical faculties needed for language.

There have been several other important projects on chimpanzee language. Each of these has used a different technique, and each has contributed immensely to our understanding of this issue. One chimp, Lana, was taught to "talk" to a computer terminal. The computer flashed symbols in response to Lana's manipulation of the keyboard (see Rumbaugh, Gill, & von Glasersfeld, 1973). That is, by pressing the appropriate keys in the right sequence, Lana could convey various messages to the computer which, in turn, caused the trainer to deliver rewards, praise, or attention. It has been shown that Lana could complete sentences based on their meaning as well as reject incomplete sentences that were grammatically incorrect. Ordering words appropriately is, of course, one fundamental feature of language.

In another experiment, language was taught to a chimpanzee named Sarah (Premack, 1976). This experiment used small plastic chips which stood for various objects such as food rewards. Sarah could manipulate the chips and arrange them appropriately on a magnetic board in order to convey her messages. Once again, correct syntax was shown in addition to other aspects of language. For example, Sarah was trained to identify the names of two colors. She then demonstrated that she had mastered the meaning of the word "on" by placing one color on top of the other when told to do so. In addition, when the trainer placed one color upon another, Sarah could correctly describe the action in language; that is, she could correctly arrange the plastic symbols (for example, "green on red") to describe the placement.

Can apes really learn language? The answer would appear to be yes. The scientists who have worked closely with these animals have done an impressive job in showing that chimpanzees have the cognitive structures that allow for language use. There are many who do not believe that apes are fully capable of human language (see, for example, Bronowski & Bellugi, 1970; Mistler-Lachman & Lachman, 1974). But the collective judgment of the scientific community at this time appears to be that chimpanzees do indeed possess language skills.

Human beings, of course, may still be unique in their facility for and dependence on language (Lenneberg, 1967; McNeill, 1970). According to McNeill, humans had a particular need for language that was not shared even by other primates. He claims that during evolution the human brain (and therefore skull) increased in size in order to accommodate the development of intellectual processes. Thus, the enlarged brain permitted human beings to be more adaptive in their environment, enabling them to cope with changing conditions, etc. At the same time, human beings evolved an upright position. Walking upon two legs gave them added mobility that could be used to avoid predators. The combination of these two developmental changes, however, produced a problem: the upright position meant that the birth canal could not enlarge beyond a certain size, a size that was too small to accommodate the larger skull. So, according to McNeill, human beings followed a different developmental strategy during evolution. Infants were born earlier relative to their nonhuman primate relatives, that is, human infants were born when their skulls were still small and when the brains that were housed in the skulls were relatively undeveloped. But the cost of this strategy meant that human infants were practically helpless. The helpless, immature human could not gesture or locomote; communication with the mother had to develop in other ways. And this was done by emitting sounds. Language developed in human beings, but not in other species, because humans had a unique and special need for language. Whether or not this particular anthropological theory is correct, it still highlights the important fact that human beings have a far greater capacity for language than any other animal. Or perhaps it is more accurate to say that human beings have a far greater capacity for *human* language than any other species so that we do not deny the remarkable communicative skills of lower animals.

Let us conclude this section by quoting Lenneberg (1967). His words not only reflect the thinking of many scientists on the topic of animal language, but they also encapsulate the thinking that pervades this entire chapter on biological specializations in learning. "It seems unlikely the genes actually transmit behavior as we observe it in the living animal because the course that an animal takes in its peregrinations through life must necessarily depend on environmental contingencies which could not have been 'programmed and prepared for' in advance. Inheritance must confine itself to propensities, to dormant potentialities that await actualization by extra-organic stimuli, but it is possible that innate facilitatory or inhibitory factors are genetically transmitted which heighten the likelihood of one course of events over another. When put into these terms, it becomes quite clear that nature-nurture cannot be a *dichotomy* of factors but only an *interaction* of factors. To think of these terms as incompatible opposites only obscures the interesting aspects of the origin of behavior [and of learning]" (p. 22).

Summary

This chapter examined directly the many biologically specialized learning abilities that many species have developed throughout evolution. Our awareness of these specialized abilities has resulted, in part, from the discovery that many of our principles of learning, designed to explain "conventional" learned behaviors, do not adequately account for certain other learned reactions. In addition, an evolutionary perspective of learning would suggest that various species should have evolved specialized learning abilities.

One important area of research deals with naturalistic action patterns. These general patterns of behavior are susceptible to modification through positive reinforcement and punishment, but the nature of the changes depends largely on the type of treatment and on which action pattern is involved. Food, shock punishment, and Pavlovian fear cues all affect naturalistic responding but in very different ways.

One phenomenon that has been studied in recent years is acquired food preferences. If an animal is given a deficient diet, then it will experience a slow-developing malaise. The flavor of that diet, in turn, becomes aversive. Moreover, new flavors, particularly if they are paired with recovery from the illness, come to be preferred.

A related phenomenon is taste aversion. In this case, malaise is induced quickly by the injection of a mild poison. The result is an aversion to the associated flavor. This form of learning has two distinct and interesting characteristics. First, tastes seem to be conditioned more easily than other features of stimuli such as their visual appearance. This is true even though the sense modality actually used by the animal to procure food helps to determine the strength and nature of the aversive conditioning. Second, flavors and illness may be associated even though the poison administration is delayed for several hours after the taste CS.

Avoidance learning is another area of research in which species-specific behavior patterns have been identified. According to one theory, animals are able to learn an avoidance reaction only if such a reaction is consistent with their natural response tendencies. Several experiments have supported such an approach to learning.

Two socially oriented learning phenomena are imprinting and animal language. In studies investigating imprinting, it was found that shortly after their birth precocial animals, especially birds, follow moving objects. Normally their attachment would be to their mother, but research has shown that they will follow even neutral objects. While some evidence suggests that this learning is unique, many recent findings indicate that imprinting is consistent with other known principles of learning.

Research on animal language has strongly suggested that chimpanzees are capable of using sign language to convey simple messages. In various studies, the subjects formed sentences which conformed to the rules of syntax. On occasion, subjects generated novel communications. While this exciting area of research holds great promise, human beings still appear to be unique in the extent to which they learn and use language.

7

Extinction

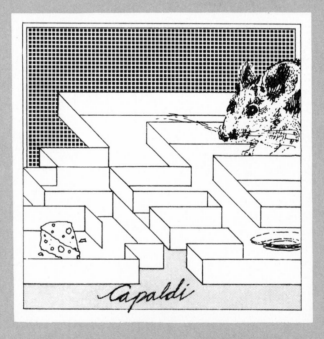

Introduction

Stimulus Extinction

Response Extinction

Procedures That Affect Extinction

General Theory of Extinction

Partial Reinforcement Effect

Summary

Introduction

Our discussion of stimulus and response learning in the previous chapters has focused almost exclusively upon acquisition. But, as mentioned in Chapter 2, there is the mirror image process called extinction. In this chapter we will discuss the factors and theories pertaining to stimulus and response extinction.

Extinction really refers to a procedure, namely, the elimination of the contingency between the stimulus or response and the US. More specifically, Pavlovian excitation is extinguished through the withholding of the US following the CS presentation. In instrumental learning, the CR is extinguished by withholding the reward following CR execution. The result of this procedure is the gradual reduction in CR strength. Several reasons may account for the reduction in responding, however. First of all, there may be a loss of the original expectancy because the stimulus or response no longer predicts the US presentation reliably. In addition, inhibitory reactions may develop. After all, during extinction the stimulus and response signal no US presentation; such a stimulus/response-outcome correlation is the essence of inhibitory conditioning. This fact, of course, means that the original expectancy is eventually replaced by a new expectancy, namely, that the stimulus or response is now followed by no US.

The notion that inhibition operates during extinction is supported by the concept of spontaneous recovery. If a subject is given a rest interval following extinction training and then placed in the apparatus once again for a test, one observes that the original CR "spontaneously" reappears. Clearly, extinction did not eliminate the behavior altogether; the potential was still present. Extinction merely caused the CR to become suppressed. Some of the processes underlying extinction, then, are inhibitory ones since, when they have had an opportunity to dissipate (during the rest interval), the behavior reappears.

A study by Burdick and James (1970) illustrates spontaneous recovery. The authors trained thirsty rats to lick a drinking tube for water. During this time, a Pavlovian CS (noise and light) and a shock US were presented. The subjects, of course, acquired a conditioned fear to the CS. After training, all the animals were given 40 extinction trials during which the CS was presented by itself without the US presentation. The fear reaction, measured in terms of the hesitancy to drink during the CS, declined as a result of the extinction procedure. Different groups of animals then received a rest interval, ranging from about 3 minutes to 72 hours.

Finally, in the fourth phase of the study, the animals were put back into the testing apparatus, and the CS was presented while they were drinking. Again, the degree to which the drinking was suppressed indicated the strength of the Pavlovian fear. The results are shown in Figure 7–1. The spontaneous recovery data show an interesting relationship between fear and the length of the rest interval. When subjects were tested after a very short rest interval, .058 hour, little fear was shown; it is only natural that immediately after extinction the conditioned reaction would continue to be suppressed; not enough time had elapsed to permit the inhibition to dissipate. However, as the interval between the extinction and test sessions increased, the suppression became gradually stronger. The subjects that were tested 24 hours after extinction showed almost as much fear of the CS as they had at the end of the training session. This suggests that the effects of the extinction procedure were not permanent. In summary, spontaneous recovery, which is obtained for both stimulus and response learning, indicates that extinction, as it is usually conducted, involves inhibition. If the inhibition were not present, that is, if the suppression of the CR were not due to a temporary inhibitory process, then spontaneous recovery would not be observed; the response would be lost forever.

Stimulus Extinction

Let us consider extinction of a Pavlovian excitatory reaction. The typical way to extinguish a Pavlovian or stimulus expectancy is to present the CS by itself. When this is done, the CR declines in strength as a function of the number of CS presentations (Kalish, 1954). One important study was

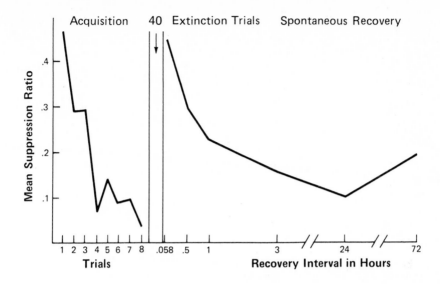

Figure 7-1. Mean CER suppression ratio during acquisition and following the postextinction recovery interval. (From Burdick & James, 1970)

done by Shipley (1974). Rats were trained to fear a tone CS by pairing the CS with a shock US. Stimulus learning was assessed with a CER technique. During extinction, both the number and the duration of the CS presentations were varied. Some subjects received a 25-second CS, while others received a CS that lasted 100 seconds. Different animals in each of those two conditions received either 200, 400, or 800 exposures. Shipley found that neither the number of extinction trials nor the duration of the CS, whether it lasted 25 or 100 seconds, alone accounted for the behavior. Rather, fear declined as a function of the combination of number and duration: the greater the total exposure to the CS, regardless of whether the exposure was due to shorter but more CS exposures or longer but fewer CS presentations, the faster the extinction of the fear. In other words, the longer the time that the animals had to discover that no US would follow the CS, the greater the extinction of the Pavlovian reaction. This study is important because it challenged a number of extinction theories that focused on either the number of extinction trials or the duration of the CS (but not both).

As we discussed in Chapters 2 and 3, classical conditioning is based on a stimulus-outcome correlation. The CS-no US technique is, therefore, not really the procedure that is opposite to acquisition. It would be more appropriate to show that the random presentation of the CS and US leads to the decline of the reaction. In other words, the elimination of the correlation, by the presentation of the CS and US in a random fashion, should be the counterpart to acquisition. Such a procedure would evaluate the effect of degrading the CS-US correlation without permitting the formation of a new inhibitory expectancy.

Ayres and DeCosta (1971) examined this question. In their study a tone was followed by shock; the fear reaction to the tone was assessed by the normal CER technique. During Pavlovian extinction, subjects were given random presentations of the CS and US. The authors found, surprisingly, that the animals continued to show suppression. That is, the random presentations of the CS and US did *not* lead to the elimination of conditioned fear (although, admittedly, some CR reduction was observed later on). This finding is interesting because it contradicts what one would predict based upon the acquisition of stimulus expectancies. To be specific, if excitatory stimulus learning involves the correlation of the CS and a US, then one would expect that the CR, once it was formed, would be abolished by degrading the CS-US correlation. The effect of this procedure is weak, however, and there are several reasons why this may be true. First of all, the US presentation during extinction may, on occasion, closely follow the CS presentation; this "accidental" pairing would, of course, tend to sustain the conditioned reaction. Second, the US presentations during extinction may continue to remind the subject about its acquisition experience. It is well known that various procedures which affect the memory of the US do influence responding (see Rescorla, 1974). Finally, it may be that subjects are quite unready to abandon a particular expectancy (even if the correlation between the CS and US is zero) when they have no alternative expectancy to take its place. For example, the subjects would continue to "want" to predict shock in this experiment; although the tone no longer is a good predictor of shock, nothing else is either. Under these conditions, when the prediction is important, extinction of a previous stimulus expectancy may be quite slow. It certainly is the case that the most predictive element becomes the strongest (see Chapter 3; Pearce & Hall, 1978). Perhaps, then, without having a better predictor, the animal continues to respond to the formerly predictive CS.

The extinction of conditioned inhibitory reactions has also been studied (see Zimmer-Hart & Rescorla, 1974). In Phase 1 of this study, all subjects underwent normal inhibitory conditioning: CS_1 was followed by the US (a conditioned excitor) and the compound stimulus, CS_1-CS_2, was never followed by shock. This procedure is one way of making CS_2 into a conditioned inhibitor (see Chapters 2 and 3). During the second phase of the study, Zimmer-Hart and Rescorla experimented to see what procedure would be most effective in eliminating the conditioned inhibition. Giving extra presentations of CS_2 alone, the conditioned inhibitor by itself, proved ineffective; it did not change the strength of the conditioned inhibitor. Moreover, failing to follow the excitor, CS_1, by a US did not affect the strength of the conditioned inhibitor, CS_2. The only procedure that reduced the conditioned inhibition elicited by CS$_2$ was the direct reinforcement of that stimulus; that is, following the conditioned inhibitor by the shock US. This finding is consistent with those reported just above. The most effective procedure to extinguish a conditioned excitor is to omit all USs following its presentation; similarly the most effective way to extinguish a conditioned inhibitor is to *present* the US following its occurrence. Conditioned excitors and inhibitors, therefore, appear to work in a similar fashion; both require the reversal of the acquisition conditions in order for extinction to take place (that is, a change from excitatory to inhibitory conditioning in order to eliminate the strength of a conditioned excitor, and a change from inhibitory to excitatory procedures in order to extinguish a conditioned inhibitor).

We have discussed two major extinction procedures: the presentation of the CS only (or, in the case of inhibitory conditioning, the pairing of the formerly conditioned inhibitor with a US) and the random presentation of the CS and US. The former procedure, reversing the CS meaning, works far more effectively than merely eliminating the correlation. But there are several other techniques as well. One is presenting the US only. This procedure has been used in a number of experiments (Colby & Smith, 1977; Rescorla, 1973a). Colby and Smith's experiment dealt with a conditioned flavor aversion. Rats were given taste aversion training in which a flavor was followed by poison during acquisition. Then, during the treatment phase, some of the animals were given 1, 5, or 10 CS-only exposures (normal extinction), while other animals were given either 1, 5, or 10 US-only exposures. The authors found that exposing the US-only was much less effective in reducing the conditioned taste aversion than the CS-only technique. Whereas the CS-only groups rapidly extinguished their aversion for the solution, extinction in the US-only exposure groups was not evident until late in the testing period. Again, we have evidence that the most effective procedure is the CS-only presentation, because it is this procedure that allows the animal to form a counterexpectancy to the one developed during acquisition. The other procedures, such as random CS-US presentation and US-only presentation, do not provide for the development of a new expectancy, or they tend to promote the conditions that extend the life of the old expectancy.

Let us summarize the evidence on the extinction of stimulus learning. By far the most widely investigated, and the most effective, means for eliminating a conditioned Pavlovian reaction is the presentation of the CS only (or the pairing of the CS and US in the case of inhibitory CR extinction). Such a procedure involves the degradation of the former CS-US correlation, *and* it also leads to the formation of a new stimulus expectancy: the animal comes to expect that no US will follow the stimulus. Even though random presentation of the two stimuli prevents acquisition, extinction (presumably the "opposite" process) is not facilitated by the random technique. This result makes sense from an evolutionary viewpoint. If an animal is to behave adaptively, it must be able to predict biologically strong USs. During acquisition, when the animal first experiences these USs, it attempts to "explain them" by associating them with the most reliable and predictive cue. During extinction, on the other hand, when the USs continue to be presented (as they are in a random CS-US procedure), the animal's need to predict them is still there; but since there is no alternative element, either a stimulus or a response, which is able to predict the USs, then the animal is slow to abandon its only choice, the former CS.

Response Extinction

The general principles concerning response extinction are similar to those described above. There are three basic techniques used in studies of response extinction: reward training techniques, omission training, and avoidance training. We will discuss each of these methods in the sections that follow.

Reward Training

When the response is no longer followed by reinforcement, then suppression of the behavior is observed. This is the usual extinction procedure. Inherent in this procedure is the degradation of the correlation between a response and the reward as well as the development of a new response expectancy: the response now predicts no US. Many hundreds of studies on reward conditioning have demonstrated that extinction takes place when the reward is omitted; the more extinction given to the animal, the greater the suppression of the response (see following sections of this chapter for factors that affect extinction).

The most important study on extinction of reward conditioning was done by Rescorla and Skucy (1969). Recognizing that virtually all the studies on extinction had simply omitted the US presentation, Rescorla and Skucy set out to show that the degradation of the R-US correlation also would lead to extinction. (Recall from our discussion above that a debasement of the S-US correlation in Pavlovian conditioning did *not* lead to rapid extinction.) In Rescorla and Skucy's experiment, subjects were taught to press a lever for food; during extinction free food was presented, thus debasing the R-US correlation. The results were remarkably similar to those observed in the stimulus learning situation: free food retarded extinction. In other words, even though the response-US correlation is paramount for acquisition, debasing that correlation, by giving free food, does not necessarily lead to rapid extinction. The extra food appeared to elicit lever pressing instead. The reasons are not clear, but superstitious responding or accidental contingency did not seem to play a role. It is more likely that the presentation of free food maintained the animals' orientation toward the area where the lever was located and therefore increased the probability that the lever would be hit. In support of that idea, it is certainly known that animals will direct their attention to and approach locations associated with food (Karpicke, Christoph, Peterson, & Hearst, 1977). In summary, Rescorla and Skucy's study as well as others supports the notion that the reduction in the response-US correlation, the ingredient most critical for acquisition, is not the most effective procedure for extinction. Animals continue to execute the formerly reinforced behavior even though the behavior no longer predicts the US presentation very reliably.

Omission Training

In addition to the basic techniques in response extinction such as withholding the reward and debasing the response-US correlation there is a third technique that has received attention in recent years. This is the omission procedure first mentioned in Chapter 2. Recall that omission training involves the withholding of reinforcement following the criterion response; reward is given at other points in the training session as long as it does *not* follow the CR. Such a procedure is similar to extinction to the extent that the criterion response is not rewarded, but it differs in that reward is given for "not responding."

Several studies have questioned whether extinction leads to greater suppression of behavior as compared to omission training (for example, Uhl, 1973; Uhl & Garcia, 1969). In the first phase of the Uhl and Garcia study, two groups of rats were trained to press a lever in order to receive food. Following this acquisition phase, extinction was given to one group of subjects and omission training to the other. The extinction procedure merely involved the withholding of reinforcement after each response; omission, however, not only involved withholding reward after a lever response, but also entailed the presentation of food if the subject refrained from pressing the lever for a certain period of time. Thus, omission-trained subjects could receive food merely by remaining still in the box and not pressing the lever. If they did press the lever, food was postponed. This type of omission training is called the differential-reinforcement-of-other-behaviors (DRO) because, in theory at least, the food presentation for "not responding" is really a reward for performing some behavior other than the criterion reaction.

As shown in Figure 7–2, the authors found that the suppression of lever pressing was less severe for the omission-trained subjects than for the regular extinction subjects. That is, extinction was more effective in eliminating the behavior than was omission. This result seems to confirm the idea that "normal" extinction, the procedure that involves no reinforcement whatsoever, involves two factors: the degradation of the R-US correlation *and* conditioned inhibition. According to the authors omission training was less effective than extinction because the omission subjects tended to treat the delivery of free food as a signal for additional lever pressing. After all, during the acquisition phase, food was a reliable signal indicating that additional lever pressing was appropriate. During omission training, then, the food continued to elicit lever responding even though pressing was counterproductive, that is, it led to the postponement of food.

There are other ways to conduct the omission training procedure. Some of them produce important differences from the one just described (see Leitenberg, Rawson, & Mulick, 1975; Rawson & Leitenberg, 1973; Rawson, Leitenberg, Mulick, & Lefebvre, 1977). In each of these studies, omission training led to *greater* suppression of the behavior than did conventional extinction. In these experiments reinforcement was not given merely for "not responding." Rather, it was presented following a *specific*

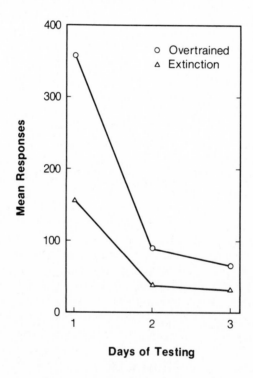

Figure 7-2.   Mean number of responses on three test sessions for the group that was extinguished in the normal fashion and the group that received omission training. (From Uhl & Garcia, 1969)

alternative response. This form of omission training is called countercondi-tioning because a new behavior, counter to the original response, is rein-forced, while at the same time the original response is not rewarded. In the Rawson et al. (1977) study, for instance, reward was withheld for pressing lever number 1, but food was provided for pressing lever number 2. Re-sponding to lever 2, of course, increased, whereas responding to the first lever declined dramatically. It is quite clear that the increasing tendency to press the second (rewarded) lever interfered with pressing the first lever. In other words, when a *specific* alternative response is strengthened at the expense of the first behavior, then omission training *does* suppress respond-ing of the first response more quickly than conventional extinction proce-dures. Giving reward at unspecified times merely for making "no CR" is a far less efficient way to suppress behavior and does not lead to any systematic increase in the alternative behavior. It is interesting to note in this connection that no differences were observed between groups that were reinforced during omission training for making a response that was similar to the original CR, and groups that were rewarded for making a different type of response (Leitenberg et al., 1975; Pacitti & Smith, 1977). One would imagine that the reinforcement of a response vastly different from the original one would hasten the suppression of the original behavior more than reinforcement of a response that was similar.

Finally, consider the work of Holland (1979a). As we discussed in Chapter 3, Holland showed that different types of CSs elicit different kinds of reactions. In this study, he was interested in how omission training suppresses these various reactions. Generally, he found that the effective-ness of omission training depended upon the kind of CR that was being suppressed. For example, explicitly failing to reinforce a magazine-approach response increased the incidence of rearing on the hind legs. It is as if the animal compensated for the loss of reward following one action with an increase in the other behavior; as one was selectively suppressed by the omission contingency, the other increased. Auditory CSs induced head jerk reactions. But when omission training was carried out for these re-sponses, *no* compensation was observed in other behaviors. Omission train-ing had different effects depending upon the kind of reaction that was being treated. Therefore while it is true in general that omission training sup-presses behavior, there are instances when certain responses may actually increase as a result of the omission contingency.

Let us summarize these discussions of reward training and omission training. The typical extinction procedure is to present the stimulus without the US (in the case of stimulus learning) or allow the response to be executed without reinforcing it (response learning). In both cases the procedure leads to the suppression of the CR. However, such a procedure really involves two distinct components: the degradation of the stimulus or response-outcome correlation and the acquisition of a new expectancy based upon the stimulus or response-no US correlation. When only the correlation is debased, responding in both the stimulus and response learning paradigms, paradoxically, does not decline appreciably (Ayres & DeCosta, 1971; Rescorla & Skucy, 1969). It may be that the lack of suppression reflects the fact that the animal does not abandon the expectancy that it first developed during acquisition unless it has a more predictive element to take its place. Finally, in both the stimulus and response learning paradigms, additional procedures in which behaviors other than the original CR are reinforced lead to the suppression of responding. Suppression is most apparent when these other behaviors are explicit competing behaviors, that is, specific reactions that interfere with the execution of the original CR, although even these techniques are not permanent (Leitenberg, Rawson, & Bath, 1970). Suppression continues as long as these alternative behaviors are reinforced, but the response returns once the omission contingency is abandoned.

Avoidance Training

Avoidance responding poses some unique problems for the analysis of extinction. The reason that avoidance is somewhat special is that omission of the shock altogether does not parallel the procedure used in appetitive conditioning (omission of the reward). To omit the shock is to withdraw the source of motivation for the avoidance reaction. Admittedly the animal continues to be fearful for a while, but if no shock is ever given, the behavior eventually deteriorates.

A number of studies have demonstrated that extinction of avoidance behavior is most effective when the avoidance contingency itself is removed. Recall from Chapters 4 and 5 that the avoidance contingency, the ability to avoid the shock, is, fundamentally, the most important ingredient in avoidance acquisition (Bolles, Stokes, & Younger, 1966). It would, therefore, be plausible to assume that elimination of the avoidance contingency would be the most effective way to suppress the responding. Such results have been found in a number of studies (Bolles, Moot, & Grossen, 1971; Reynierse & Rizley, 1970). In the Reynierse and Rizley study, animals were trained to make an avoidance response during a tone CS by running from one side of a shuttlebox to the other and back again on the next trial. After receiving 200 acquisition trials, all animals were given extinction using various procedures. Some animals received the conventional extinction procedure: the CS sounded but no shock was ever delivered. A second group of animals was given essentially the same treatment, but the CS offset was delayed for 5 seconds following each response. Four additional groups were used. Each received CS and US presentations during extinction in a random fashion. For one of these groups, the CS offset was delayed following a response, although the shock went off immediately. A second of these groups received the opposite treatment: an extinction response terminated the CS immediately, but the US offset was delayed for 5 seconds. A third group got a delay in the offset of both the CS and the US, while a fourth random CS-US group experienced no delay in either of the offsets. All the extinction procedures were effective in reducing the level of avoidance responding below that achieved during acquisition, but delaying the US termination reduced avoidance responding faster and more completely than any other procedure. In other words, not being able to control the shock offset was the most important factor in this study. It has also been found that punishing the avoidance reactions with shock, in addition to delaying the shock offset, facilitates suppression of the avoidance reaction (Bolles et al., 1971).

Extinction of avoidance behavior has been considered from a different vantage point. Several investigators have tried to reduce the R-US avoidance correlation by presenting "free" shocks during extinction (Coulson, Coulson, & Gardner, 1970; McKearney, 1969). In the Coulson et al. study, rats learned to bar press in order to postpone shock (Sidman avoidance learning). By maintaining a rather consistent level of responding, the subjects experienced only a few shocks. During extinction, animals in one group were given no shocks at all (regular extinction), whereas other animals were given the same pattern and number of shocks that they had experienced on the last acquisition session. These "free" shocks were designed to weaken the response expectancy. The results were surprising because they disagreed with those mentioned just above: even a few free shocks during extinction on a noncontingent basis were sufficient to maintain responding. Extra shocks should have eliminated the avoidance contingency, and thus caused extinction, but they did not; avoidance reactions were instead maintained by the free shocks probably because the subjects couldn't discriminate the acquisition conditions from those of extinction.

Powell (1972b) also discovered that free shocks during Sidman avoidance extinction maintain responding, but after closely examining the behavior pattern, he developed a different theory about this phenomenon. He noticed that most of the responses during extinction were given shortly after the free shock had been presented. Furthermore, the rate of pressing was related to the frequency and intensity of the free shocks. This led Powell to speculate that these responses were actually shock-elicited aggression reactions. It is well known that shock, a highly aversive stimulus, is capable of eliciting aggressive behavior (see Chapter 11). Moreover, aggression is one form of species-specific defense reaction. When bar pressing no longer works effectively, because the response-no shock correlation has been debased, and when there are no other responses that are effective, then aggression often is observed. In this case, Powell reasoned that the aggressive reaction was directed toward the lever because that was the only distinctive feature in the environment and because the lever had been associated with the shock schedule.

In summary, avoidance responding declines when the principle factor, the avoidance contingency, is eliminated; the normal procedure of extinction, presenting the CS only with no US, is not really the correct way to study the problem. And, as Reynierse and Rizley showed, CS-only presentations are not as effective in suppressing the avoidance CR as is eliminating the US contingency itself. Second, under certain conditions, Sidman avoidance is maintained during extinction by free shocks. Such an outcome could occur because animals do not discriminate between acquisition and extinction conditions adequately or because of a species-specific aggressive reaction directed toward the lever.

Procedures That Affect Extinction

Some of the major factors that affect extinction are the size of the reward, the number of acquisition trials, delay of reward, and the physical effort that the response requires of the subject. We shall consider each of these variables in the sections that follow.

Magnitude of Reward

One of the major parameters that affects resistance to extinction is the magnitude of the reward used during acquisition. The general finding is that the larger the reward during acquisition, the faster the extinction (Campbell, Batsche, & Batsche, 1972; Roberts, 1969; Wagner, 1961). This principle contradicts that found for acquisition. Although larger rewards tend to increase CR strength as measured by acquisition scores, larger rewards have the opposite effect in extinction: they lead to the rapid elimination of the behavior.

In the Roberts study, rats were trained to run down an alleyway in order to receive 1, 2, 5, 10, or 25 food reward pellets. During extinction, food, of course, was never presented in the goal box. The mean running speed out of the start box during the last 30 extinction trials was expressed as a percentage of the terminal performance level during acquisition (see Figure 7–3). It is clear from the figure that performance declined as a function of the reward magnitude during acquisition; the larger the acquisition reward, the faster the extinction.

The variability in reward magnitude has also been shown to influence resistance to extinction. In one study (Sytsma & Dyal, 1973), rats were trained to run down two alleyways. One group was the constant group; animals in this group were given 10 food pellets in both goal boxes following their response. Animals in a second group, the variable group, received 10 pellets in the goal box of one alley, but a variable amount of pellets (either 1 or 10) in the goal box of the other alley. During extinction all the rats

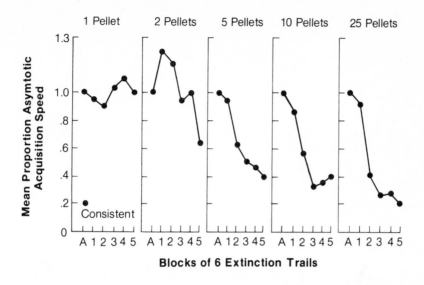

Figure 7-3.   Mean proportion of terminal acquisition speed in the start box during
extinction as a function of blocks of extinction trials for groups receiving
different amounts of food reward during acquisition. Terminal acquisition
performance given at point "A." (From Roberts, 1969)

were tested in the alley in which they had received consistent reinforcement
(that is, the 10-pellet alley). Sytsma and Dyal found that the variable reward
affected resistance to extinction. Specifically, the speed of running down the
alley was greater for subjects in the variable group than for those in the
constant group. Even though the variable subjects had always received 10
pellets in the alleyway used during extinction, the fact that they had received a variable number of pellets in the other alleyway caused their
response speed to increase.

Variability of reward magnitude has been considered from another
viewpoint (Wike & Atwood, 1970; Wike & King, 1973). In these experiments, the magnitude of the reward was systematically increased, decreased, or randomized over a series of trials. For example, in the Wike and
King study, an increase in magnitude over the 3 daily trials, from 0 to 45
to 500 milligrams, caused the animals to run at a progressively faster rate
on each successive trial. Similarly, decreasing the reward level, from 500 to
45 to 0 milligrams, caused those animals to slow down progressively on each
trial. Interestingly, a third group in the study, for whom the amount of
reward was randomized across the 3 daily trials, ran as fast on all trials as
the other two groups ran on their fastest trials. One might expect that

animals in the random group would have run at an intermediate speed, but instead they ran at the fastest rate on all trials.

Wike and King's study suggested two important principles. First, the degree to which varied amounts of reward are randomized during training is critical to the subject's resistance to extinction. Randomization produces faster performance and slower extinction than fixed patterns. This is consistent with the principle cited in Chapter 4 in connection with contrast. An animal's response expectancy not only involves the presence or absence of subsequent reward, but also the characteristics (for example, magnitude) of that reward. The second point noted in the Wike and King experiment was that resistance to extinction increases if the reward pattern is one in which a small reward is followed by a more preferred, that is, larger, reward.

Training Level

The number of acquisition trials is known to influence resistance to extinction. Traditionally, it was thought that the greater the number of acquisition trials, the greater the resistance to extinction (Williams, 1938). Such a result suggested that extinction was a good measure of habit strength: the stronger the response (due to a greater number of reinforcements during acquisition), the more persistent the habit.

Recent evidence, however, indicates that this positive relationship occurs only when small rewards are used during acquisition (Ison, 1962; Ison & Cook, 1964; Sperling, 1965a, 1965b). For instance, Ison and Cook (1964) rewarded rats during acquisition for running down an alleyway. They administered 15 trials per day for a total of either 30 or 75 trials. Half the subjects in each of those groups received 1 pellet in the goal box, while the other half received 10 pellets. In Phase 2 of the experiment, 15 extinction trials were given. The authors found that the response strength during acquisition increased as a function of the magnitude of reward and the number of training trials. Such a finding, of course, is no surprise (see Chapter 4). During extinction, though, the outcome was more complicated. For the subjects that received the small reward during acquisition, extinction was a positive function of the number of training trials: the more trials they had received, the more persistent they were during extinction. On the other hand, for the subjects that received the large reward during acquisition, extinction speeds were inversely related to training: extinction was more rapid with the greater number of training trials. It is interesting to note that a similar result is obtained when a low deprivation level rather than a small reward is used: resistance to extinction increases with training, but only if the experimental animals are maintained at a low drive, or motivational, level (Traupmann, 1972).

Delay of Reward

Delay of reward is an extremely important variable. Although it has a pronounced effect on the acquisition of behavior (the greater the delay, the slower the rate of acquisition), it has various effects on resistance to extinction (see Tarpy & Sawabini, 1974). Some early studies showed that delay of reward during acquisition increased later resistance to extinction (Crum, Brown, & Bitterman, 1951). More recently, however, constant delay of reward during acquisition has been shown to have no effect (Habley, Gipson, & Hause, 1972; Renner, 1965; Sgro, Dyal, & Anastasio, 1967), or, actually, to cause a decrease in later resistance to extinction (Tombaugh, 1970; Tombaugh & Tombaugh, 1969). The reason that such discrepant results have been found is that delay of reward interacts with a variety of other parameters such as the magnitude of reward, the food deprivation level, and the training level.

One important study that helped to clarify how constant delay of reward affects later resistance to extinction was done by Tombaugh (1966). Groups of rats were trained to run down an alley for food reward in the goal box. Different groups of rats received various delays of reward: either 0, 5, 10, or 20 seconds. Later, extinction was given. On these trials, the animal would enter the goal box and be confined for a period of time before being removed. For some of the animals, the length of confinement during extinction was the same as the length of their acquisition delay; that is, they received a confinement duration of 0, 5, 10, or 20 seconds depending upon which was used during acquisition. Other animals, however, received a 0-second confinement period during extinction regardless of their previous delay condition. The results showed that, when the confinement period following an extinction response equaled the previous delay value, then delay of reward did not affect resistance to extinction. Tombaugh therefore concluded that although delay of reward affects the terminal performance levels during acquisition, it does not change the rate of extinction.

The effect on resistance to extinction of delaying the reward only occasionally during acquisition has also been studied (see P. E. Campbell, 1970; Knouse & Campbell, 1971; Shanab & Birnbaum, 1974; Tombaugh, 1970). Most studies have confirmed that partial delay during acquisition increases resistance to extinction. This is in sharp contrast to constant delay, which, under most circumstances, does not increase resistance to extinction. In the Knouse and Campbell experiment, rats were trained to run down an alleyway for food. On 50 percent of the acquisition trials the reward was delayed for either 0, 8, 16, 24, 32, 40, 48, or 56 seconds. In a second part of the experiment, animals were given extinction. On these trials, they were merely confined in the goal box for 15 seconds without receiving reward. The results, as shown in Figure 7–4, indicated that resistance to extinction increased as a function of delay duration. With greater delays on a portion of the acquisition trials, extinction speed increased and responding was prolonged.

A somewhat different procedure was used by Wike, Mellgren, and Wike (1968). During acquisition, they delayed the reward for 20 seconds; they also used a 20-second goal box confinement period during the extinction trials. The difference was that, during acquisition, they varied the percentage of trials that involved delay. That is, different groups of rats received their 20-second delay on either 0, 33, 67, or 100 percent of the acquisition trials. Wike et al. found that resistance to extinction declined as a function of the percentage of delay trials during acquisition: the greater the number of acquisition trials that involved a delay of reward, the faster the subjects extinguished.

Let us summarize this work on delay of reinforcement and extinction. Constant delay during acquisition appears to decrease resistance to extinction somewhat, if it has any effect at all. This is especially true if the delay duration and the goal box confinement duration during extinction are equal; if the confinement period on extinction trials is shorter than the delay period used during acquisition, then resistance to extinction increases (Tombaugh, 1966). Partial delay of reward, however, increases resistance to extinction: the longer the delay, the greater the increase. If

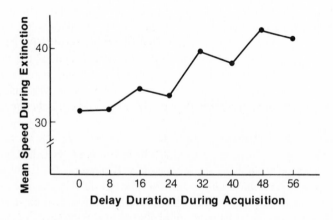

Figure 7-4.   Mean running speed during extinction as a function of the partial delay of reward during acquisition. (From Knouse & Campbell, 1971)

the percentage of partially delayed trials approaches 100 percent, then the increase is not observed.

Response Effort

The physical effort required by the subjects also affects extinction. Specifically, extinction occurs more rapidly when the response during extinction (Fischer, Viney, Knight, & Johnson 1968) or acquisition (Johnson & Viney, 1970) requires more effort. In the Fischer et al. study, for instance, rats were given 60 acquisition trials in an alleyway. For some of the trials the alley was horizontal, but for other trials the alleyway was tilted as much as 40 degrees. Here, effort was defined in terms of the angle of slant of the alleyway: the greater the slant, the more difficult the response. During extinction, Fischer et al. found that resistance to extinction was an inverse function of the response effort. The animals that had to climb a steeper alleyway during extinction stopped running sooner during extinction than subjects that were permitted to run down a horizontal alley. It is interesting to note that in this study, running speed during acquisition was not affected by response effort.

General Theory of Extinction

Let us begin by reviewing the two basic types of expectancies that are developed in any learning situation (see Chapter 5). We can represent them schematically as shown in the top of Figure 7–5. A stimulus complex, because it is associated with reinforcement, comes to elicit a stimulus expectancy. At the same time, the response-reward relationship leads to the acquisition of a response expectancy. Normally the response is executed only in the context of the stimuli because the stimuli also serve as discriminative cues; they signal when reinforcement is available. In other words, responses invariably are performed in the presence of a particular stimulus complex because, while the response predicts reward, it does so only in the appropriate situation, that is, in the presence of a stimulus context that also predicts reward. Animals develop separate response and stimulus expectancies, but it is their *joint* occurrence, as discussed in Chapter 5, that is important to our theory of behavior.

If the organization of a learned reaction may be depicted as shown in the top of Figure 7–5, then extinction theories must deal with the effect of removing the reinforcer. Quite clearly, the removal of reinforcement in this model (corresponding to the procedure of extinction) would have important consequences on the two types of expectancies. Namely, the organism would no longer expect to receive reward follow-

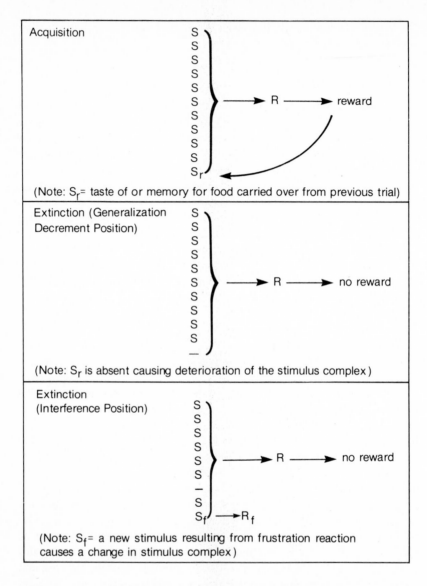

Figure 7-5. Schematic diagram depicting how the generalization decrement (middle panel) and interference (bottom panel) theories explain extinction.

ing the stimulus complex, and eventually it would expect to receive no reward following its response. This is precisely the theme we have emphasized in this chapter. Extinction alters the expectancies that were developed during acquisition and causes the development of new expectancies. The gradual decrease in CR strength during extinction, therefore, can be considered a result, in part, of the change in stimulus and response expectancies.

Stimulus Generalization Decrement

We can analyze this process in somewhat more concrete terms by considering the precise consequence of the omission of reinforcement during extinction. During acquisition, reward not only leads to the formation of the stimulus and response expectancies but it also is itself a stimulus; the aftereffects of reward (for example, taste or memory of the food) become part of the stimulus complex. Imagine a rat that is placed in the start box of an alleyway. It remembers having received reward previously; it may even have in its mouth the lingering taste of food from those occasions. Such a stimulus, called a reward aftereffect, is an important part of the stimulus complex.

Now consider the middle portion of Figure 7–5. When food is *not* provided for the response, the very important and salient stimulus aftereffect, S_r, resulting from the reward during acquisition is absent. The stimulus complex has suffered a loss of one important element. If, as we claim, the stimulus complex undergoes such a dramatic change during extinction, then it can no longer serve as an effective discriminative stimulus for the response. The signaling properties that were established during acquisition no longer "work" effectively because not all of the appropriate stimuli are present. In other terms, in acquisition, the original stimulus complex, which included the reward aftereffects, served as a source of information about the reward. In extinction, however, there is essentially a "new" stimulus complex, one that lacks the reward aftereffects, one that has no history of predicting reward.

If the response declines during extinction in part because the controlling stimulus complex differs noticeably from the complex to which the subjects were conditioned during acquisition, then the extent of the decrease in responding during extinction would be related to the discrepancy between the acquisition and extinction stimuli. In other words, if the stimulus complex during extinction changes radically, then the conditioned response will decline quickly; if, on the other hand, the stimulus complex changes only slightly during extinction, then there will be somewhat greater resistance to extinction. Normally, withholding reward causes a CR decline because it eliminates the important reward aftereffect, S_r. But the omission of other elements in the stimulus complex may also produce a decrement in responding.

Evidence for this relationship between the rate of extinction and the degree of change of the stimulus complex is abundant. First of all, recall that extinction is more rapid when acquisition is conducted with large rewards. Going from large reward in acquisition to no reward in extinction is a more dramatic change in the reward aftereffects, that is, in the stimulus complex, than going from a small reward to no reward. Similarly, with more training trials, extinction is faster. This occurs because the stimulus complex is more developed with greater training and therefore any change during extinction, due to the absence of reward aftereffects, is more noticeable. Third, investigators have shown that an extinguished behavior will recover if an animal is given a food-related cue prior to the test. Such a procedure facilitates performance by helping to recreate the original stimulus complex (Homzie, 1974). Finally, several experiments have shown that extinction is more rapid if it is conducted in a different environment. Not only is the reward aftereffect eliminated, but other elements in the stimulus complex are changed as well, for example, apparatus cues and the like (Bouton & Bolles, 1979; Welker & McAuley, 1978).

Let us illustrate this last point. Welker and McAuley (1978) gave rats 10 1-hour sessions in a lever box during which time they could press a lever to receive food on a VI 1-minute schedule. All animals achieved a stable rate of responding. During this period, the actual handling procedure and apparatus characteristics were noted. In particular, the transportation of the subjects from their home cage to the apparatus involved removing the water bottles from the home cage, placing the cage on a cart lined with newspaper, positioning a wire mesh lid on the cages, and wheeling the cart into the testing room. The apparatus, that is, the experimental context, also was carefully noted: a circular disk above the lever was illuminated, a hissing noise was continually present, and the tray beneath the grid floor was filled with wood shavings.

In the second phase of the study, all animals were given five extinction sessions. The purpose of the experiment was to determine how changes in the animals' environment would affect extinction. One group of subjects received the same conditions as before (that is, the same transportation procedure and apparatus). However, the transportation procedure was changed for a second group, and the testing context was changed for a third. Finally, a fourth group experienced a change in both the transportation procedure and context. The changes in the transportation method involved leaving the water bottle on the cage but merely sliding the cage out so that the animals could be picked up and placed in a plywood box containing wood shavings. The new context (apparatus) consisted of the same lever box but the circular disk was now dark, opaque black plastic inserts covered the ceiling and walls, and a paper liner was placed underneath the wire mesh floor.

The results of the study are shown in Figure 7–6. Extinction, measured in terms of the percent of the terminal acquisition responding rate, declined for all groups as would be expected. The group that experienced no change, of course, declined at the slowest rate whereas the other groups extinguished much more quickly. Those subjects that experienced the new context were affected the most, but even a change in the method of handling made a significant difference. In summary, this experiment clearly shows that a potent factor in extinction is the change in the stimulus context. Such changes take place under normal conditions; for example, the fact that food is omitted during extinction deprives the subject of one very salient stimu-

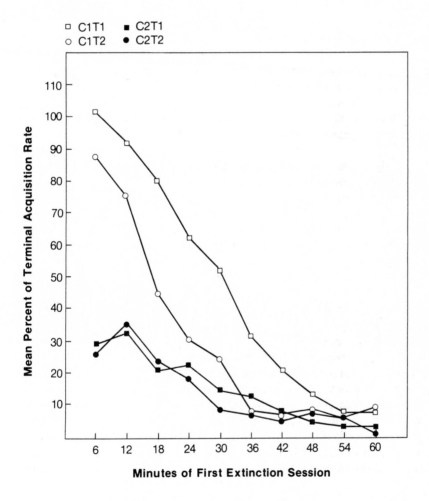

Figure 7-6. Mean percent of terminal acquisition response rate as a function of minutes on the first extinction session for groups that received the same transportation and context experience as during acquisition (C1T1), a change in the transportation method (C1T2), a change in the apparatus (C2T1), or a change in both the transportation and the context (C2T2). (From Welker & McAuley, 1978)

lus, namely, the aftertaste of the food. But other cues matter too; and if these cues are changed, extinction takes place more rapidly.

Interference

A second major mechanism for extinction is the creation of interfering behaviors by the sudden withdrawal of reinforcement. This mechanism or process is depicted in the bottom portion of Figure 7–5. The basic argument is as follows: the occurrence of no reinforcement, when reward is expected, results in a new stimulus. This new stimulus does two things. First of all, it changes the stimulus complex substantially. After all, the new stimulus was never a part of the original stimulus complex that predicted reward during acquisition. Second, the addition of this new stimulus elicits its own unconditioned reaction. This new unconditioned reaction competes with the original conditioned response. That is, the subject begins performing this new reaction at the expense of the original CR.

The interference factor can be explained more easily by referring to frustration theory (see Amsel, 1958, 1962, 1972). According to this theory, subjects come to expect reward during acquisition; during extinction, when no reward is given, the subjects experience frustration. The experience of frustration is the element S_f shown in the bottom portion of Figure 7–5. It is a new stimulus that results from receiving no reward; it is the feeling of frustration.

As mentioned above, frustration does two things. For one thing, this new feeling, this S_f, alters the stimulus complex; the conditioned reaction, of course, never took place in the context of the frustration stimulus. Second, and more important, frustration naturally elicits frustration reactions (the R_f shown in Figure 7–5). It is presumed that the performance of these frustration behaviors competes with the performance of the CR. Animals cannot do both at the same time. For this reason, the conditioned reaction declines.

We now have a great deal of evidence indicating that frustration is a motivational state. Consider the classic study by Amsel and Roussel (1952). Rats were trained to run down an alleyway. They were fed upon entering the first goal box, located halfway down the alleyway. Then they were allowed to continue their running in the second half of the alley to a second goal box, where they were fed once again. Throughout acquisition, the animals came to expect food in both goal boxes. After giving a sufficient number of training trials, the authors then omitted reward in the first goal box. Omission of reward, when reward is expected, produces frustration. In this case, the frustration experienced in the first goal box energized the behavior; the motivation was heightened by this feeling of frustration, which, in turn, made the animals run more quickly in the second part of the alleyway.

The motivating characteristics of nonreward have been documented in many other studies. The amount of frustration, of course, should reflect the discrepancy between what is expected and what is received. Larger rewards during acquisition should lead to a greater sense of frustration during extinction when no reward is given. Consistent with this is the finding that response speed in the second half of the double alleyway is increased when the omitted reward was large (Bower, 1962; Daly, 1968; Peckham & Amsel, 1967). Other experiments have demonstrated that frustration is produced when the reward in the first goal box is delayed (Sgro, 1969; but see Sgro, Glotfelty, & Moore, 1970). A third factor that affects frustration level is the degree of previous training. The longer the acquisition period, the greater the expectation of reward and, thus, the greater the frustration when nonreward is experienced (Stimmel & Adams, 1969; Yelen, 1969).

The research discussed above indicates that frustration has an energizing function. Let us clarify that important argument. When frustration is experienced it is aversive; aversive states naturally produce unconditioned escape behaviors. For example, an animal will run or jump over a hurdle to escape an aversive shock. It has also been shown in numerous instances that animals will do the same in order to escape from frustration (Daly, 1970). The greater the reward and therefore the greater the discrepancy between the expected reward and no reward, the greater the escape from frustration (Brooks, 1975). Likewise, the more training an animal receives, the greater the escape from frustration cues (Senkowski & Vogel, 1976).

It appears from these experiments that frustration is essentially like fear (Bertsch & Leitenberg, 1970; Brown & Wagner, 1964; Linden & Hallgren, 1973; Wagner, 1969a; but see Fallon, 1971). One study that showed this was done by Daly (1970). Two groups of rats were repeatedly placed in a white box and given 15 food pellets; two other groups were simply placed in the box the same number of times but were not given food reward. All animals then received a series of placements in this box without receiving food. The first two groups, of course, were expected to experience frustration at this time since they had come to expect food in the box. The latter two groups should not have experienced frustration because they never had received food in the box. In the third phase of the study, these four larger groups were divided in half. One group in each of the frustration conditions received fear conditioning in a separate apparatus; a hissing noise was paired with shock, thus creating fear of the noise. Subjects in the other group in each of the frustration conditions were placed in the fear apparatus, but they received backward conditioning (that is, shock followed by noise); they were not expected to develop conditioned fear. Finally, in the last phase of the study, all subjects were once again placed in the original white box and the noise was sounded. One group was expected to feel both frustration and fear, a second group frustration but not fear, a third fear but no frustration, and a fourth group neither fear of the noise nor frustration.

The actual test of the effects of fear and frustration involved escaping the frustration (white box) and fear cues (noise) by hurdling over a small barrier into an adjacent compartment. The more motivated or energized the animals were, the faster they should learn the hurdle-jump response.

As shown in Figure 7–7, frustration and fear combined to affect the overall performance level. While both frustration alone and fear alone increased hurdle-jump performance relative to the control subjects that experienced neither emotion, the fastest group was the one that experienced fear and frustration simultaneously. Thus it appears that frustration is functionally equivalent to punishment. Both lead to aversive states that energize behavior.

It is a matter of assumption that the frustration reaction elicited by nonreward induces competing responses which then interfere with the ongoing behavior. But there are several lines of evidence indicating that this is true. In one experiment (Adelman & Maatsch, 1955) rats were trained to run down an alleyway to get food in the goal box. During extinction

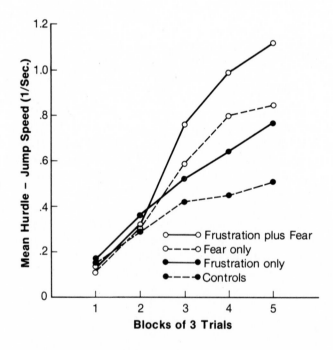

Figure 7-7.   Mean hurdle-jump speed as a function of blocks of three trials for groups that were frustrated, fearful, both frustrated and fearful, or neither frustrated nor fearful. (From Daly, 1970)

subjects were divided into three groups. Animals in the "normal" group remained in the goal box without food for 20 seconds following each response. Recall that these subjects are presumed to experience frustration which, in turn, induces reactions that are incompatible with running down the alleyway (the CR). Subjects in the second group, the "jump" group, were allowed to jump out of the goal box into a separate cage. The aversive nature of the frustration experience, of course, would energize such a behavior. It would be like the hurdle-jump reaction described above. Rats in the third group, the "recoil" group, were allowed to retrace their steps back into the alleyway after they discovered no food in the goal box. Returning to the alleyway, like jumping out of the goal box, is a response that is energized by frustration.

The investigators argued that all the groups would experience frustration due to no reward and therefore that they all would perform competing behaviors. The competing behaviors were visible to the psychologists in the jump and recoil groups; they were explicit behaviors that could be observed. In one of those groups, the jump group, the frustration-induced behavior was compatible with continued running in the alleyway. Jumping out of the goal box is a forward-going motion consistent with approaching the goal. This was not true for the recoil group, however. Retracing into the alleyway from the goal box was incompatible with the CR of running down the alleyway. The recoil animals could not run *to* the goal box and, at the same time, be reinforced (through a reduction in the frustration drive) for running *away from* the goal box. For this reason, the authors predicted that the jump group would continue responding during extinction whereas the recoil group would not. These predictions were confirmed, indicating that frustration does energize behavior and that the behavior may compete with the original reaction if the two are incompatible. It is *presumed* that such competition also operated in the "normal" group, although it was not possible to observe such an effect directly.

In conclusion, the frustration theory suggests that extinction results from the action of competing responses, energized by frustration following nonreward. The frustration-induced reaction competes with the criterion CR, but also the feeling of frustration, the S_f in Figure 7–5, constitutes a change in the stimulus complex.

Counterexpectancy

Finally, there is a third mechanism that is responsible for the reduction of the CR during extinction, in addition to the generalization decrement and interference factors discussed above. This third factor is an obvious one: animals suppress their conditioned reaction during extinction because they are developing new stimulus and response expectancies. To be specific, the "new" stimulus complex, which includes frustration stimuli but not reward aftereffects, predicts the nonoccurrence of reinforcement. Similarly, the formerly correct response now is correlated with no reward. The formation of new expectancies, coupled with the deterioration of the old stimulus complex and interfering responses, all combine to suppress the behavior during extinction.

Partial Reinforcement Effect

One of the most interesting and durable phenomena in learning theory is the partial reinforcement effect. It has received perhaps more attention than any other learning phenomenon. For this reason, and because it highlights some of the theoretical ideas concerning extinction discussed above, we will treat it separately here (see Robbins, 1971 for a review).

The partial reinforcement effect is defined as the increase in resistance to extinction following intermittent reinforcement during acquisition. If an animal is given reinforcement on only a portion of the acquisition trials, resistance to extinction, in terms of speed and persistence of the CR during extinction, is increased relative to subjects that received reward on every trial during acquisition. Traditionally, the partial reinforcement effect was considered paradoxical. If reward strengthened a response during acquisition and if the strength of the CR depended upon the number of rewards or the percentage of rewards, then why wouldn't the 100 percent animals persist longer than partially reinforced animals? After all, persistence should be, according to some theoretical accounts, an alternative measure of response strength. The partial reinforcement effect, however, is no longer viewed as a paradox primarily because it can be explained quite adequately by the general theories discussed earlier.

Stimulus Learning

The partial reinforcement effect traditionally has been limited to response learning situations. Indeed, numerous experiments have failed to show a partial reinforcement effect using classical conditioning techniques (for example, Thomas & Wagner, 1964; Vardaris, 1971; Vardaris & Fitzgerald, 1969). It was claimed by many of these investigators that partial reinforcement during stimulus learning, that is, following the CS only occasionally with the US, weakened the Pavlovian CR; this in turn led to *quicker* extinction. Moreover, the concepts that were successful for explaining the partial reinforcement effect in instrumental conditioning, such as frustration, for some reason, did not seem applicable to stimulus learning.

Recently, however, several investigators have confirmed that the partial reinforcement effect does occur in classical conditioning (Hilton, 1969; Leonard, 1975). One explanation, by Leonard, was that subjects couldn't discriminate between the acquisition and extinction conditions. Quite simply, partially reinforced subjects persisted in executing their Pavlovian CR during extinction because the stimulus complex was similar to the one established during acquisition, at least more so than for the continuously reinforced subjects. Partially reinforced animals continued to expect the US (and therefore continued to emit their CRs during extinction) because the stimulus complex for both acquisition and extinction contained frustration stimuli. In other words, frustration, arising out of nonreinforcement, became part of the stimulus complex during acquisition; if frustration already were a part of the stimulus complex during acquisition, then it makes sense that the animal would continue to expect reward (and continue to perform its CR) during extinction.

Conditions Affecting Extinction Following Intermittent Reward Training

Many of the training procedures mentioned previously as important for extinction behavior affect extinction responding differently if acquisition has been carried out using intermittent reinforcement. One such factor is the magnitude of the reward during acquisition. Recall that larger rewards given on 100 percent of the trials decrease resistance to extinction. In contrast, larger rewards given intermittently during acquisition *increase* resistance to extinction (for example, Capaldi, Lanier, & Godbout, 1968; Leonard, 1969; Ratliff & Ratliff, 1971; Wagner, 1961). Consider the experiment by Ratliff and Ratliff. Rats were trained to run down an alleyway for food. During acquisition, the percentage of trials on which they received reward in the goal box was either 25, 50, 75, or 100. The animals in each of those conditions were further divided into separate groups; on the rewarded trials these groups received either 2, 4, 8, or 16 food pellets in the goal box. Thus the experiment combined four levels of reward magnitude with four levels of partial reward training. Figure 7–8 illustrates the performance during extinction (the numbers on the ordinate refer to rate of extinction; see Anderson, 1963). First, consider the continuously reinforced animals. Larger rewards, from 2 pellets to 16 pellets, decreased resistance to extinction. In contrast, resistance to extinction increased as a function of reward magnitude for the partial reinforcement conditions. Note the changes particularly at the 50 and 75 percent levels. In summary, reward magnitude has a differential effect on behavior depending upon, among other things, the schedule of reinforcement. Resistance to extinction is low following continuous large reward, but resistance to extinction is high following partial large rewards.

The same results have been found when the number of training trials has been considered. Specifically, resistance to extinction following partial

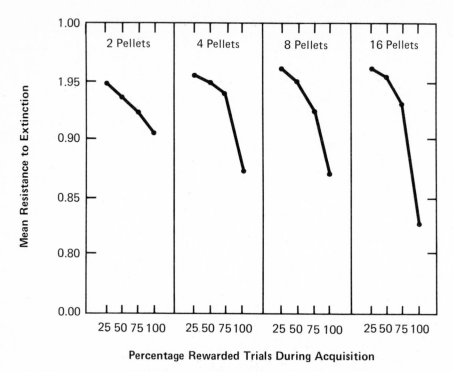

Figure 7-8. Mean resistance to extinction as a function of the percentage of rewarded trials during acquisition for groups that received different amounts of reward. (From Ratliff & Ratliff, 1971)

reinforcement increases as a function of the number of partially reinforced trials during acquisition (for example, Hill & Spear, 1963; Wilson, 1964). This is in contrast to the findings cited earlier, that resistance to extinction decreases as a function of the number of continuously reinforced trials. In the Hill and Spear study, five groups of rats were given partial reinforcement training for either 8, 16, 32, 64, or 128 trials. During extinction, running speed in the alleyway was a positive function of the degree of training. That is, subjects that received 128 trials continued to run fast during extinction, while subjects that received fewer trials were slow. It should be noted that increased resistance to extinction following partial reinforcement does not require many training trials. Indeed, partial reinforcement subjects are more resistant to extinction than continuously reinforced animals even after experiencing only a few acquisition trials (Capaldi & Deutsch, 1967; McCain, 1966).

The percentage of rewarded trials during acquisition is, of course, one parameter that has a strong effect on later resistance to extinction. It has been shown that resistance is an inverted U-shaped function of percentage of reward during acquisition (Bacon, 1962; Coughlin, 1970; see also Lewis, 1960). More specifically, extinction occurs rapidly if very low or very high percentages of reinforced trials were used during acquisition. Animals that receive 0 percent reinforcement during acquisition, of course, could not be expected to have much resistance to extinction; a response that was never reinforced could hardly be expected to be persistent. Similarly, animals that receive 100 percent reinforcement during acquisition extinguish rapidly; this is, essentially, the partial reinforcement effect. The most persistent subjects should be those animals that receive reward on about 50 percent of the acquisition trials. This appears to be the case.

The pattern of reward during acquisition also influences extinction behavior. Normally, an alternating pattern, that is, the recurring sequence of reward followed by nonreward, produces less resistance to extinction than a random pattern of reward and nonreward even though the two procedures use the same number of trials and percentage of reward (E. J. Capaldi, 1958). Such a result suggested to early investigators that animals had trouble distinguishing between the acquisition and extinction conditions. It is easier to discriminate when an alternating pattern has been discontinued than when a random pattern has ceased. Interestingly, with a small number of training trials, this principle is reversed: an alternating pattern leads to greater resistance to extinction than a random one (Capaldi & Hart, 1962).

The notion that animals discriminate between acquisition and extinction more easily when the acquisition pattern alternates between reward and nonreward was confirmed in a study by Rudy (1971). In several of his groups, Rudy replicated the well-known fact that compared to a random pattern of reinforcement and nonreinforcement, alternation of reward and nonreward during acquisition decreases later resistance to extinction. Rudy, however, included another group; subjects in this group received the random schedule, but also a light in the alleyway was turned on during the nonrewarded trials. It was believed that such a light would help the subjects to discriminate rewarded from nonrewarded trials. During extinction, these subjects stopped performing just as quickly as the alternation subjects. In summary, if the pattern itself is easy to discern (alternation), then the subjects readily perceive the onset of the extinction trials and stop responding rather quickly; if the subjects cannot discriminate, but a cue is provided to help them make the discrimination, then extinction also takes place quickly.

Finally, an important factor that affects extinction is the order of intermittent schedules, that is whether a series of continuously reinforced trials is given before or after a series of partially reinforced trials. Most recent research indicates that resistance to extinction is greater when the partially reinforced trials follow the continuously reinforced trials (Dyal & Sytsma, 1976; Mellgren, Seybert, & Dyck, 1978; Theios, 1962; Theios &

McGinnis, 1967). In the Dyal and Sytsma study, subjects were given three phases of training. First, half the animals received partial reinforcement training, and the other half received continuous reinforcement for running down an alleyway. In Phase 2, half of each of these groups continued with this same training procedure (partial or continuous reinforcement), whereas the other half was switched to the opposite reinforcement procedure. In other words, some animals received partial reinforcement throughout Phases 1 and 2 (Group P-P), and some received continuous reinforcement throughout (Group C-C). Other animals, however, received continuous followed by partial reinforcement (Group C-P), or partial reinforcement training followed by a series of continuously reinforced trials (Group P-C). All animals were then extinguished in Phase 3.

As shown in Figure 7–9, the usual partial reinforcement effect was demonstrated (compare the P-P and the C-C groups). Note that the group that received the partial reinforcement just prior to extinction, Group C-P, was more resistant to extinction than the group that received continuous reinforcement following partial reward training, Group P-C. It is as if such a block of continuous reward trials helps the animal to discriminate the acquisition from the extinction conditions. It is also interesting to note that some partially reinforced trials given initially (Group P-C) increased resistance to extinction relative to the C-C treatment. Finally, there was an increase in resistance to extinction for groups that received partial reinforcement just prior to extinction (Group C-P versus C-C), a difference that is magnified by increasing the number of continuously reinforced trials initially used (see Mellgren et al., 1978).

There seem to be two principles in operation. First, extinction is increased when partial reinforcement follows continuous; this would suggest that the ease of discriminating between acquisition and extinction is at least part of the explanation for the partial reinforcement effect. Second, the effect of partial reinforcement on extinction persists even though a block of continuously reinforced trials is given between the partial reinforcement and extinction treatment. This finding suggests that, although discrimination between acquisition and extinction may be important, other factors are also involved. Some of these factors are discussed in greater detail below.

Theory of the Partial Reinforcement Effect

In this section we will take the three major factors discussed previously —generalization decrement, interference or frustration, and the formation of new expectancies—and show how they may be used to explain the partial reinforcement effect. First of all, recall that any given trial, reward or nonreward, involves aftereffects. Quite simply, animals remember whether they were rewarded on previous trials. The aftereffects may have particular emotional and motivational significance, such as the feeling of frustration

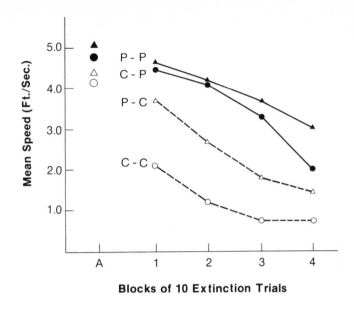

Figure 7-9. Mean running speed in the alleyway as a function of blocks of extinction trials for groups that were given continuous reward throughout acquisition (C-C), continuous followed by partially rewarded trials (C-P), partial followed by continuous reward (P-C), or partial reinforcement throughout acquisition (P-P). (From Dyal & Sytsma, 1976)

upon the receipt of no reward. We argued previously that when these nonreward aftereffects, emotional or not, suddenly occurred during extinction, the stimulus complex was disrupted and competing behaviors occurred.

The disruptive effect of these events is depicted in the left half of Figure 7–10. During acquisition, for continuously rewarded subjects, the stimulus complex and response are followed by reinforcement on 100 percent of the trials. During extinction, the new stimuli, S_n and S_f (memory and frustration aftereffects from no reward respectively), occur and disrupt performance; in addition, previous food-related cues are absent. Extinction takes place very readily because the stimulus complex is markedly degraded.

Now consider how these factors operate for a partially reinforced animal (refer to the top right half of Figure 7–10). During acquisition, the reward produces food-related aftereffects (S_r) which, admittedly, are absent during extinction. But during acquisition, the lack of reinforcement on the nonrewarded trials results in stimulus aftereffects too (memory of nonreward—S_n) as well as emotional or frustration stimuli (S_f). These stimuli,

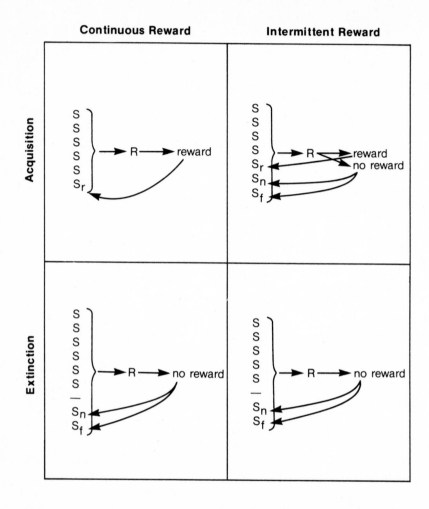

Figure 7-10. Schematic diagram representing how added cues from nonreward, S_n and S_f, disrupt extinction responding following continuous reward but become part of the stimulus complex during partial reward and therefore continue to elicit the expectancy for reward during extinction. Note the absence of the S_r cues during extinction.

in fact, become part of the stimulus complex that signals reward during acquisition.

Imagine the reaction of a partially reinforced animal that experiences nonreward for the first time during acquisition. It feels frustration (S_f) and remembers, while in the start box for the next trial, that it was not rewarded on the previous trial (S_n). These stimuli are now part of the animal's environment for that trial. And when the animal runs down the alleyway and receives reinforcement, the stimulus complex, which contains the S_n and S_f stimuli, gets conditioned. In other words, an animal learns to run in the presence of nonreward cues because those cues predict reward.

Now consider the bottom right portion of Figure 7–10. This is the schematic diagram of extinction following partial reinforcement. Notice that the nonreward cues, S_n and S_f, are a part of the stimulus complex stemming from no reward. But they are *already* a part of the stimulus complex because they occurred during acquisition. That is, subjects were previously reinforced for responding in the presence of those frustration cues. During extinction, therefore, the partially rewarded animals persist much longer because, in one sense, they developed a tolerance for frustration and nonreward during acquisition.

The specific version of such a theory of the partial reinforcement effect was proposed by E. J. Capaldi (1966, 1967). He stressed that nonreward trials produce stimulus aftereffects which are present during acquisition and therefore become part of the stimulus complex that elicits the expectancy for reward. Continuously reinforced animals, of course, never experience the nonreward aftereffects during acquisition and so their occurrence during extinction, for the first time, causes disruption. In particular, Capaldi specified three crucial conditions that determine later resistance to extinction. If we designate a nonrewarded trial as N and a rewarded trial as R, then we can refer to a series of N trials or a series of R trials. Using this terminology, we can state Capaldi's three principles as the following. Resistance to extinction is a function of (1) the N-length, that is, the number of consecutive nonrewarded trials prior to an R trial; (2) the number of times a particular N-length occurs, that is, the number of N-R transitions; and (3) the variety of different N-lengths. As shown by Capaldi (1964), resistance to extinction is greater if the animal has experienced longer N-lengths during acquisition. In addition, if an animal receives many N-lengths (extended training) or a variety of different N-lengths, then resistance to extinction also increases. These three principles appear to account for much of the data on this phenomenon.

What about some of the major parameters of training and their effect on resistance to extinction following partial reinforcement? These are perfectly consistent with Capaldi's theory. If the reward during acquisition is large, then the stimulus expectancy, based on the stimulus complex that includes nonreward aftereffects, is stronger. This means that partially rewarded animals should persist longer during extinction with large reward than with small reward, but the opposite should be observed for continuously rewarded subjects; this is indeed what is found (Capaldi & Capaldi, 1970; Capaldi & Lynch, 1968). Similarly, a greater number of training trials during acquisition merely strengthens the role of the nonreward aftereffects in the stimulus complex. This should and does lead to greater resistance to extinction for partially reinforced animals (Capaldi, 1964).

Let us now consider a specific version of the interference position as it relates to the partial reinforcement effect. The frustration theory, in fact, was designed to explain the partial reinforcement effect and it does so quite well. Refer once again to the left part of Figure 7–10. For continuously reinforced animals, the sudden appearance of frustration stimuli during extinction has a disruptive effect on performance; the aversive frustration state elicits unconditioned reactions which interfere with the criterion response (see Figure 7–5). Partially reinforced animals, on the other hand, as indicated on the right half of Figure 7–10, have already experienced frustration during acquisition. In fact, they were reinforced for performing while experiencing frustration cues. Thus, frustration stimuli became part of the stimulus complex that predicts reward. Specifically, on the nonreinforced trials during acquisition, frustration stimuli occur; because the animal is then later reinforced for running in the presence of these stimuli, the stimulus complex that includes the frustration stimuli, is reinforced; it is followed by reward and therefore comes to elicit the expectation for reward. During extinction, then, when no reward is presented, the frustration stimuli that arise from nonreward are already part of the stimulus complex. The behavior continues to be elicited during extinction because the stimulus complex contains the frustration stimuli that were established earlier. In essence, then, there is very little difference between the memory aftereffects emphasized by Capaldi and the frustration aftereffects emphasized by Amsel, at least in terms of how they become a part of the stimulus complex. Both result from nonreinforcement, and both can become part of the stimulus complex during partial reward acquisition. The main difference is that the nonreward aftereffects are explicitly emotional in nature for Amsel but are merely memories for Capaldi.

Frustration theory also accounts for many of the variables discussed previously. For example, the strength of the frustration is a function of the discrepancy between the expected reward and no reward. Larger continuous rewards during acquisition produce greater frustration during extinction. This fact accounts for the rapid extinction of the continuously reinforced subjects. However, larger rewards during acquisition for partially reinforced animals strengthen the stimulus expectancy based on the frustration cues. The opportunity for intense frustration to develop during acquisition, and consequently for strong conditioning of the frustration cues, is greater with larger rewards.

The same argument can be made for the degree of training: the more opportunity the animal has had during acquisition to have the frustration stimuli become part of the stimulus complex, the more persistent the animal will be during extinction. In short, the frustration theory, as well as Capaldi's aftereffects theory, explains well the effects of many variables on extinction. In particular, it explains why certain variables, such as reward magnitude and degree of training, influence extinction of partially reinforced animals differently than continuously reinforced animals.

The notion that nonreward produces an emotional aftereffect is, of course, critical to frustration theory. Several researchers have attempted to confirm that this is, indeed, the case. We have already discussed several of the experiments that show that animals will learn new behaviors in order to escape from frustration cues. This, by definition, indicates that frustration is an aversive drive state. Other investigators have gone in the opposite direction. Specifically, they have tried to reduce or eliminate the emotionality stemming from nonreward by administering tranquilizing drugs to the subjects (for example, Gray, 1969; Ison & Pennes, 1969). In Gray's study, rats were trained under either a continuous or partial reinforcement schedule to run down an alleyway for food. Prior to the acquisition trials half the subjects in each group were injected with amobarbital, a drug that reduces emotionality. The other half received an inert placebo injection. During extinction, all the subjects received this placebo. Gray found that the partial reinforcement effect was greatly reduced in the subjects that had been given the drug. According to the frustration theory, these partially reinforced but drugged animals did not experience frustration, and, therefore, the frustration aftereffects stemming from nonreward did not get incorporated into the stimulus complex. Continuously reinforced subjects, of course, never experienced nonreward (frustration) either and so when both groups encountered frustration for the first time during extinction, the CR decline was rapid in each group.

Although evidence from these drug studies provides strong support for the frustration theory, others have failed to obtain similar efforts (Ziff & Capaldi, 1971). In addition, a number of investigators have aruged that frustration takes time to develop during acquisition. If this were the case, then how could frustration account for the partial reinforcement effect that is observed after only a very small number of acquisition trials? This point is not thoroughly resolved; several studies have even attempted to show how the frustration theory could deal with such experiments (Amsel, Hug, & Surridge, 1968; Brooks, 1969). It is, perhaps, best to note that *both* the aftereffects and the frustration theories are applicable to the problem. The aftereffects stemming from nonreward certainly can be emotional and frustrative in nature. On the other hand, it may be that the mere memory of nonreward, a memory aftereffect, is sufficient too (Capaldi & Waters, 1970).

The above discussion indicates that a number of factors are responsible for extinction behavior: the loss of important stimuli like reward-related cues, the inclusion in the stimulus complex of new unfamiliar stimuli like nonreward memories or frustration cues, the conditioning of those nonreward aftereffects during partial reinforcement training, and so forth. The third major factor cited earlier in this chapter that helps to account for extinction behavior is the formation of a new expectancy. This factor can not be overlooked: the former CR is being suppressed but, at the same time, the animal is learning new expectancies about the relationship between its stimulus complex, responding, and no reward. The formation of these new expectancies in the case of the partially reinforced animal is much slower because some of the salient stimuli in the stimulus complex, notably the nonreward aftereffects and emotional reactions, are already part of the stimulus complex. Those cues have come to be associated with reward and therefore the new expectation that no reward will ever be given is slow to form. This is tantamount to saying that partially reinforced animals have a greater difficulty in discriminating between the acquisition conditions and the extinction conditions.

In summary, if animals come to expect particular outcomes based on their stimulus complex and their behavior, then the environmental conditions prevalent during extinction (that is, the stimulus complex) must certainly elicit an expectation that is somewhat similar to the one elicited during acquisition for partially rewarded subjects. For continuously reinforced animals, on the other hand, there is a glaring difference between the stimulus complex during acquisition and extinction. Therefore, the expectation for reward during extinction is weak.

Summary

Extinction is a procedure whereby the contingency between the stimulus (in the case of Pavlovian learning) or the response (in instrumental learning) and the outcome, usually a US, is eliminated. The result is a decrease in the CR. Interestingly, however, merely giving the stimulus or response and the US in a random fashion does not cause a very large decline in behavior. Evidence indicates that actually reversing the predictive relationship established in acquisition is the more potent technique.

The procedures used in the reward training technique of response extinction include withholding the reward and debasing the response-US correlation. Studies using the reward training method have shown that extinction occurs when the reward is omitted. However, debasing the response-US correlation does not necessarily lead to rapid extinction.

Omission training is a second method used to eliminate conditioned responding, although under some conditions it is not as effective as normal extinction. If, however, specific competing behaviors are reinforced, then suppression of the CR is more rapid.

Avoidance extinction represents yet a different problem for analysis. Merely giving the CS by itself appears to be less effective than removing the avoidance contingency. If shocks are given during extinction, then responding is maintained much as it is in appetitive conditioning when food is presented during extinction.

Several factors affect resistance to extinction. Among the most important is the magnitude of the food reward experienced during acquisition. Large rewards lead to faster extinction; variable magnitude increases resistance to extinction. Training level operates in a similar fashion: the more extensive the training, the faster the extinction. Delay of reward is a third important factor. Generally, constant delay does not appear to affect resistance to extinction although changes are observed if the duration of confinement in the goal box on the extinction trials differs from reward delay during acquisition. Finally, greater response effort during acquisition or extinction causes animals to stop responding more quickly during extinction.

The first of the three general theories of extinction discussed was the stimulus generalization decrement theory. According to this theory, the environment changes during extinction and such changes disrupt performance. Essentially, the complex of stimuli that come to elicit the expectation of reward during acquisition is incomplete; the aftereffects of reward, an important stimulus, are no longer present during extinction.

The second theory, the interference theory, claims that new stimuli stemming from nonreward during extinction elicit unconditioned responses that interfere with the learned behavior. They do so because these new stimuli induce an aversive state of frustration.

Finally, extinction may take place because, like in acquisition, new expectancies develop. In this case it would be the expectancy that the response will not lead to reward.

The partial reinforcement effect refers to the finding that animals that were intermittently rewarded during acquisition persist longer and run faster during extinction than subjects that received constant reinforcement during acquisition. The conditions that affect this phenomenon include the magnitude of the acquisition reward, the number of training trials, the percentage and pattern of the reward during acquisition, and the order of reinforcement schedules.

The stimulus generalization decrement, interference, and counterexpectancy may be used to explain the partial reinforcement effect. Specifically, if aftereffects from nonreward are experienced during acquisition, as they are in partially rewarded subjects, then they become part of the stimulus complex that elicits the expectation for reward. Accordingly, during extinction, when the aftereffects stem from continuous nonreward, the expectation for reward, and the accompanying behavior, is more durable. Such an increase in resistance to extinction will occur when the aftereffects are memories of nonreward or an aversive feeling of frustration.

8

Memory

Introduction

General Theories of Memory

Short-Term Memory

Intermediate-Term Memory

Long-Term Memory

Summary

Introduction

In many respects, to study learning is to study memory. Learning would be impossible without memory because each execution of a learned reaction requires memory of the previous trial. Indeed, improvement from one trial to the next reflects the gradual buildup of the memory for that particular response. One could even say that learning is the strengthening of memories —stimulus and response expectancies—and that their persistence over some period of time (a feature that is integral to the notion of memory) is part of the definition of the learning process.

Why then separate the material on memory from material on learning? If learning is merely the accumulation of memories, then why have a special category labeled "memory" at all? The answer to these questions is complex, but certainly one reason is that psychologists have only recently begun to study animal memories directly. The study of memory, historically at least, has been mainly concerned with human verbal memories. Human beings, of course, have verbal memories of incredible complexity, and good memory is an important talent. So although it has long been recognized that animals can remember, it was within the area of human learning and memory that the technology and theory developed, rather than in the field of animal learning and behavior.

As shown in Figure 8–1, memory actually involves three different stages. In the first stage, learning takes place (left portion of Figure 8–1). At this stage the expectancies are formed; they become embodied in the information pool within the subject. During this encoding stage, the otherwise naive subject is changed in a fundamental way; it incorporates a new learned unit of behavior.

The second stage is the retention or storage phase of the memory process (middle portion of Figure 8–1). It is here that the memory is fixed or simply stored; it persists over time in some fashion. Indeed, it is this stage, the retention interval or storage phase, that defines memory: memory is the evidence of learning after some period of retention.

Finally, as shown in the right portion of Figure 8–1, the third basic stage of memory is the retrieval or performance stage. At this stage the subject actually performs the learned reaction, indicating that it has retained the CR over the storage interval. If performance is as good as it was during acquisition, then memory is said to be perfect; no forgetting has occurred. On the other hand, if performance is below the level shown at the end of the learning phase, then forgetting is said to have occurred.

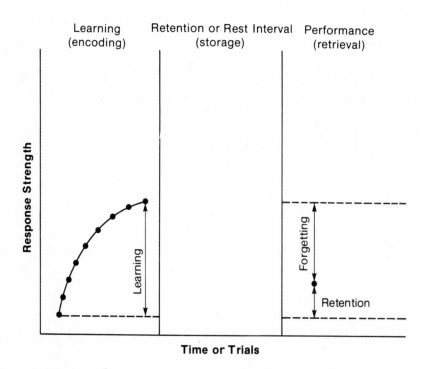

Figure 8-1. Hypothetical curves showing the three major stages of memory.

General Theories of Memory

The major task of memory theorists, of course, is to explain forgetting, or, alternatively, retention. For many years, it was believed that forgetting occurs because of a decay process during the retention interval. That is to say, memories were encoded during learning and then, like footprints in sand, faded during the retention interval because they were not used. When the subject was asked to retrieve the memory, little retention was evident because the memory had deteriorated merely with the passage of time. Although some evidence has been provided for this point of view, memory theorists currently believe that it is largely incorrect to describe forgetting as a consequence of memory decay.

Another general theory of forgetting is the interference position. This theory claims that competing memories develop during the retention interval which interfere with the memory. In other words, the learned unit of behavior does not deteriorate through time; rather it is susceptible to disruption by the appearance of other competing memories. Unless a given memory is periodically strengthened by rehearsal, these other memories encroach upon the original memory, causing poor retention. There is no doubt that memories can be interfered with. The acquisition of new memories certainly causes problems for the retention of old memories. In addition, the presence of certain memories can cause problems for the retention of future memories. This is true in both humans and subhuman animals and we shall discuss some of this work in more detail below.

A final theory focuses on the third phase of memory (right portion of Figure 8–1). This position claims that forgetting results from a failure in retrieval. That is, a memory survives the retention interval intact, without decaying or without being interfered with. However, due to various factors which we shall discuss, the subject simply cannot perform the memory accurately. The behavior is "stored" somewhere in the subject, but it cannot be retrieved. It is like trying to find a library book that has been misplaced on the shelves. The book is there, that is, the memory is intact, but the subject cannot find the book; it cannot retrieve the memory. Recent work on memory suggests that retrieval failure is a major problem of forgetting. This is true for long-term memories in humans as well as in animals.

Consolidation Theory

Let us consider in detail some of the research over the last decade that has confirmed the notion that retrieval failure is a major source of forgetting, beginning with an examination of some early work on memory and electroconvulsive shock. Electroconvulsive shock (ECS) is a technique that is used in clinical practice to help treat depression. It also has effects on memory; namely, it causes retrograde amnesia, that is, the forgetting of events that occurred just prior to administration of electroconvulsive shock. The ECS is administered to animals in much the same way as it is to human beings: a very large electric current is applied to the head region of the animal, causing it to lose consciousness and go into a seizure. The seizure lasts only a few minutes. Afterward the subject appears to have recovered fully. No one knows precisely how ECS works, but its apparent therapeutic and memory effects are somehow caused by the massive firing of brain neurons (see Lewis, 1969).

There is an explicit and well-known relationship between memory and ECS administration. One of the earliest studies on this was done by Thompson and Dean (1955). Animals were taught to run from a start box into a goal area to avoid shock. They did so by going through either one of two doors outside the goal box. After this basic response was thoroughly learned, the rats were required to learn a discrimination. Specifically, cardboard targets, comprised of either horizontal or vertical stripes, were placed on the doors and the subjects had to choose the correct one. Failure to pick the appropriate door resulted in shock. The animals learned to do this quite readily too. Following the last trial of this discrimination learning, Thompson and Dean gave a retention interval of either 10 seconds, 2 minutes, 1 hour, or 4 hours, and then an ECS treatment (a control group was not given ECS). Finally, two days later, after the animals had fully recovered, they were given a relearning test. The prediction was that if ECS had interfered with the memory of the original discrimination, then relearning should be rather difficult. On the other hand, if the ECS had had no effect on memory, then the subjects should relearn the horizontal-vertical discrimination almost immediately.

The data from this study are shown in Figure 8–2. Subjects that received ECS immediately after learning the discrimination (the 10-second animals), showed very poor relearning; their relearning curve is nearly the same as it was for original conditioning, indicating that they remembered very little from their previous experience. In contrast, animals that received ECS 1 or 4 hours after learning showed very quick relearning. This would indicate that they remembered the original task well and, therefore, had no trouble relearning it.

This relationship, between memory and the delay of the ECS administration, supports the traditional consolidation theory of memory. According to this theory, memories are created at a particular point in time and, if they are to endure, they must undergo a period of consolidation. During this time the memories become fixed; the neural trace that results during learning requires some time in order to "jell." Imagine, for a moment, that memories are like impressions in concrete. After the impression is first made (the learning phase), one must allow a certain period of time for the concrete to "set" or become consolidated. If this is done, the impression (memory)

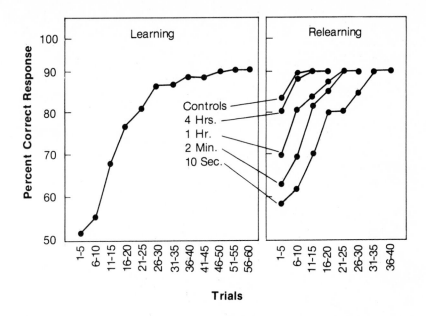

Figure 8-2. Percent correct discrimination choice on learning and relearning as a function of training trials for groups that received ECS treatment 10 seconds, 2 minutes, 1 hour, or 4 hours after learning (controls were not given ECS). (From Thompson & Dean, 1955)

is permanent. If, however, something is done to interrupt this consolidation process, for example, if water is splashed on the impression, then the impression fades; it is damaged and recovery is impossible. According to consolidation theorists, ECS operates in this way. Learning creates a neural trace that must become fixed before it is permanent. The ECS disrupts this fixing, or consolidation, process. In particular, the shock scrambles the brain neurons in such a way that the neural trace established during learning does not consolidate. But the more time one allows for the consolidation process, the stronger the memory. This is how we can interpret the Figure 8–2. The longer the learning-ECS interval, the greater the consolidation and, therefore, the greater the memory. Conversely, the shorter the learning-ECS interval, the greater the disruption of consolidation and the greater the forgetting.

Retrieval Theory

Many theorists now believe that the consolidation position is inaccurate. The inability to perform properly after ECS does not reflect the *preven-*

tion of memory formation, which is the consolidation position, but rather reflects an inability to retrieve the memory at the time of the test (see Miller & Springer, 1973; Spevack & Suboski, 1969). There is considerable evidence to support such a retrieval theory. One of the strongest predictions from consolidation theory is that if ECS prevents the memory from being consolidated, then it should never be possible to retrieve that memory. In other words, if ECS operates to eliminate the fixing process, then the memory is never properly formed; it should be lost forever. Recent research has shown, however, that memories disrupted by ECS administration are not, in fact, lost (Koppenaal, Jagoda, & Cruce, 1967; Lewis, Misanin, & Miller, 1968; Miller & Springer, 1972). For example, in the Lewis et al. (1968) study, animals were given a single passive avoidance learning trial. The procedure used in such a trial is this: First the subjects are allowed to step off a safe platform onto an electrified grid. Later, upon being placed on the platform again, they avoid the shock by remaining passively on the platform. Such an avoidance reaction, of course, requires that they remember having been shocked on the grids previously. However, if an experimental animal receives ECS soon after it steps onto the shock grids, it "forgets"; when tested the next day, it will step off the safe platform as if it had never encountered shock on the grids before. This is the retrograde amnesia phenomenon. In their study, Lewis et al. trained the passive avoidance reaction and administered ECS immediately. Just prior to testing on the next day, they gave the animal a single mild foot shock in a different apparatus. This experience was enough to remind the subject about the previous learning episode, because when it was then tested for its fear of the grid bars (that is, the passive avoidance test), it showed perfect memory. In other words, Lewis et al. showed that a memory, which had apparently been disrupted by ECS administration, actually could be retrieved if the animal were given a reminder about the early learning. The reminder trial did not in any way "teach" the animal the memory; control groups that got *only* the reminder trial showed no fear of the original grid bars. Therefore, it can be safely concluded that the reminder trial did precisely that: it reminded the animal about the previous day's learning experience so that it could retrieve the memory of that experience properly. Without such a reminder, the animals perform as if they have forgotten; indeed, they have forgotten since they can not retrieve the original learning. In summary, it appears that ECS affects the retrieval process during the test rather than abolishes the consolidation of the memory just after learning.

This general finding, that memories which are apparently lost can be retrieved if the animal is reminded about the learning conditions, has been duplicated in many experiments. For instance it has been shown for appetitive memories based on food reinforcement (Miller, Ott, Berk, & Springer, 1974), and it has been shown to occur when the reminding agent is not the same as the original training agent. An example of this was a study by

Springer and Miller (1972). They punished control animals with ice-cold water and then administered ECS. These subjects showed no memory for the location where they were immersed in water; that is, they showed the normal retrograde amnesia finding. However, experimental animals that were given a mild shock in a different apparatus before testing recalled their previous water immersion experience; they showed an avoidance reaction. Here, the retrieval of a memory was facilitated by a reminder trial even though that reminder experience was different from the original learning conditions. Apparently, the fact that both experiences were highly aversive was sufficient to "jar" the animals' memory. If a reminder trial is dissimilar in nature, then this process does not work.

In summary, then, the administration of ECS immediately after learning produces retrograde amnesia; the animals perform on memory tests as if they had never experienced the original learning. However, the reason for this poor performance is not that the memory was prevented from being fixed or consolidated. Lack of consolidation suggests that the memory would never again be present, but the reminder-trial studies unequivocally demonstrate that the amnesia occurs, fundamentally, because of a lack of retrieval capability. The reason that the animals perform poorly on the test is that they cannot retrieve the memory. However, if the animals are reminded of their previous learning, even though that reminder experience itself is unable to teach the animals the memory, then performance is restored.

The notion that forgetting is due to retrieval failure has been developed extensively in recent years in areas of research other than the ECS literature (see Spear, 1973, 1978; Spear & Parsons, 1976). What is important is that the general concept of retrieval failure is nearly identical to the generalization decrement theory we discussed in Chapter 7. Let us review that theory briefly. A learned response occurs in the presence of a stimulus complex. The stimulus complex and the response both elicit expectations of reward because both are correlated with the reward presentation. When new stimuli, resulting from either internal or external sources, encroach upon the stimulus complex, then the behavior declines; the new stimulus complex is no longer the one that signaled reward. In addition, the loss of certain salient stimuli in the complex can also cause a disruption of the eliciting process.

Consider now how the retrieval theory is virtually the same as the generalization decrement theory of extinction. Extinction involves a change in the environment (stimulus complex) due to the elimination of reward. Like extinction, ECS also causes deliberate alterations in the stimulus complex. But the mere passage of time, such as experienced during the retention interval, can cause the stimulus complex to be different in the future too. New stimuli that arise during the retention interval are added to the stimulus complex and stimuli are lost during the retention interval. The net effect is an inability for the stimulus complex to function efficiently during the memory test.

One well-known phenomenon that illustrates this is the warm-up decrement. If an animal is given training, such as Sidman avoidance learning, and tested each day at approximately the same time, then the animal will perform poorly at the beginning of each session. Later, performance improves. In a study by Spear, Gordon, and Martin (1973), rats were trained to make a response to avoid shock. A flashing light and light noise accompanied the shock. Twenty-four hours later the animals were retested. The average number of seconds required to relearn this response was 554.1. However, if the animals were placed in a different apparatus and given a brief shock, then they required only 413.8 seconds to relearn the response. Finally, if the animals were placed in a different apparatus and given CS-US pairings, they only took 366.0 seconds. In other words, within a single day, animals showed a warm-up decrement; it took them some practice before they were able to perform efficiently once again. However, if they were reminded of their previous training, then the warm-up decrement was eliminated. The better the reminder, such as getting CS-US pairings instead of just shock, the better the relearning. In summary, the passage of time can cause a deterioration in the stimulus complex; when the animals are reminded of the previous learning context, then the decrement is eliminated. Such evidence strongly points to retrieval failure as an explanation for the warm-up decrement as well as for forgetting (see Deweer, Sara, & Hars, 1980; Feldman & Gordon, 1979; Gordon & Feldman, 1978; Gordon & Spear, 1973; Hickis, Robles, & Thomas, 1977; but also Powell, 1972a).

Short-Term Memory

Research on human memory has traditionally distinguished between at least two types of memory processes: short-term memory and long-term memory. It is believed that the principles governing these memories are different. The literature dealing with animal memories is not nearly so divided. Most of the work has focused on memories that persist for hours or longer, that is, long-term memories. However, a good deal of research recently has focused on memories that last for a shorter period of time (see Honig & James, 1971; Medin, Roberts, & Davis, 1976). The material discussed above in connection with the retrieval theory of memory essentially focused on long-term memories. We shall return to that topic in a later section. First, however, we would like to discuss the recent research on animal short-term memory.

Delayed-Matching-to-Sample

The most common procedure for investigating short-term memory in animals is the technique called delayed-matching-to-sample, or DMTS (see D'Amato, 1973; D'Amato & Cox, 1976; Roberts & Grant, 1976). Here is how a typical DMTS study is conducted. The animal, usually a monkey or pigeon, is placed in a box containing three plastic disks behind which different stimuli can be illuminated. For example, each disk, or key, has a back light that can illuminate different colored patches, various geometric shapes, or even shapes of various sizes that differ in color from their background. When the trial begins, the center key is illuminated, showing the "sample" stimulus. Subjects must demonstrate that they recognize the sample by pressing (for monkeys) or pecking (for pigeons) at the key. Following this, a delay period is given during which the animals must wait. After the delay interval, both of the side keys are illuminated. One of them contains the same stimulus as was just presented on the center key (the sample). The other side key is a different stimulus. The animal's task is to remember the characteristics of the sample over the delay period (short-term memory) and demonstrate this memory by choosing the correct stimulus. If the animal is correct, that is, if it matches the sample after the delay period, then it receives reinforcement; if the animal is incorrect in its choice, it receives no reward.

The basic finding is that performance declines as a function of the delay interval. The longer the animals have to remember the characteristics of the sample, the less efficient they are on the DMTS task. Striking differences, however, are observed between different species of animals. For instance, monkeys and dolphins can match the sample even with delays as long as 2 minutes (for one monkey, DMTS was possible even with a delay of 9 minutes; see D'Amato, 1973). With pigeons, on the other hand, short-term memory seems to be much less; they can match the sample with delays of only 10 to 15 seconds.

In the typical DMTS experiment, simple stimuli like colored patches or geometric figures are used. The animal simply has to remember the sensory characteristics of the sample. However, as shown by D'Amato and Worsham (1974), monkeys are capable of far more complex memories. In that study, the monkeys demonstrated conditional matching, that is, matching based upon some arbitrary association between the sample and a comparison stimulus rather than matching based upon simple sensory characteristics. For instance, the animals might be given a red disk as the sample and required to choose an inverted triangle on the test; choice of the alternative stimulus on the test would not be rewarded. Similarly, on other trials, the animals might be given a vertical line which, after the delay interval, they would have to match by choosing a dot pattern. Clearly, the correct performance on this task is based on the memory of the sample, not simply a lingering visual aftereffect. D'Amato and Worsham found that monkeys could, indeed, perform this task.

Using the DMTS technique, many investigators have attempted to manipulate conditions so that the variables that affect memory can be identified. One of the important variables that affect DMTS performance is the duration of the sample presentation. One phenomenon noted in a number of investigations is that the longer the sample presentation prior to the delay, the better the matching after the delay (Devine, Jones, Neville, & Sakai, 1977; Herzog, Grant, & Roberts, 1977; Roberts & Grant, 1974). In the Roberts and Grant experiment (1974), for example, pigeons were presented with a sample stimulus (a red or green key) for either .5, 1, 2, 4, or 8 seconds. Then all animals received a delay of 1 second, followed by a choice between the sample and an alternative stimulus. As shown in Figure 8–3, the percent correct on the DMTS task increased as a function of the sample duration. Whereas a sample that lasted only .5 seconds produced about 56 percent correct choices, performance was nearly 80 percent correct with sample durations of 8 seconds.

A second variable which has an interesting effect on short-term memory in the DMTS paradigm is the number of dimensions that define the sample stimulus. If the sample contains both a geometric form and color, but the subject is required to match only on the basis of the geometric figure (that is, ignore the color), then performance is inferior relative to the

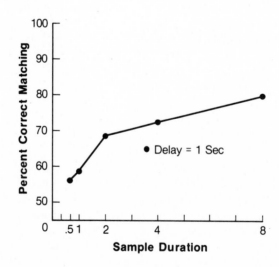

Figure 8-3. Percent correct on the DMTS task as a function of the sample duration. (From Roberts & Grant, 1974)

condition when the sample contains only one dimension. Several investigators have claimed that the reason for this decrement in performance is that the animal must "share" its attention between two dimensions. When this is the case, the animal has an inferior memory for either of the dimensions (see Maki & Leith, 1973; Maki, Riley, & Leith, 1976).

There is another theory, however, that could explain this phenomenon. When the sample contains two elements or dimensions but the choice stimuli contain only one dimension (recall that this leads to inferior matching performance), it could be simply a matter of a stimulus generalization decrement. In the discussion above we stated that performance declines when the stimulus complex is changed. If one can conceive of the sample as essentially defining the stimulus complex, then it is plausible to attribute the lowered performance to the fact that the complex has changed; the choice stimuli contain only one dimension and this is substantially different from the original sample stimulus which contained two. At this time, it is not possible to resolve this theoretical issue. Roberts and Grant (1978b) investigated these two hypotheses and discovered that *both* factors contributed to the inferior performance. That is, the fact that the subjects must "share" their attention to two dimensions *and* the fact that the sample is not accurately given on the choice trial (causing a stimulus generalization decrement) are both contributing factors to the reduction in DMTS performance.

If memory decreases as a function of the delay interval, then the memory loss must be due to certain factors. One very important contributor to forgetting in the DMTS paradigm is interference. Retroactive interference occurs when some event, occurring between the original sample and the choice stimuli, interferes with the memory of the sample, causing inferior retention. The most important source of retrograde interference is the illumination cues during the delay interval. D'Amato and O'Neill (1971) were the first to observe that DMTS performance improved if the lights in the apparatus were turned off during the delay interval. In fact, performance was better if the lights were off all the time except during the intertrial interval. The presence of lights in the apparatus apparently has some unique interfering effects. Moreover, interference is observed for both monkeys and pigeons, even when the source of light is merely the illuminated response key or disk (Grant & Roberts, 1976). Other sources of interference, specifically auditory stimuli like a hissing noise or monkey vocalizations, do not have the same interfering effect (Worsham & D'Amato, 1973). In a recent analysis of this phenomenon, Roberts and Grant (1978a) found that interference occurred even when the light presentation was extremely brief. As shown in Figure 8–4, the longer the interfering light, the greater the memory decrement. Also, interference was greater if the light occurred toward the end of the delay interval rather than at the beginning. These findings confirm that memory is decreased by the presentation of an interfering light, but the exact reason, or mechanism, for the interference is still unclear. It is possible that the presence of the light causes competition with the visual memory; alternatively, the afterimage or trace might be better preserved in the darkness (but see Cook, 1980).

In one study by Maki, Moe, and Bierley (1977), illumination caused interference for the memory of the sample *as well as* for the memory of a response. In this experiment, the pigeon had to peck at the key either 1 or 20 times; the correct match later depended upon which response requirement had been in effect. If the requirement had been 1 peck, then one stimulus was appropriate. However, if the requirement was 20 pecks, then a different stimulus was appropriate. The authors found that illumination caused interference with the response memory as well as the stimulus memory. To be specific, turning on the lights caused the birds to forget whether they had just pecked 1 or 20 times. Maki et al. argued that the illumination had disrupted the rehearsal of the response memory. This theory, perhaps, is the most plausible of all. For whatever reason (for example, fewer distracting stimuli), darkness facilitates rehearsal of the sample or of the response, whereas illumination disrupts the rehearsal.

It is interesting to note that a different kind of interference has also been investigated. Several researchers have shown that events occurring *prior* to

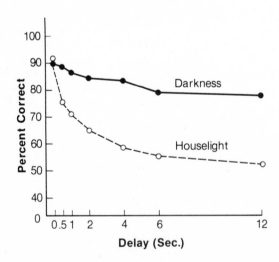

Figure 8-4.   Mean percent correct performance on the DMTS task as a function of the delay between the sample and choice stimuli for groups that spent the delay in darkness or with the houselight on. (From Roberts & Grant, 1978a)

the sample also may interfere with the memory of the sample (see Grant, 1975; Zentall & Hogan, 1977). In these experiments, two stimuli were used. The first was the interfering stimulus and the second was the official sample. After a delay interval, it was observed that memory for the sample was reduced because of the presence of the competing stimulus prior to it.

Theories of DMTS

A good deal of effort has been expended in explaining DMTS performance. One of the simplest theories is the trace decay notion. According to this theory, a stimulus establishes a neural trace that slowly decays. Progressively poorer retention reflects the lowered intensity of the neural trace when the comparison stimulus is chosen. This theory, clearly, is not a very good explanation for short-term memory in this context. The reason is primarily that longer sample durations improve DMTS performance. If it were merely a decaying trace that were responsible for forgetting, then the duration of exposure to the sample should make no difference; the decay process should start at the stimulus offset regardless of how long the subject had been exposed to the sample.

A more viable theory was proposed by Roberts and Grant (1976). Their simple modification of the trace decay theory is as follows: the memory trace increases as a function of the sample duration; the longer the exposure, the greater the strength of the memory trace. Second, the strength of the trace decreases during the retention interval, that is, as a function of the delay. Third, the probability of a correct match depends upon the interaction of these two tendencies, that is, on the *momentary* trace strength. In other words, the longer the duration of the sample, the greater strength the memory trace attains; the shorter the delay or retention interval, the less strength is lost. This theory is simple yet elegant, and it accounts for much of the data (see Reynolds & Medin, 1979), especially the finding that DMTS performance improves with longer duration samples.

There is a third theory about DMTS performance that also seems to have merit (D'Amato, 1973; D'Amato & Cox, 1976). According to this temporal discrimination hypothesis, declining performance as a function of the retention interval does not reflect a deterioration of the memory trace. Rather, it is claimed to reflect some sort of confusion. To be specific, animals are exposed to a limited number of stimuli on these tests. The problem is not really in remembering the characteristics of the stimulus (so that it can be matched), but rather trying to recall, on any given trial, *which* of the stimuli in fact had just been presented. In one sense, the animals become confused about which stimulus was most recently used as the sample.

Although this theory is, perhaps, deficient in some respects, it does account for some important facts. First of all, animals can improve their

DMTS performance with sufficient training. D'Amato (1973) showed that one of his monkeys was able to tolerate only 9 seconds delay between the sample and the choice when first trained. However, after several years of training, and many thousands of trials, this monkey was able to match the sample correctly after a delay of 9 minutes. According to the notion of trace decay, such a dramatic improvement in performance could not have taken place because the neural memory trace should decay at approximately the same rate regardless of the experience of the animal. These results, therefore, suggest that the animal improves in its ability to discriminate which stimulus was used most recently as the sample.

A second piece of evidence that supports the temporal discrimination hypothesis is the finding that the use of many, rather than just a few, stimuli as samples, improves DMTS performance (Worsham, 1975). With a greater variety of stimuli, animals find it less difficult to discriminate which one was presented most recently; when only a few stimuli are used, however, confusion increases.

In conclusion, the precise mechanisms for short-term memory using the DMTS task are still unspecified, although it appears that trace intensity as well as temporal discrimination factors are important. This field of research shows some remarkable parallels to the research on human short-term memories and, therefore, it has helped to enrich our understanding of memory mechanisms in general.

Intermediate-Term Memory

As the research discussed above indicates, short-term memory refers to memories that last only for seconds or minutes. Long-term memory, in contrast, as the name implies, refers to memories that last for days or even years. Here, we would like to discuss animal research pertaining to inter-mediate-term memories, that is, memories that last for hours. This category is not usually recognized as being distinct since intermediate-term memories normally are thought of as being long-term in nature. But the label is a convenient one because there is a significant block of literature dealing with an important problem in memory that is neither unequivocally short-term or long-term in nature.

Research in this area began with a study by Kamin (1957c). He was interested in the retention of a partially learned avoidance reaction after an intermediate-term retention interval. He trained rats to perform an avoid-ance response (learning was only partially completed in the 25 trials he administered) and then placed the animals back in their home cages for a retention interval. Specifically, the animals waited either 0, .5, 1, 6, 24, or 456 hours (19 days). After the retention interval, the animals were returned to the shuttlebox and given 25 relearning trials. The results were dramatic: the number of avoidance trials during relearning, that is, the efficiency of the animals' performance, declined from 0 to 1 hour and then increased again to 24 hours. In other words, those animals that were given relearning immediately after the original training and those that were given relearning at least 24 hours later performed very well on the relearning task. In contrast, animals that were given the relearning task after an intermediate-term retention interval (in this study it was 1 hour) relearned very poorly. Since Kamin's experiment, many studies have replicated this phenomenon (see Brush, 1971; Spear, 1971).

One of the earliest theories concerning this relearning decrement suggested that fear of the apparatus, acquired during the original learning task, spontaneously increased over time. That is to say, once the animals were put back in their home cage, the general fear that they experienced during training gradually increased. The increase in fear, in turn, caused greater and greater disruption of the behavior on the relearning test. If the retention interval was long enough, however, the fear had a chance to dissipate; no disruption was observed after 24 hours. In support of this theory, it has been shown that the decrement in relearning is eliminated if the animal is allowed to spend its retention interval in the training apparatus (Bintz, 1972). Presumably, this detention permits the apparatus fear to extinguish; once extinguished, the fear does not cause competing behaviors and disruption of the relearning process.

Other versions of a fear hypothesis were offered. For example, some experiments suggested that fear, rather than increasing over the retention interval, was actually decreasing (Bintz, Braud, & Brown, 1970). In this study, an avoidance reaction was conditioned to a location in the apparatus by shocking the subjects there. The authors found that the avoidance CR decreased after a 3-hour retention interval, suggesting that fear was decreasing rather than increasing.

All this research focused on changes in the motivational state of the animal following avoidance learning. Other investigators, however, developed an alternative theory (see Anisman, 1975). This general theoretical position was based on the idea of motor response competition. Basically, the hypothesis claimed that hormones, resulting from stress, change the reactions of the animal and, thereby, cause competition with the relearning performance. In particular, these time-dependent hormonal changes induced the animal to freeze; freezing, of course, is a behavior that would compete with active avoidance relearning. After a few hours, the hormonal changes subside, and the tendency to freeze is reduced, providing for more efficient relearning at the long-retention intervals.

There has been some support for this position. First, it has been shown that a decrease in activity does follow shock stress (Pinel, Corcoran, & Malsbury, 1971). That is, foot shock induces progressively less activity up to about 8 hours. In addition, the reactivity to future shocks, the sensitivity to shock after already having received shock stress, first increases then decreases as a function of time (Pinel & Mucha, 1973). Conceivably, a combination of changes in freezing or activity, and alterations in the animals' reactivity or sensitivity to shock, could account for the phenomenon.

These ideas provoked several investigators to test the motor competition theory. In one study (Anisman & Waller, 1971a), animals were given shock stress (inescapable shock) followed by a retention interval of 5 minutes, 4 hours, or 24 hours, and then avoidance learning in Phase 3. The learning task that these authors used was a one-way avoidance response (animals always ran from one box to a discriminably different box). The authors found that performance was poor in the animals that had received the 4-hour retention interval. Furthermore, the decrement in performance did not depend on whether the animals were running to or away from the box previously used for shock stress. In other words, the original shock stress induced freezing behaviors; these competed with the avoidance learning.

Other experiments have also supported the motor competition theory of this avoidance decrement (Barrett, Leith, & Ray, 1971; Steranka & Barrett, 1973; but see Bryan & Spear, 1976). In the Steranka and Barrett experiment, animals were trained to choose the bright side of a Y-maze; if they did not run to the correct side, then they were shocked. After learning and then a retention interval (short, intermediate, and long), the animals were given relearning trials in Phase 3 of the study. Some of them, however, had to learn the opposite habit during these relearning trials; that is, they had to run to the dim side of the Y-maze. In general, these authors found a performance decrement following an intermediate retention interval when they considered the speed of the reactions. However, they did *not* find such a performance decrement when they considered the choice data. Although the animals performed more slowly after the intermediate retention intervals (suggesting that they were freezing or crouching—behaviors that compete with the active avoidance relearning), they had no difficulty in learning to choose the correct alley. If the animals had actually had a memory loss during the intermediate retention interval, then they would also have shown a decrement in their choice behavior.

The other major alternative theory of the avoidance relearning decrement phenomenon centers upon memory and retrieval processes. Recall our discussion about the stimulus complex and its importance in retrieval. During original learning, the organism responds in the presence of a particular stimulus complex; during the retention interval, this complex changes somewhat because of the change in the internal hormonal state of the organism. Then, during the retention test, the performance declines because the stimulus complex is unable to elicit the proper response. But when tested after 0 or 24 hours, animals have no difficulty retrieving the correct response; the stimulus complex is either the same as in training (after 0 hours) or has been restored to a state very similar to original learning (after 24 hours) even though a change has taken place in between those times.

The evidence in favor of the memory-retrieval theory is substantial. First of all, consider the transfer studies (Bintz, 1970; Klein & Spear, 1970a). Klein and Spear trained animals to avoid shock by running from a light-colored compartment to a black compartment. The subjects learned this original avoidance task to the point where they made five consecutive avoidance responses. Different retention intervals were then used (5 minutes, 1 hour, 4 hours, or 24 hours) followed by a new task, namely, the acquisition of a passive avoidance response. Here, the animals were shocked in the black box and then were permitted to remain passively on the white side, thereby avoiding the black side. If the animals remembered their original *active* avoidance reaction, then they should have great difficulty in learning this new *passive* avoidance reaction; on the other hand, if the animals forgot (couldn't retrieve) their previous active avoidance learning, then they should easily learn this new task, namely, the passive avoidance response. As shown in Figure 8–5, naive subjects learned the new passive avoidance response quite easily. Animals that were tested after a 1- or 4-hour retention interval also learned the new response easily. For those subjects there was no competition; they learned it as if they had forgotten their original active avoidance learning response entirely. However, the animals that were given passive avoidance training 5 minutes or 24 hours after original learning had great difficulty in learning this new response; apparently, they remembered their original response quite well and therefore suffered great competition in learning this new response. In summary, this experiment provides powerful evidence for the memory theory by showing that animals, given an intermediate-term retention interval, learned a new response as easily as naive subjects; if they had remembered the first active avoidance task, then they should have experienced interference with the learning of this new task as the 0- and 24-hour groups did.

It should be noted that Klein and Spear (1970a) also did the reverse experiment: animals were first given passive avoidance training, followed by a retention interval, and then active avoidance learning. The same reasoning applied. If animals forgot their original passive avoidance learning, then active avoidance learning should be easy. The animals that received the intermediate-term retention interval indeed learned the new active avoidance response as easily as naive animals. On the other hand, if the animals remembered their original passive response, then they should learn the new active response with great difficulty; this was true for the 0- and 24-hour groups.

What could be the mechanism for this retrieval failure after 1 to 4 hours? The answer is related to the work discussed previously. It is believed that stress induces hormonal reactions; these are actually internal cues that

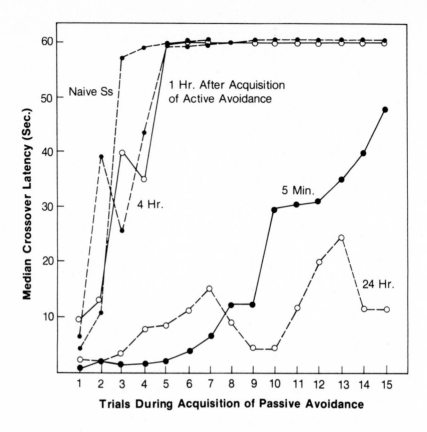

Figure 8-5. Median latency to cross back into the black side as a function of passive avoidance training for groups that differed in terms of the retention interval between their active avoidance and passive avoidance. (From Klein & Spear, 1970a)

change the stimulus complex. It is known that the hormonal changes following stress parallel the relearning deficits shown in the above studies (Brush & Levine, 1966). And if after an intermediate-term retention interval animals are given a drug supplement to overcome the changes that took place during that interval as a consequence of shock stress, then no deficit in relearning is observed (Klein, 1972; Singh, Sakellaris, & Brush, 1971). In other words, the avoidance relearning decrement can be eliminated by the administration of drugs such as adrenocorticotrophic hormone; these drugs are designed to correct the hormonal "imbalance" created by the shock stress. By correcting the imbalance, the original stimulus complex is restored and no retrieval problems (relearning decrement) are observed.

Not only does the administration of the appropriate drugs eliminate the problem, but the stimulus complex can be restored by giving shock stress again just prior to the relearning test (Klein & Spear, 1970b).

In summary, animals have trouble relearning an avoidance reaction if they are not in the same "state" (Spear, Klein, & Riley, 1971). Shock stress seems to induce biochemical stimuli that are time-dependent; these are important internal cues to which the animal is sensitive. These biochemical changes, in turn, alter the stimulus complex, which causes problems for retrieval of the original memory. If the biochemical state is altered but the original stimulus complex is restored through drug administration, then no decline in relearning is observed.

Long-Term Memory

Comparatively little is known about the long-term memories of subhuman animals. There are reports of pigeons remembering discriminative stimuli years after training, and, of course, anecdotes concerning the keen memory of elephants are legendary. But memories do not have to be retained for years in order to qualify as long-term. Most literature on human memory considers retention to be long-term when material is retained for as long as 24 hours, although there is nothing special about that time period.

There is one area of animal memory research that bears upon long-term retention. This area deals with the permanence of memories formed at a young age. Conventional wisdom would assert that traumas experienced at an early age have a disproportionate effect on adult behavior. According to this thinking, many neuroses that become apparent during adulthood can be traced back to childhood traumas.

This theory concerning the memory for early traumas has not been confirmed in the animal research literature. Despite the appealing nature of such a theory, empirical data deny that this is how memories operate. Rather, it appears from a large number of studies that early memories are, in fact, forgotten more easily than memories formed during adulthood (see Campbell, 1967; Campbell & Coulter, 1976; Campbell & Spear, 1972). Although most of the research investigates memories for aversive events, for example, responses or stimuli that were associated with painful trauma like shock, even memories for appetitive tasks can be lost if they were learned during an early age (Campbell, Jaynes, & Misanin, 1968).

Age-Retention Phenomenon

Some of the most convincing evidence indicating that traumas learned early in life are forgotten more easily than traumas learned later comes from studies on the retention of fear as a function of age (Campbell & Campbell,

1962; Coulter, Collier, & Campbell, 1976). In the experiment by Coulter et al., rats were placed in a box and given four tone-shock pairings. Control subjects received the same treatment except that the tone and shock were presented randomly, thus eliminating a conditioned reaction to the CS (see Chapter 3). The fear training took place at various ages. Some animals were given training when they were 11 to 13 days of age, others experienced fear training at 14 to 16, 17 to 19, or 20 to 22 days of age. (It should be noted that rats are weaned at the age of 21 days and reach sexual maturity by about 40 to 50 days of age.) The animals were then returned to their home cage and allowed to mature for about a month. At this point, all animals were taught to press a lever in order to get food. Once this was established, they were given a fear test, specifically a CER test using the tone CS. This test consisted of the presentation of the CS while they were lever pressing. Fear was assessed through the usual CER suppression ratio (see Chapter 2). The greater the fear, the greater the suppression (disruption of lever pressing).

As shown in Figure 8–6, all the subjects that received unpaired CS and US presentations early in life did not show any fear to the tone CS on the CER test; their suppression ratios were approximately .5 on all the trials. However, various degrees of fear were shown by the experimental subjects, and the amount of fear changed as a function of the age at training. Animals that experienced fear training at a very early age (11 to 13 days) failed to show any suppression at all. Apparently, they forgot completely. Fear retention increased, however, in the older groups. The best retention was shown by the oldest subjects (20 to 22 days at original training); suppression was nearly complete on the first CER trial and never recovered even after four presentations. In summary, this study by Coulter et al. clearly indicates that aversively motivated memories are forgotten if they were learned at a young age.

This age-retention phenomenon is by no means limited to Pavlovian fear training. Several investigators, for example, have shown that escape (Smith, 1968) or avoidance reactions are also forgotten if they were learned early in life. And it made no difference whether the original avoidance reaction was an active running response in an alleyway (Kirby, 1963), in a small shuttlebox (Feigley & Spear, 1970), or whether the original learning was a passive response (Schulenburg, Riccio, & Stikes, 1971). In all cases, animals trained at a young age showed poorer retention than those trained as adults.

In summary, research supports the conclusion that memory capacity increases with age. Although young animals are able to learn various responses perfectly well, they seem to have trouble remembering them. The development of memory, moreover, is gradual. There is evidence that mice can remember their responses for a few hours when they are 5 to 7 days of age (Nagy, Misanin, Newman, Olsen, & Hinderliter, 1972), and that they

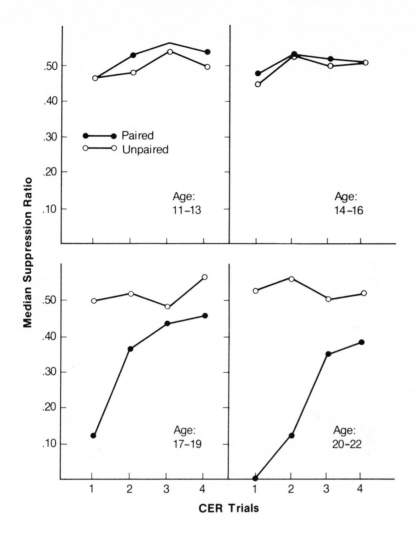

Figure 8-6. Median suppression CER ratios as a function of trials for groups that received fear training or random CS-shock presentations at ages 11–13, 14–16, 17–19, and 20–22 days. (From Coulter, Collier, & Campbell, 1976)

are capable of retaining material for 24 hours by 9 days of age (Misanin, Nagy, & Keiser, 1971; Nagy & Mueller, 1973). But it is not until sometime later, around weaning at the age of 21 days, that the ability to remember over a longer period of time develops.

There are two ways to theorize about the age-retention phenomenon. First, one can note that there are numerous biological conditions that are developing early in life. These could account for the improvement in memory as a function of age. For example, myelinization (fatty insulation around the nerves) in the brain develops early in life. Similarly, various brain cells are becoming differentiated, neural transmitters are developing, and RNA (a biochemical often associated with memory) is developing in the immature animal. Any one or all of these conditions could contribute to the neural mechanisms that permit memory to occur.

The second type of theory for this phenomenon focuses on memory processes directly. It is believed (Spear, 1973) that the age-retention phenomenon, in particular, reflects a memory retrieval failure. This kind of theory is similar to that used to explain the forgetting after ECS and at the intermediate retention intervals following avoidance conditioning. The reasoning is the following: although animals learn at an early age (the appropriate stimulus complex is formed at that time), the changes that are endured during maturation cause a dramatic and substantial change in the stimulus complex. Thus, when the animals are tested as adults for their infantile memory, forgetting is observed because the stimulus complex that was appropriate to the infantile memory has changed. Since the stimulus complex is so radically different after maturation, the memory cannot be retrieved effectively.

There are numerous experiments that support such a retrieval theory. For example, Parsons and Spear (1972) used environmental enrichment as a source of interference. They trained young and old rats to make an active avoidance response. The control animals were detained in their home cage during the retention interval, while experimental subjects were placed in a special cage containing a platform with a ladder, a running wheel, other rats, food dish, and so forth. After 60 days, all animals were tested for their retention of the avoidance reaction. The authors found that the young animals did, indeed, forget their avoidance, and more importantly, that the enriched living conditions enhanced this forgetting. This latter result was found even in adult rats. In other words, enrichment provided a source of interference for the old memory. The maturational changes that took place in the enriched rat, presumably, were greater than those in the control animals.

There are, of course, other ways to change the stimulus complex and thereby cause a retrieval failure. One way is to fail to present all of the relevant cues on the retention test. When this is done, forgetting is increased (McAllister & McAllister, 1968). Unfortunately, there is no good way to go in the other direction, that is, to restore the stimulus complex that was experienced during an early age. The reason is that those changes which are said to have taken place during maturation can never be reversed in the adult animal; quite simply, we cannot convert an adult animal back into a young subject. Nevertheless, sufficient information on retrieval theory in general and on the age-retention effect in particular confirmed that memory retrieval failure is the cause for this phenomenon.

Reinstatement

Let us return to our original paradox. It does appear that some memories from childhood *are* disproportionately important during adulthood. Some memories do, indeed, persist. But how do we remember these early experiences if, as the research discussed above suggests, such memories are uniformly forgotten? One answer was suggested in a paper by Campbell and Jaynes (1966). According to those authors, memories normally are forgotten, but if the organism is periodically "reminded," then the stimulus complex will be "updated" and no forgetting will occur. This procedure is called reinstatement. It involves placing the animal back into the apparatus and giving it a reminder trial every so often. The reminder itself is not sufficient to teach the animal the original response. Rather, it is merely a periodic reminder to the animal concerning its original learning. When young organisms are given reinstatement (periodic rehearsal), they do not suffer the amnesia normally observed. Reinstatement has been shown to occur for appetitive discrimination behaviors (Campbell & Randall, 1976) or when animals are merely exposed to the fear cues that comprised original training (Silvestri, Rohrbaugh, & Riccio, 1970).

The age-retention effect is important not only because it illustrates an interesting memory phenomenon, but also because it supports the notion that retrieval failure accounts for forgetting: amnesia results from ECS, from changes in the stimulus complex that are based on stress-induced hormonal reactions, or from changes in the stimulus complex resulting from longer-term maturational changes. In each of these cases, some agent is responsible for inducing an internal change in the subject. In all these areas of research, various techniques, such as reminder trials, which are designed to restore or preserve the original stimulus complex, all improve memory by eliminating the retrieval failure. Similarly, procedures that produce a change in the animal, or conditions that facilitate those changes that occur naturally, cause even greater forgetting.

Summary

Memory is a topic that is closely linked to learning. In fact, some of the processes that control memory are the same as those that control learning. In particular, the retrieval of a response from memory is believed to involve the elicitation of the behavior by the stimulus complex present during acquisition. Changes in the complex result in forgetting (retrieval failure) just as they result in faster extinction. The literature on electroconvulsive shock and memory highlights this notion. The shock treatment normally causes amnesia for events that occurred just prior to its administration, although if the animal is "reminded" of the previous learning, no memory deficit is later observed.

Short-term memory has been studied in animals using the delayed-matching-to-sample technique. In this procedure, a sample stimulus is presented and, after a delay, two comparison stimuli are given. The subject must choose the comparison that matches the sample in order to get food reward. Success at this DMTS task presumably reflects memory for the sample over the delay interval. Memory improves as a function of sample duration but deteriorates when lights are turned on during the delay. The ability to perform the DMTS task efficiently may be based on the decay of a neural memory trace or upon remembering which sample, from a very limited number of possible stimuli, was actually just given.

There is a significant block of research on intermediate-term memories that last for 1 to 4 hours. If an animal is trained to make an avoidance reaction, good memory (relearning) is displayed either 0 or 24 hours later, but poor performance occurs from 1 to 4 hours afterward. The reason appears to be that stress initiates time-dependent hormonal changes. These internal cues change the stimulus complex, causing a decrease in the elicitation of the behavior.

One main topic considered in the research on long-term memory is infantile amnesia: the finding that young organisms can learn well but cannot remember their learned behaviors. Again, the source of forgetting appears to be the changes in the stimulus complex that occur during maturation. If procedures are taken to restore the complex, that is, if reminder trials are given, then memory retrieval is improved.

9

Generalization
and Discrimination

Introduction

Generalization Gradients

Factors Affecting Generalization

Factors Affecting Discrimination

Theories of Discrimination

Discrimination Phenomena

Summary

Introduction

Up to this point we have emphasized the idea that CSs can become powerful through Pavlovian training; they can come to elicit expectations for events that follow, like USs. These expectations, coupled with the expectations based upon responses, provoke or control responding. In this sense, we can say that all behavior is under stimulus control; that is, responding occurs in the proper context, not in the presence of inappropriate stimuli. Responding, however, is not restricted to the precise CS used in training. Other stimuli, similar to the original CS, that were not used in the original training task, also can elicit CRs. This phenomenon is called stimulus generalization. Two points are fundamental. First, the degree to which a novel stimulus will elicit the CR is related to the similarity between this novel stimulus and the original CS. The more similar the two stimuli are, the greater the conditioned reaction on the generalization test. Second, generalization may take place along virtually any dimension of the stimulus. For example, if the original CS were a tone, animals might react to novel tones that were similar in terms of their pitch, intensity, or any other characteristic.

The fact that animals respond to stimuli other than the original CS indicates that certain stimuli, defined by the dimensions over which generalization occurs, control behavior. Consider the example of a tone CS. The tone can be characterized in terms of its pitch and intensity. By presenting various tones that differ in terms of both pitch and intensity, one can discover which stimulus dimension actually controls behavior. If responding is observed to novel tones that are similar in pitch to the original CS, then it is the frequency of the tone that controls behavior. Conversely, if intensity were more important, novel tones that were similar to the original CS in loudness but not in pitch would elicit the CR.

It is important to note that generalization allows animals to behave adaptively in their environment. It would be a calamity if they were constrained to respond to only those stimuli used during training. In the natural environment, for example, it would be impossible to identify new sources of food by exploring those locations that were similar to the ones that previously contained food if generalization could not take place. That is, the animal would have to learn from the start how to identify each food source even though all of them were similar along some dimension. A species could not possibly survive for very long if it had to learn the significance of each and every stimulus separately. Given that generalization from one stimulus to another can take place, the animal can make appropriate and adaptive responses to novel stimuli and, thereby, take advantage of past learning experiences.

Discrimination is the opposite of generalization. Whereas generalization is responding in the same way to two different stimuli, discrimination is responding differently to two stimuli. This means that an organism performs a CR to one stimulus, the S+, because it is followed by a reward, but not to another stimulus, the S−, because that stimulus is followed by no reward, less reward, or punishment (see Gilbert & Sutherland, 1969). The ability to react differently to stimuli, of course, is also a highly adaptive capacity. Without it, organisms would be unable to solve problems, to differentiate between stimuli and make the necessary response. Discrimination is the perception of differences between stimuli which permits differential responding. Because discrimination and generalization are "companion" processes—mirror images—discrimination too is a measure of stimulus control. That is, by observing generalization, we can understand the dimension along which animals equate stimuli; correspondingly, by observing discrimination behavior, we can assess the basis on which animals distinguish between stimuli. Although the two processes are highly complementary, we shall deal with each of them separately in this chapter.

Generalization Gradients

One of the most striking aspects of generalization is the orderly relationship between the strength of the CR to the generalized stimuli and the similarity of those stimuli to the original CS. This graded relationship is called the generalization gradient: the generalized CR becomes weaker as the training and test stimuli become more dissimilar (see Kalish, 1969;

Mednick & Freedman, 1960; Mostofsky, 1965; Prokasy & Hall, 1963). Let us illustrate this phenomenon by discussing a classic experiment on generalization by Guttman and Kalish (1956). They trained pigeons to peck a small plastic disk located on the side of the cage to gain access to food reward. A light was placed behind the disk so that, with the use of filters, the color of the light could be controlled accurately. Four groups of animals were used. The only difference in the treatment was the wavelength of the CS (color of the disk). For some animals, the CS wavelength was 530 nanometers; for the other three groups, the color of the CS was 550, 580, or 600 nanometers. Acquisition consisted of training the animals to peck at the disk when it was illuminated with the CS color. The generalization test took place later. Here, the original CS plus 10 other stimuli that were both higher and lower on the color spectrum were presented for 30 seconds. Each set of 11 stimuli was then repeated 12 different times.

The results of the study are shown in Figure 9–1. Mean total number of responses to each stimulus is plotted separately for the four groups. Maximum responding in each group of subjects was, of course, to the CS used originally in training. However, it is clear that other novel stimuli also elicited the pecking response, and, more importantly, that the magnitude of responding was a positive function of the similarity between the training and test stimuli. Although Guttman and Kalish obtained symmetrical generalization gradients (the rate of decline on either side of the original CS was approximately equal), such symmetry is not always observed.

The generalization gradients drawn in Figure 9–1 show the "spread" of excitatory conditioning to stimuli that are similar to the original CS. But, as discussed in Chapters 2 and 3, inhibitory strength can also develop to stimuli. Accordingly, several studies have shown that conditioned inhibition also generalizes (see Hearst, Besley, & Farthing, 1970; Jenkins & Harrison, 1962; Weisman & Palmer, 1969). Recall that a conditioned inhibitor is a CS that is paired explicitly with no US; its strength is assessed usually by the summation technique: the presumed CS− is combined with a known excitatory CS+; if excitation is reduced, then the CS− is said to be a conditioned inhibitor (see Chapter 3). This summation technique is used on generalization tests too. The original CS− subtracts from the excitatory tendency of a CS+ more than other stimuli; but as one measures the inhibitory strength of stimuli similar to the CS−, one observes generalized conditioned inhibition. In short, conditioned inhibition generalizes just like conditioned excitation; an inhibitory generalization gradient, which is the mirror image of an excitatory gradient, shows the same orderly relationship between strength of conditioned inhibition and similarity of the CS− to the novel test stimuli.

The Weisman and Palmer study provides a good example of generalization of conditioned inhibition. Pigeons were given discrimination train-

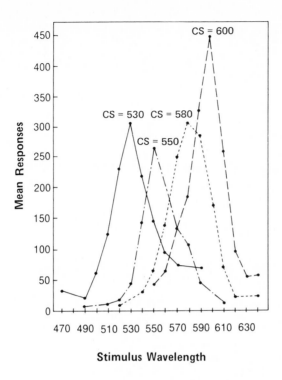

Figure 9-1. Mean responding as a function of the stimulus wavelength on a generalization test for four groups for which the original S+ wavelength was 530, 550, 580, or 600 nanometers. (From Guttman & Kalish, 1956)

ing in which pecking a green disk was reinforced and pecking the green disk with a vertical white line projected onto it was not rewarded. The green color was, of course, the CS+, whereas the vertical white line was the CS− (because it uniquely predicted no reward). After sufficient training was provided, each subject was given a generalization test. This consisted of the presentation of the original vertical white line (the CS−) plus 6 other lines that departed from vertical by −90, −60, −30, +30, +60, and +90 degrees. If the vertical line elicited conditioned inhibition, then the other lines that varied in their slant should elicit generalized conditioned inhibition.

The individual inhibitory gradients for five of the pigeons are shown in Figure 9–2. Number of responses is shown as a function of the stimulus.

It is clear that the original CS−, the line that deviated 0 degrees from vertical, elicited the greatest conditioned inhibition; responding was most depressed during that stimulus. The other stimuli also elicited conditioned inhibition (responding was depressed relative to the level for the S+ alone shown at the right of each gradient), but, of course, the degree of suppression was reduced. That is, tilted lines caused less suppression; the greater the tilt, the less the suppression of responding.

Factors Affecting Generalization

In this section we will review the effect of several training variables on generalization. Bear in mind that a steep gradient reflects little generalization; even stimuli that are quite similar to the original CS evoke only a

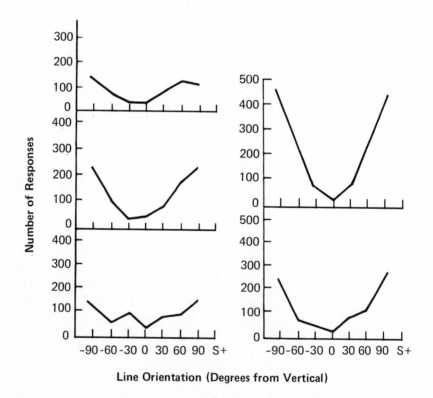

Line Orientation (Degrees from Vertical)

Figure 9-2. Number of responses given to the S− and six other generalized stimuli by five different pigeons. (From Weisman & Palmer, 1969)

marginal reaction. Flatter gradients, on the other hand, reflect more generalization to novel stimuli.

Degree of Acquisition

Several studies have shown that generalization is influenced by the amount of acquisition training to the CS. Specifically, the more extended the training, the less the generalization (the gradient becomes sharper). In one study, Hearst and Koresko (1968) trained pigeons to peck at a vertical line on a colored disk (S+). Groups differed in terms of the number of sessions of training: 2, 4, 7, or 14 50-minute sessions using a VI-1 minute schedule of reinforcement. After training, generalization was measured using 6 stimuli that differed from the S+ in terms of the slant of the line. The results are shown in Figure 9–3. The overall height of the generalization gradient increased as a function of training level. This is consistent with material discussed in Chapters 3 and 4: response strength increases with training level. In addition, Hearst and Koresko found that the gradients became steeper with more acquisition. Proportionately more of the total responses were given to the S+ with proportionately fewer given to the generalized stimuli. In other words, the *absolute* level of responding to generalized stimuli increased as a function of training, whereas the *relative* response rate to generalized stimuli decreased. Similar results have been found in other studies using different measures of generalization (see B. L. Brown, 1970; Friedman & Guttman, 1965; Kalish & Haber, 1963).

Motivation Level

It has been found that the motivational level of the organism affects its tendency to generalize. Kalish and Haber (1965), for example, trained pigeons that had been deprived to 90, 80, or 70 percent of their normal body weight to peck a plastic disk for food. During the generalization test the subjects were also maintained at that same weight level. The authors found that the slope of the generalization gradient increased as a function of deprivation. That is, the 90 percent group (low deprivation condition) displayed a relatively flat gradient, whereas the 70 percent group (high deprivation group) showed a much steeper gradient.

In a study by Newman and Grice (1965), acquisition was conducted while all subjects were deprived of food for 24 hours, but generalization was measured while the subjects were either 12- or 48-hours food-deprived. Those authors observed that the overall response level during generalization, as well as the steepness of the gradient, was higher for the hungrier subjects. In short, it appears that the greater the drive level, the less likely the subject is to generalize to novel stimuli (but see Coate, 1964; Sidman, 1961).

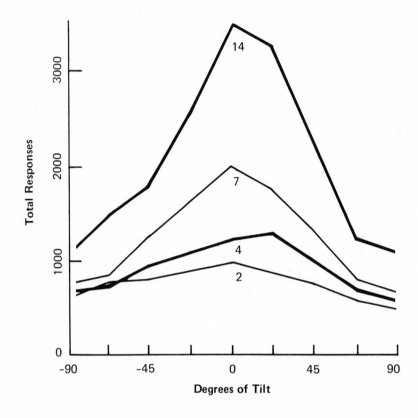

Figure 9-3. Mean total responses as a function of the tilt of the line stimulus for groups that received 2, 4, 7, or 14 days of training. (From Hearst & Koresko, 1968)

Training-Test Interval

The tendency to generalize is not static; it increases with time (the gradient becomes flatter). This finding has been observed using appetitive tasks (Burr & Thomas, 1972; Perkins & Weyant, 1958; Thomas & Lopez, 1962) as well as aversive tasks such as avoidance (Desiderato, Butler, & Meyer, 1966; McAllister & McAllister, 1963).

Consider the experiment by Thomas and Lopez. Pigeons were first trained to peck at a plastic disk. The S+ was illumination of the disk by colored light (550 nanometers). Generalization testing then was administered either immediately, 1 day, or 1 week after training. Although all the animals received the same amount of training and the same generalization test, the degree of generalization to the 11 stimuli varied. Those tested immediately showed a rather steep gradient, whereas generalization 1 day or 1 week later was much greater. The gradient was flatter. The relative number of responses to generalized stimuli, even those quite discrepant in wavelength from the original S+, increased with the passage of time.

We should point out that this finding is extremely similar to those reported in Chapter 8 on memory. Indeed, the predominant theory for this result (that generalization increases with time) is that the animals forget the exact features of the S+, although they do remember the general task of pecking the disk. Another way to express that idea is to say that the stimulus complex changes over time, causing a failure in memory retrieval.

Early Sensory Experience

An important problem in research on generalization has been to discover the effect of restricting sensory experience right after birth. The basic question is the following: is experience with various stimulus dimensions important for generalization, or do animals discriminate between stimuli even if they have never experienced the dimensions by which they differentiate them?

Several studies have found that early experience with sensory stimuli is critical (Ganz & Riesen, 1962; Peterson, 1962), but more recent research suggests that animals can discriminate stimuli even without experience with those stimuli (Mountjoy & Malott, 1968; Riley & Leuin, 1971; Rudolph, Honig, & Gerry, 1969; Tracy, 1970).

In Riley and Leuin's study, chicks were raised in an environment which was illuminated by a light of one wavelength only (589 nanometers). All hues other than 589 were filtered out. At 10 days of age, all the animals were trained to peck a plastic disk that was illuminated by this same color (589). After seven days of this training they were given a test for generalization. Here, the birds were presented with the S+ (the original 589 nanometer light) plus two similar colors (569 and 550 nanometers). The authors found that these new generalized stimuli elicited the pecking response, but not as much as the original S+. That is, a normal generalization gradient was observed. This suggested that some animals, birds at least, have an innate ability to distinguish between colors; they do not have to experience those colors early in life in order to be able to discriminate them later on. If the ability to discriminate colors depended upon experiencing them, then those subjects whose sensory experience had been severely restricted should not show any differential responding; they should show complete generalization by responding to all wavelengths equally.

It is plausible to assume that sensory discrimination in many animals is innate. However, early experience certainly does influence animals' generalization, at least to some degree. Kerr, Ostapoff, and Rubel (1979), for instance, investigated the influence of acoustic experience on frequency generalization gradients. They used a special technique to assess generalization in 1- to 4-day-old chicks. The birds were presented with an 800-Hz tone immediately after they emitted a distress call; usually this made them stop calling and orient to the stimulus. After the tone went off, however, vocalizations were resumed. In fact, the authors found that, over training, the speed to resume distress calling increased progressively. The animals were then given a generalization test using tones of frequencies ranging from 800 Hz (the original CS), to 825, 850, 900, and 1000 Hz. Here, the authors also measured the latency, or speed, of emitting the calls following the offset of the tone. As shown in Figure 9–4, animals that were 1 day old at training showed a very flat gradient of generalization; they resumed their distress calling rather quickly after the offset of almost all the tones. In contrast, animals that were 3 to 4 days old during initial training had a much steeper gradient. They began vocalizing quickly once the original stimulus was terminated but took much longer to resume distress calling when tones of different frequencies were used.

In another portion of the Kerr et al. study, the animals were temporarily deprived of much acoustic input by means of earplugs. After spending the first few days of life in this condition, the ones that were 3 to 4 days of age behaved as the 1-day-old subjects did in the previous part of the

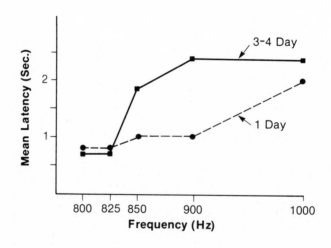

Figure 9-4.   Mean latency in seconds to resume distress calling as a function of tone frequency in Hz for groups habituated to the 800-Hz tone at 1 and 3–4 days of age. (From Kerr, Ostapoff, & Rubel, 1979)

experiment: they demonstrated a very flat generalization gradient. The authors concluded that experience with acoustic sensory input "sharpened" the animal's perceptual acuity. This perceptual sharpening occurs during normal maturation, but only when it is deliberately manipulated, as in this study, can its effects be clearly seen. In general, then, this study demonstrates that early sensory experience does, indeed, affect the perceptual development of the organism. Although some sensory systems may be essentially innate, others are more influenced by experience.

One of the reasons that this work is so important is that one theory of generalization claims that experience with the stimulus dimension is essential for discrimination (Lashley & Wade, 1946). According to this hypothesis, generalization is, essentially, confusion; animals cannot discern the difference between stimuli and so they respond to all of them. This theory also claims that it is only through experience with the stimuli that animals can come to discern the difference. Therefore, if an animal were deprived of experience with a particular stimulus, like color or pitch, then, presumably, it would generalize completely along those dimensions; it would have had no opportunity to learn about those stimulus dimensions and, therefore, would not be able to discriminate between stimuli. As discussed above, there is evidence both for and against such a theory. Some animals are able to discriminate along certain stimulus dimensions innately, but experience does play a role in other instances.

Prior Discrimination Training

Now let us consider the important effect that prior discrimination training has on generalization. The basic finding is that discrimination training given before the generalization test sharpens the generalization gradient; that is, novel stimuli produce less generalized responding if the subject has undergone discrimination training previously (Friedman & Guttman, 1965; Hanson, 1959, 1961; Jenkins & Harrison, 1960; Thomas, 1962). One of the clearest examples of this phenomenon was the study by Hanson (1961). Pigeons were trained to peck a disk for food that was delivered on a VI schedule. The S+ was the illumination of the disk by a colored light (550 nanometers) which signaled when reinforcement was available. After five days of acquisition, the control subjects continued with this same treatment, while experimental subjects were given discrimination training consisting of the presentation of the S+ (reinforced) as well as two S− cues (light of 540 and 560 nanometers). The subjects were reinforced for responding to the S+ but not for responding during either S− stimulus. In Phase 3 of the study, after reaching a criterion of discrimination, both groups of pigeons were given generalization tests. Specifically, the S+, both S− cues, and 8 other generalized stimuli were presented randomly. As shown in Figure 9–5, both groups of subjects showed a generalization gradient, but the gradient for the control subjects was relatively flat compared to the gradient for the experimental subjects. Discrimination training had produced not only a higher rate of responding to the S+, but also a much lower rate of responding to the generalized stimuli, that is, a steeper gradient.

This effect of discrimination training on generalization occurs not only when the discrimination training involves the same stimulus dimension as in Hanson's study, but also when the discrimination stimuli and the generalization dimensions are entirely different. Mackintosh and Honig (1970), for example, showed that discrimination training using tilted lines as the stimuli later influenced the generalization gradient for wavelength. Specifically, the gradient for wavelength was steeper for subjects that were given

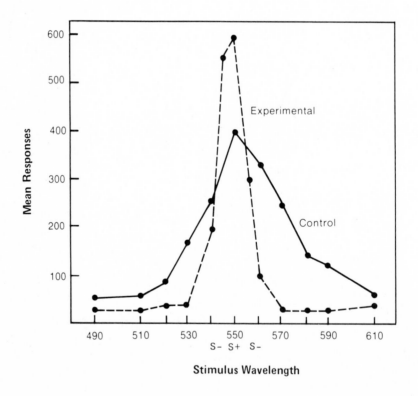

Stimulus Wavelength

Figure 9-5. Mean responding as a function of stimulus wavelength for the experimental and control groups. Both groups received training to the S+ (550 nanometers), but the experimental group also received discrimination training using two S− cues (540 and 560 nanometers). (From Hanson, 1961)

prior line-tilt discrimination training than for subjects that did not receive this prior training.

Inhibitory-Excitatory Interactions

As we have mentioned, discrimination involves the reinforcement for responding during one cue (the S+) but not during another cue (the S−). According to conventional theory, as outlined in Chapters 3 and 4, the S+ becomes an excitatory cue, whereas the S− becomes an inhibitory cue. Furthermore, as noted earlier in this chapter, both excitation and inhibition generalize to neighboring stimuli that lie along the same dimension. We therefore can interpret the findings on generalization after discrimination training in terms of interacting inhibitory and excitatory tendencies. That is, the gradient following discrimination training should result from the interaction of the inhibitory and excitatory generalization gradients.

The postdiscrimination gradient does, in fact, reflect the interaction of excitatory and inhibitory tendencies. Specifically, the inhibitory strength for the stimuli subtracts from the excitatory strength; the resulting behavior is the algebraic difference of the two competing tendencies. When this is computed one observes an interesting phenomenon called the peak shift. The peak shift is defined as a displacement of the peak of the excitatory gradient away from the original S+ in the direction opposite to that of the S− (see Purtle, 1973). Consider a study by Hanson (1959). Pigeons were trained to peck a disk during an S+ (a colored light with a wavelength of 550 nanometers). Then, different groups were given discrimination training for which the S+ continued to be the 550 light and the S− was either 555, 560, 570, or 590 nanometers in wavelength. A fifth control group did not receive discrimination training. In the third phase of this study, generalization tests were given which consisted of 13 different stimuli ranging from 480 to 600 nanometers.

The results of the generalization test are shown in Figure 9–6. First, consider the performance of the animals in the control group that did not receive any discrimination training. The peak of their responding was, predictably, at the original S+; and an orderly decrease was observed to occur to the various generalized stimuli. Now, consider the results for the groups that received discrimination training using various S− cues. In each case, the peak of the curve, that is, maximum responding, was not given to the original S+; rather, the peak was displaced away from the S+ in a direction opposite to the side of the S−. Furthermore, the amount of displacement was a function of the difference between the S+ and S− cues. The closer the S− was to the S+, the greater the peak shift. If their S− cues were located further away from the original S+, as in the group that received a light of 590 nanometers as the S−, the animals did not show as great a shift in the peak of their responding.

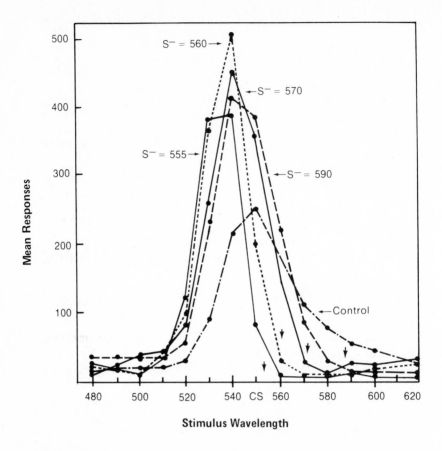

Figure 9-6. Mean responding as a function of the stimulus wavelenth for groups for which the S+ was 550 nanometers and the S− was either 555, 560, 570, or 590 nanometers. The control group did not receive any S− training. (From Hanson, 1959)

How can we account for this unusual phenomenon, that maximum responding during generalization is not to the original S+ cue? One theory is that the inhibitory gradient represents an emotional reaction. The S− becomes aversive during discrimination training because it is associated with no reward. That such emotional reactions occur is entirely consistent with our discussion about inhibitory conditioning and emotion (see Chapters 3 and 5). During generalization testing, then, when the aversive S− is presented, the subject tries to "avoid" that cue; this results in overshooting the S+. The closer the S− is to the S+, the greater the tendency to avoid the aversive S−, and thus, the greater the tendency to overshoot the S+.

This emotional theory of the peak shift has been supported in a variety of ways. Terrace (1963), for example, found that when the S— was introduced into the experiment very gradually, so that subjects made no errors during its presentation, the peak shift was not obtained. The interpretation of this experiment was that the S— did not become aversive because, in fact, it was never associated with no reward. If the S— was not aversive, then it should not produce the overshooting. This is what Terrace found. Other experimenters have supported the emotional theory of the peak shift too. For instance, Lyons, Klipec, and Steinsultz, (1973) gave a tranquilizing drug to their subjects during the generalization test. It was predicted that the tranquilizer (chlorpromazine) would reduce the emotional reaction to the S— and, therefore, would eliminate the peak shift. This prediction was confirmed.

Other investigators have preferred a different theory of the peak shift. For one thing, not all of the results discussed above have been replicated. Hoffman (1969), for instance, reported that the peak shift could, indeed, be obtained following the special "errorless" discrimination training previously used by Terrace. Perhaps more important was the finding that the postdiscrimination generalization gradients can be predicted from normal gradients of excitation and inhibition (Marsh, 1972). That is, the responding rate to any particular stimulus can be conceived of as an algebraic sum of the excitatory and inhibitory tendencies for that stimulus. If the excitatory and inhibitory gradients are measured separately and the algebraic sums are computed, then the resulting gradient shows a peak shift away from the S—. Finally, the peak shift quite simply may reflect the incompatibility between two response tendencies. Excitatory and inhibitory reactions, of course, are incompatible, and it may be the incompatibility that underlies the peak shift (Blough, 1973).

Factors Affecting Discrimination

As discussed previously, discrimination training involves the presentation of reinforcement following a response to one stimulus, but the withholding of reinforcement (or the presentation of punishment or a lower magnitude of reward) following a response to a different stimulus. The fact that the animal responds differently to the two stimuli indicates not only that it can perceive the difference (it does not generalize), but also that the stimulus and response expectancies combine. That is to say, the joint occurrence of a stimulus expectancy, elicited by the stimulus complex, and a response expectancy, based upon a response-outcome correlation determine the animal's discriminative reaction. In particular, in a discrimination experiment a response is given only when the animal has developed both a response and a stimulus expectancy; when this is the case, then the animal performs the response only in that context (that is, when the appropriate stimulus complex is present). In the following sections we will outline many of the variables that affect discrimination learning and discuss various theoretical treatments concerning discrimination.

Problem Difficulty

It is certainly not surprising to discover that discrimination learning occurs more rapidly when the two stimuli are easily distinguished from one another than when they are not. Animals that must distinguish between a black and white alleyway in order to receive a reward will learn faster than animals that must differentiate between two shades of gray. Presumably, the stimulus complex that comes to elicit the stimulus expectancy is more distinct when those stimuli in the complex which uniquely predict reward are salient.

Stimulus salience, or distinctiveness, however, is not the only factor. Experience with the stimuli, or related stimuli, can affect discrimination learning too. This effect was first shown by Lawrence (1952). Rats were trained to jump from a start box into one of two goal boxes. Their task was to choose, on the basis of the brightness, the rewarded box; jumping into the incorrect goal box resulted in no reward. Animals in one group were given 80 trials with a very difficult discrimination (each goal box was a different shade of gray). Subjects in a second group also received this same discrimination problem, but, in addition, they were first given 30 trials with goal boxes that were easily contrasted (one was black, the other white). In other words, the basic question addressed in this study was: what effect would mastering an "easy" problem have on later discrimination that used a "difficult" problem?

The results were unambiguous: the experience of solving the easy problem facilitated mastery of the difficult problem later on. In other words, prior training with easily distinguished stimuli facilitated later discrimination of hard-to-distinguish stimuli. This phenomenon, known as the easy-to-hard effect, has been found in many studies. The explanation favored by Lawrence, and more recently by Mackintosh and Little (1970), is based on the idea that animals learn to attend to the relevant stimulus dimension during the easy trials, and this facilitates their discrimination along this same dimension later, when the stimuli are more difficult to distinguish. Stated somewhat differently, the attention theory claims that an animal must learn about the relevant stimulus dimension (for example, goal box brightness in Lawrence's study) if it is to perform accurately; attending to the proper dimension, and thus performing the reaction, is easy when the two stimuli are highly discriminable. Later, the animal already has learned to attend to the relevant dimension and so performance is easier than it otherwise would be with stimuli that are so difficult to distinguish. Control subjects, that never experienced the easy problem, must learn *both* the instrumental response as well as the stimulus characteristics with the hard-to-distinguish stimuli.

A second theory of this phenomenon maintains that the facilitation of discrimination stems from a general improvement in discrimination behavior. Merely experiencing one discrimination problem facilitates the solution of other problems. According to this theory, *any* experience with discrimination problems, regardless of what stimulus dimension was used, would promote some general problem-solving skill that would aid the organism when it later confronted the difficult problem. This hypothesis was not supported by Marsh (1969). In that experiment, pigeons were given a difficult discrimination based upon the color of the cue. Some of the animals had received a much easier color discrimination problem previously, and these animals, predictably, showed the easy-to-hard effect described above. Other animals, however, that were first given a discrimination problem involving brightness, did not show the improvement in wavelength discrimination later on. Thus, the facilitation of discrimination occurred only when the stimuli on both the easy and difficult problems were from the same category; when the stimuli were from different dimensions, brightness and color in this case, the facilitation did not occur. Such a finding tends to support the attention theory mentioned above. If the animals learn to attend to one stimulus dimension (for example, brightness), then they certainly will be less likely to perform a discrimination based upon a different dimension (for example, color).

Quite a different theory of this phenomenon was proposed by Logan (1966) based on the notion that both excitatory and inhibitory conditioned reactions generalize. During the easy problem, an excitatory reaction is developed to the S+, whereas an inhibitory reaction is developed to the S−. These reactions then generalize to novel stimuli, namely, the new difficult stimuli used later in the experiment. It would appear that the ability to make the second, difficult discrimination would be influenced by the amount of generalization from the former cues. In fact, according to Logan, the superiority of the animals that received the easy discrimination first can be explained in terms of the differential generalization of the excitatory and inhibitory tendencies.

Recent support for this theory was provided by Turney (1976). During the easy discrimination problem, the S+ was a vertical line (0 degrees tilt), whereas the S− was a horizontal line (90 degrees tilt). Rats readily learned to discriminate between these two stimuli. During the second phase of the study, a more difficult discrimination was used: one line was tilted −60 degrees and the other +60 degrees. What is important in this design is the fact that both of these tilted lines in the hard task are equally different from the vertical 0-degree tilt line used previously. That is, the excitatory strength that developed to the 0-degree tilt line should generalize equally to the

−60- and +60-degree tilt lines. Therefore, learning to distinguish between these two lines should not be differentially influenced by the tendency, established previously, to respond to a vertical 0-degree tilt line. Turney found that learning the easy task did not facilitate learning the hard task. The easy-to-hard effect was not observed when the generalization factor was eliminated in the experimental design. Obviously, this experiment fails to support the attention theory; were the attention theory correct, then improvement in the difficult discrimination should have been observed because the animals would have learned to attend to line tilt during Phase 1 of the study.

Recently, a study by Seraganian (1979) has shown that both attention and generalization factors are important. Improvement in learning a difficult discrimination arises from exposure to stimuli of the same dimension; this is termed intradimensional transfer. Specifically, the gradients of excitation and inhibition interact to produce facilitated discrimination learning later on. However, the second factor is general attentiveness, based upon having undergone discrimination training previously. In summary, it appears from the work of Seraganian that both a general effect (attention) and specific effects (interaction between gradients) combine to produce this interesting phenomenon.

Stimulus Information

As we discussed in Chapter 3, the information value of the stimulus is the most important characteristic in conditioning. Therefore, it is not surprising that discrimination should be influenced by this factor. One study by Olton (1972) was designed to assess whether the choice behavior during discrimination depended upon the animal's learning about the rewarded stimulus or about the nonrewarded stimulus. Olton trained thirsty rats in a maze containing three different alleyways. On each of the 24 days of the experiment, the subjects were given two forced-choice trials (one to the rewarded alleyway and the other to the nonrewarded alleyway) followed by a free-choice trial. In other words, on the third trial, the animal could express its preference for the rewarded alley versus its avoidance of the nonrewarded alleyway. Olton found that the animals avoided the nonrewarded alley on the choice trials, but, interestingly, they chose the previously rewarded alley and the "neutral" alleyway about the same number of times. The animals distinctly avoided the nonreward alley, but they were more "indifferent" between the rewarded alley and the third neutral alley. Olton concluded that the information the animals derived from the nonreinforced trial influenced their choice behavior more than the information they received on rewarded trials.

A related study was done by Hearst (1971). Pigeons in one group were rewarded for pecking an illuminated disk on which a short vertical line was projected. The S— cue, the disk to which a peck was not rewarded, contained no line; it was blank. Animals in another group were also given discrimination training. For them, the S— was an illuminated disk on which a long vertical line was projected. Their S+ cue was blank. Stated differently, during the pretraining phase Group 1 was given reward for responding to a short vertical line, whereas the Group 2 experienced nonreward for responding to a long vertical line. In a second phase of the study both groups were given discrimination training. Here, the S+ cue was the short line and the S— cue was the long vertical line. The principal question was the following: which group would learn the discrimination better; the first group that had experienced the S+ or the second group that had experienced the S—?

Hearst found that the first group, the one that had experienced the S+, required only about 1.5 sessions to reach the discrimination criterion during this second phase. The second group, on the other hand, that had experienced the S— during the preliminary phase, took about 4.0 sessions. According to this study, then, experiencing the rewarded cue is far more important in terms of later discrimination learning than is experiencing the nonrewarded cue. That is, excitatory consequences of reward are more important for later discrimination learning than are the inhibitory consequences of nonreward.

A different but related phenomenon is the feature-positive effect (see Jenkins & Sainsbury, 1970). In one study by Sainsbury (1971), pigeons were trained to peck at one plastic disk (S+) but not at another (S—). Both these disks were divided into four areas and each area was illuminated by either a red or green circle projected from behind the apparatus. Thus, Sainsbury could manipulate the degree to which the S+ and S— stimuli differed merely by altering the combination of colors on each target stimulus. On any given trial, the S+ and S— were identical except for a single color difference in one of the areas of the stimulus. That is, all four areas for both stimuli were illuminated with the same color except for one area on one of the stimuli. This area had the opposite color. This distinctive feature, of course, could be located on *either* the S+ or S— display. The basic question was the following: do animals learn to discriminate between these two general targets when the distinctive feature, the single discrepant color on an otherwise identical stimulus, is located on the S+ or the S— target?

Sainsbury found that discrimination learning was quite poor when the distinctive feature was located on the S— disk. In contrast, the subjects readily learned to discriminate between the two targets when the distinctive circle was part of the S+ display. In other words, animals learn to discriminate between stimuli only when that single distinctive feature which makes the stimuli discriminably different is associated with reward. When the distinctive feature is part of the nonrewarded display, discrimination learning is poor.

The explanation for this phenomenon encompasses two different concepts. First, the feature-positive effect would seem to be related to attention. In order to make the correct reaction, the animals have to attend to the distinctive element. Failing to do so means that they would not be able to differentiate between the stimuli. When a distinctive element is part of the S+ display, attending to that element is rewarded immediately; this results in the strengthening of the attention reaction to this distinctive feature. When the distinctive feature appears on the S− stimulus, however, reward is never given for attending to that element; even though the animals may attend to that distinctive feature, they are never reinforced for doing so because reward never follows a response to that stimulus.

A second, and related factor, is a concept called sign-tracking (see Hearst, 1978; Hearst & Jenkins, 1974). Sign-tracking refers to the motor reactions which an animal directs toward stimuli that are associated with reward or nonreward. Specifically, it has been shown that birds and other species will direct their motor responses (for example, body position or pecking behavior) toward signals for food; they will even position themselves at some distance from cues that are associated with nonreward. This concept of sign-tracking has been applied to the feature-positive effect (Crowell & Bernhardt, 1979; Looney & Griffin, 1978). Specifically, animals literally peck at or respond to the distinctive element. Not only do they respond during the S+, but they actually come into physical contact with the very element that differentiates the S+ from the S−. Directing their responding to the distinctive feature, of course, would facilitate discrimination learning because such a reaction would be rewarded immediately.

Differential Outcomes

A third condition that affects discrimination learning is the degree to which the response leads to a unique type of reward. Consider the following example: during S_1 the animal is required to make R_1, whereas during S_2 the animal is required to make R_2 in order to get reinforced. Experimental subjects are perfectly capable of learning this conditional discrimination, that is, learning when to make R_1 or R_2 based upon which stimulus is being presented. However, learning this discrimination is facilitated if the outcomes for each stimulus-response sequence differ (Brodigan & Peterson, 1976; Peterson, Wheeler, & Armstrong, 1978; Trapold, 1970). The study by Peterson et al. used a technique that closely resembles the delayed-matching-to-sample procedure described in Chapter 8. The center key was illuminated with either green or red light; if the pigeons pecked at it, thus giving evidence that they had observed the light, two side keys or disks were illuminated. One side key had vertical lines on it and the other side key had horizontal lines on it. If green had been presented first, then the vertical-lined side key was appropriate; if red had been presented, then the horizontal side key was correct. That is, the rewarded sequence was green-vertical and red-horizontal. Animals were run under two different kinds of reward conditions. In the consistent condition, the green-vertical sequence was reinforced with food; the red-horizontal sequence was rewarded with water. Thus, in the consistent condition there was a differential or unique outcome for each type of reaction. In the inconsistent condition, on the other hand, both food and water rewards were used for each type of trial; there was no unique relationship between the type of response and the type of reward. This experiment, like the others, showed that choice behavior was more accurate when the response was correlated with a unique reward. When only a single type of reinforcer was used for both reactions, or when two reinforcers were used but in an inconsistent fashion, then performance was much lower.

The theory about this phenomenon (Trapold, 1970) is consistent with the concepts discussed in Chapters 3 and 5. Animals learn expectancies based upon the correlation between stimuli and reward or upon their responding and reward. When the relationship is consistent, that is, when the various stimuli in the environment each lead to a unique and consistent reward, then the expectancy is learned more quickly and more accurately. When the stimulus is presented in the future, it will elicit a stronger expectancy because it was associated with a distinctive outcome. Inconsistent relationships between stimuli and rewards, however, do not permit the same type of expectancy to develop.

The advantage of having distinctive feedback following each response in a conditional discrimination task is not limited to experiments in which different kinds of reinforcers like water and food are used. Other experiments have shown that nonreward versus food (Peterson, Wheeler, & Trapold, 1980) or immediate versus delayed reward (Carlson & Wielkiewicz, 1972) also are differential or distinctive outcomes that improve discrimination learning. For example, in the Carlson and Wielkiewicz study, animals pressed lever 1 following a tone or lever 2 following a click noise; both levers produced food reward. However, groups differed according to the delay experienced between the lever press and the reinforcer. Some animals received immediate reward following both responses (no delay group). A second group received a 5-second delay of reward following both lever-pressing reactions (all delay group). A third group received a 5-second delay of reward on half of the trials (random delay). Finally, a fourth group of subjects received a delay of reward following one response but not following the other (correlated delay). Subjects in this last group, thus, had a distinctive outcome following each reaction; for one lever press, reward was presented immediately, whereas for the other lever press response, reward was delayed. As shown in Figure 9–7, the authors found that this correlated procedure improved the performance of those animals. Over the training sessions, animals in the correlated delay group showed a much greater

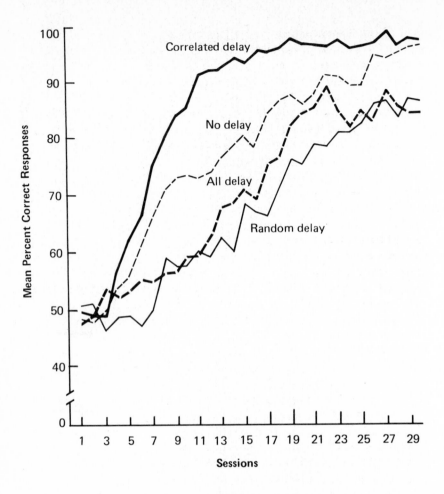

Figure 9-7.   Mean percent correct responses as a function of training sessions for groups that received immediate reward after both responses (no delay), delay of reward after both responses (all delay), delay of reward after one or the other response (random delay), or delay of reward after only one response (correlated delay). (From Carlson & Wielkiewicz, 1972)

accuracy in their behavior (pressing the correct lever to the appropriate stimulus) than did the animals in the other groups. Receiving delay following both responses, or receiving random delay following half of them, produced much poorer performance.

All these experiments suggest that expectancies are stronger when the relationship between the stimulus (or response) and the outcome is unambiguous. If a particular stimulus complex is correlated with a unique outcome, then the expectancy that that stimulus complex will produce the outcome is easier to learn than if the stimulus complex is correlated with several types of outcomes (even though all of the outcomes may be forms of reinforcement).

Observational Learning

A fourth factor that may affect discrimination learning is the opportunity for a subject to observe another animal performing the response. There are many experiments showing that observation of another subject's behavior does facilitate later learning, both on appetitive and aversive tasks (Corson, 1967; Del Russo, 1975; John, Chesler, Bartlett, & Victor, 1968; Kohn, 1976; Kohn & Dennis, 1972).

In the study by Kohn and Dennis, rats were placed in a compartment where they could observe other animals choosing between a vertical-striped door or a horizontal-striped door in order to escape shock. Subjects in one group, the Observe group, always saw the model choose the pattern which they would later experience as the S+ on their discrimination task. Subjects in the second group, the Observe-Reverse group, watched the model choose the pattern that was later used as their S−; in other words, these rats were observing the reverse of the discrimination they would later have to perform. Two control groups were used. Subjects in one control group, the No-Observe group, were not given this observation learning experience. Subjects in the other control group, the Observe-No Discrimination group, saw the model animals leave the start compartment but they could not see which target (patterned door) they chose.

Following this pretraining phase, all the subjects were given discrimination training. Their task was the same, namely, to escape shock by running to the goal box via the correct door. Kohn and Dennis found that discrimination learning was facilitated by having observed other animals perform the correct response. The control groups performed the worst: 51.6 trials were required to reach discrimination criterion for the No-Observe group and 55.4 trials were required by the Observe-No Discrimination group. In contrast, the Observe group, the subjects that saw the correct door being chosen, required an average of only 31.6 trials to meet the discrimination criterion. Interestingly, the Observe-Reverse group took significantly longer to learn the discrimination (63.0 trials). Such a result suggested that facilitation resulting from observation learning was specific; if subjects observed animals performing the reverse discrimination, then no facilitation occurred. In summary, observational experience appears to affect later discrimination learning in a rather specific fashion. Merely observing animals perform, or even observing them make a discrimination which is different from the one later required, does not facilitate later discrimination learning.

Theories of Discrimination

In this section we shall describe two major theories of discrimination learning.

Hull-Spence Theory

One of the earliest, and simplest, theories of discrimination learning was formulated by Hull (1943) and Spence (1936). The theory, in general, claims that discrimination learning involves an interaction between excitatory and inhibitory generalization gradients. More specifically, the theory has three basic principles. First, reinforcement leads to conditioned excitation to the S+. Second, nonreinforcement leads to conditioned inhibition to the S−. Finally, excitation and inhibition generalize to other stimuli and, more importantly, these conflicting tendencies algebraically summate for any given stimulus. In other words, an S+ has predominately excitatory strength but it also has some inhibitory strength (because it is a generalized stimulus to the S−). Similarly, the S− stimulus is predominately inhibitory in nature, but it also has some excitatory strength (because excitation generalizes from the S+ to the S−). The same argument can be made for any other stimulus along the dimension. Each is similar to both the S+ and S−; accordingly each evokes generalized excitation *and* inhibition. During discrimination learning itself, the S+ is reinforced and therefore it becomes predominately excitatory; the S− is not followed by reinforcement and it becomes inhibitory. Differential performance, the actual evidence that a discrimination has been made, is therefore a result of this differential reward experience.

The Hull-Spence theory has received support in a number of experiments (see Spiker, 1970). In one study by Gynther (1957), human subjects were given eyeblink conditioning. The control subjects received excitatory conditioning only; that is, a light CS+ was followed by an air puff US. A second group, however, received discrimination training using two stimuli: the CS+ was followed by the US, but a different light stimulus, the CS−, was not. According to the Hull-Spence theory excitatory conditioning to the CS+ should be lower in the second group of subjects. The reason for this is that the inhibitory tendency should generalize from the CS− to the CS+ and cause a reduction in excitatory strength (recall that performance is an algebraic summation of excitation and inhibition according to this theory). Since the control subjects did not receive any inhibitory training, their performance to the CS+ should be at full strength. This prediction was confirmed; performance to the CS+ by the control subjects significantly exceeded the response level of the second discriminative group.

The Hull-Spence theory also has been supported in other research endeavors, notably in studies of the peak-shift phenomenon discussed previously. Those experiments demonstrate that gradients of excitation and inhibition do summate. That is, postdiscrimination performance reflects an interaction between the excitatory and inhibitory gradients (see Marsh, 1972). However, the Hull-Spence theory has not been supported in other areas of research. Various phenomena, discussed below, cannot be explained simply by the interaction of two generalization gradients. Other principles seem to be involved in many of these findings. Moreover, the attention theory of discrimination, discussed in the next section, has been a more successful explanation for many of the phenomena.

Attention Theory

The attention theory was outlined most clearly in a book by Sutherland and Mackintosh (1971). According to their theory, discrimination learning involves two separate processes. First, an attention response to one or more relevant stimulus dimensions is strengthened during discrimination learning. Attention in this sense is a process that takes places centrally, in the brain; it is not simply an orientation behavior to a stimulus. Second, the subject learns to make a particular response to the relevant stimulus. Only by first attending to the proper stimulus dimension, and then by responding to the particular stimulus, can an animal perform a discrimination reaction accurately.

The brain is said to have mechanisms, or analyzers, that receive and process sensory information. When an animal is confronted with any given stimulus, various analyzers are involved. That is, each stimulus dimension that characterizes a particular stimulus (its size, color, shape, odor, brightness, etc.) has a separate stimulus analyzer which receives and processes that information. At the start of training, the strength of any given analyzer is related to the strength of the incoming signal. If a stimulus has a particularly potent feature, such as its brightness or its color, then the brightness or color analyzers operate to draw the subject's attention toward that dimension. This accounts for the fact that performance is related to CS intensity; CSs that are stronger evoke greater attention.

The analyzers, as noted above, provoke attention responses; these are dispositions to respond to a stimulus along one dimension but not others. The animal, in essence, "decides" which stimulus along the relevant dimension is correct. If that decision is followed by reinforcement, then the attention analyzer is strengthened. That is, if the animal pays attention to a particular stimulus dimension, predicts or expects to receive reward following one stimulus along that dimension, and, in fact, does receive reward, then the response of paying attention to that stimulus dimension is increased in strength. For example, if the tilt of a line were correlated with food presentation, but the brightness or color of the line were not, then the strength of the orientation analyzer would increase, but the strength of the brightness and color analyzers would decrease. Paying attention to the tilt of the line is followed by reward, whereas paying attention to its color or brightness is not. In short, attention mechanisms, called analyzers, increase in strength. They do so because they provoke an attention response. If the attention response is reinforced, then the analyzer is strengthened. If the attention response provoked by the analyzer is not reinforced, then the analyzer strength is weakened; the subject gradually stops attending to that particular dimension.

The second process specified by the attention theory is the acquisition of a particular response. A bond or attachment develops between a specific response and an analyzer. For example, the analyzer may provoke the attention response "vertical is the correct orientation" which becomes attached to the specific response "press the lever." A reward, therefore, strengthens both the attention reaction as well as the specific response. When the subject is reinforced for responding to, say, a particular line tilt, the connection between that response and the analyzer is strengthened; at the same time, the analyzer itself, the subject's attention to the relevant dimension, increases in strength.

Let us summarize. The attention theory claims that subjects learn two things during discrimination experiments. First, they learn to attend to a particular dimension of the stimulus on the basis of whether their attention reaction correctly predicts reward; this corresponds to the strengthening of the analyzer or attention mechanism. Second, subjects learn to attach a particular response to their analyzers based on whether they receive reinforcement for that response.

We have already seen considerable support for the attention theory of discrimination. For example, the easy-to-hard effect, discussed previously, appears to be based, at least in part, on attention reactions. The subjects that receive an easy discrimination first thoroughly learn about the relevant stimulus dimension; when they are confronted with the difficult problem later on, solution is relatively easy because they already have mastered the proper stimulus dimension. Control subjects, on the other hand, that received the difficult discrimination right from the start, make many errors because they have not learned to attend exclusively to the proper dimension; their analyzer strength is low.

Another important study that supports the attention theory was performed by Waller (1973a). Rats were trained to run down an alleyway in order to receive food. Four conditions were used during this first phase. Two of the groups received reward in the goal box on every trial; the other two groups were rewarded on only 50 percent of the trials. For one group in each of the above conditions, the alleyway was painted gray; for the other group in each of those conditions, the alley was covered with black and white vertical stripes. Therefore, Waller had four groups: 100 percent reward, gray; 100 percent reward, stripes; 50 percent reward, gray; and 50 percent reward, stripes.

In the second phase of the experiment, all subjects were given discrimination training. Specifically, they were required to choose between two goal boxes for food reward. One box was covered with stripes that slanted 45 degrees to the right; the other goal box was covered with stripes slanting 45 degrees to the left. Waller made several predictions. First of all, he predicted that the 50 percent-stripe group would learn the discrimination more slowly than the 100 percent-stripe group. The reason is that the 50 percent-stripe group would not pay attention to the appropriate dimension (stripes) since, in Phase 1, stripes were not highly correlated with reward. The 100 percent-stripe group, on the other hand, should have a very strong analyzer for the dimension stripes. After all, during Phase 1, the stripes always predicted reward in the goal box. Thus, during Phase 2, the animals in the 100 percent-stripe group should pay attention to stripes more than the partially rewarded group. Second, Waller predicted that the two control groups, the groups that were trained in the gray alleyway during Phase 1, would learn equally well during the discrimination phase. Animals in those groups never experienced stripes and so their attention to stripes, or analyzer strength, should be about equal.

As predicted, discrimination performance did not differ between the two control groups. The subjects that were trained in the gray alley during Phase 1 learned the discrimination response at approximately the same rate. However, the 50 percent-stripe group required 118.4 trials to meet the discrimination criterion, whereas the 100 percent-stripe group took only 93.7 trials. Thus, the partially reinforced animals, the group that performed less well during Phase 2 of the study, had not developed a sufficient attention to the proper dimension, that is, the stripes. Subjects in the continuously reinforced group, on the other hand, learned better; they had developed a stronger analyzer for stripes during Phase 1. Waller's experiment, therefore, clearly supports the attention theory of discrimination learning. The analyzer for stripes, the attention to that particular stimulus dimension, was weakened by partial reinforcement (stripes were not highly correlated with reward) but strengthened by consistent and continuous reward.

There are a number of other studies that support the attention theory of discrimination learning. For example, the presence of irrelevant cues makes it difficult to attend to a single stimulus dimension; accordingly, discrimination learning is slower, at least in rats, when irrelevant cues are present (Waller, 1971), although such an effect has not been found for cats (Warren, 1976).

Discrimination Phenomena

The following section outlines a number of important phenomena related to discrimination learning. None of these findings can be explained easily by the Hull-Spence theory, but at least two of them can be explained quite easily by the attention theory.

Overlearning-Reversal Effect

Consider the following experiment. Two groups of rats are trained to make a discrimination response. Specifically, they are given a choice between a black alley and white alley and are reinforced for running down the white alley but not the black alley. After all the animals learn to perform this discrimination, rats in one of the groups are given additional practice, say, 100 more trials; rats in the control group are merely placed in the home cage during this period. After the overlearning trials, all the animals are given reversal training: they now must choose the black alleyway to be rewarded and avoid running down the previously correct alleyway, the white side. According to the Hull-Spence theory of discrimination, the group that received overtraining should reverse much more slowly than the group that was merely given discrimination training to the criterion. The reason is that overtraining should have strengthened the original habit of choosing the white alleyway; the stronger that habit, the more difficult it should be to extinguish that habit and replace it with a new habit, namely, choosing the black alleyway. In reality, however, the opposite result is typically found (see Denny, 1970; Sperling, 1965a, 1965b). That is, overtraining on a discrimination problem makes it *easier* to reverse that discrimination than it is if the animal is not given the extra practice.

After the discovery of the overlearning reversal effect by Reid (1953) and Pubols (1956), many investigators believed that the effect was unreliable; in fact, many could not replicate the results (D'Amato & Schiff, 1965). However, more recent studies have determined that reversal learning in rats *is* facilitated by overtraining, but only when a difficult visual problem is used (Mackintosh, 1969), *and* when a large reward is given for the correct response (Hooper, 1967; Theios & Blosser, 1965). When a simple spatial discrimination problem is used, such as left versus right turn in a maze, or when small rewards are used, the phenomenon is rarely observed (Mackintosh, 1965, 1969).

Several different theories have been proposed to account for the overlearning reversal effect. Perhaps the most successful has been the explanation derived from attention theory (see Lovejoy, 1966; Mackintosh, 1969). First, the attention theory assumes that the increase in the strength of the analyzer—the attention response—develops more slowly than the strength of the specific motor response. That is, a discrimination may be learned without the subject's attending exclusively to the relevant stimulus dimension. Although the subject must obviously pay attention to the relevant dimension to some degree, analyzers for other dimensions still may be present when the original discrimination criterion is met. The theory then claims that overtraining continues to strengthen the analyzer for the relevant dimension; analyzers for other dimensions, of course, extinguish during the overtraining trials. During reversal learning the animals that received training only to the normal criterion are not giving their maximum attention to the proper dimension; their analyzer for the relevant dimension is not exclusively guiding their attention reactions. The overtrained subjects, on the other hand, maintain their attention to the relevant dimension during reversal learning; other analyzers do not compete for their attention because those analyzers extinguish during the overtraining trials. Overtrained subjects' strong and exclusive attention to the proper dimension facilitates their learning this new task which, of course, involves the same dimension as the original task. In summary, the overlearning reversal effect is explained by attention theory in terms of analyzer strength: during overlearning trials, the attention to the relevant dimension continues to increase; continued attention to that dimension is required in order to learn the reversal task.

The attention theory of the overlearning reversal effect has been supported in a number of different studies. Mackintosh (1963b) found, for example, that when irrelevant stimulus dimensions were added for the first time during reversal learning, the behavior of the normally trained subjects was disrupted; this treatment, however, did not affect the reversal learning in the overtrained subjects. This result suggested that the normal subjects were not attending exclusively to the appropriate dimension; the new dimensions competed somewhat for their attention and, therefore, disrupted their performance. The added extraneous dimensions, however, did not affect the overtrained animals because, during overtraining, those animals solidified their attention to the relevant dimension; behavior was controlled exclusively by the attention analyzer for the relevant dimension, and therefore it was not disrupted by additional extraneous dimensions.

The attention theory also accounts for the fact that a difficult problem must be used if one is to demonstrate the overlearning reversal effect. Spatial discriminations, that is, choosing right versus left in a maze, are so easy for rats to learn that attention to this dimension (the analyzer strength) is maximum within a few trials; overtraining does not have the effect of strengthening that analyzer. As the discrimination problem becomes more difficult, however, many more trials are needed to develop a strong analyzer; overtraining has a greater impact upon the analyzer strength and, thus, on reversal behavior.

Although the attention theory would appear to be successful in accounting for the overlearning reversal effect, there have been challenges to the theory. For example, several investigators have suggested that animals learn general observing or orienting strategies during overtraining; these have the effect of facilitating reversal because the animal is better at solving problems, not because the animal has increased its attention to a specific stimulus dimension (see Hall, 1973, 1974; Waller, 1973b). In the Hall (1974) study, rats were given a problem in which they had to choose between a horizontally striped target and a vertically striped target. After regular training or overtraining, the animals were given a new problem, namely, choosing between a black target and a white target. Finally, the animals were given reversal of this black-white discrimination. The attention theory would predict that the original overtraining on the horizontal-vertical discrimination would have no effect on the black-white discrimination; that is, the increase in the analyzer strength for the orientation dimension should not affect later attention to the brightness dimension (Mackintosh, 1963a). Hall, in contrast, found that overtraining, while it did not affect the learning of the new problem—the black-white discrimination —did facilitate reversal of the black-white discrimination. Thus, overtraining could involve the development of special orientation or observing responses that transfer to any new discrimination problem.

Finally, a more recent challenge claims that the response expectancy, based on the response-outcome correlation, is responsible for the overlearning reversal effect (Purdy & Cross, 1979). In this experiment, animals were taught to discriminate between horizontal and vertical targets. Then, the important experimental group received overtraining, but the overtraining was accomplished in a unique way. Specifically, animals in the experimental group were given extra trials to a single target (a gray card). Obviously, their attention to the horizontal-vertical orientation dimension could not be increasing during this period. And yet this overtraining experience produced the typical overlearning reversal effect: reversal of the horizontal-vertical discrimination was facilitated by having experienced additional practice with a single, gray stimulus. Quite clearly, this experiment does not support the attention theory. Rather, it suggests that the response expectancy, based on continued experience with the response-food correlation independent of the stimulus expectancy (target-food correlation), contributes to the phenomenon.

In conclusion, the factors contributing to the overlearning reversal effect are numerous and they interact in a complex fashion. It would appear that animals are capable of learning general orienting reactions that would facilitate future discrimination learning, but, at the same time, attentional factors to relevant stimulus dimensions are also extremely important. Whichever combination of factors proves to be the best description of this phenomenon, it is certainly true that the overlearning reversal effect cannot be explained by the simpler theory of Hull and Spence.

Intradimensional and Extradimensional Shifts

Let us consider in more detail this issue just raised: whether discrimination learning facilitates learning a new discrimination problem using the same stimulus dimension (intradimensional shift), and/or whether it facilitates the acquisition of a new discrimination using a different stimulus dimension (extradimensional shift). In one study (Mackintosh, 1964), rats were trained to respond to one of two targets for food. For one group of subjects, the targets were either black and gray squares or white and gray squares; animals in this group had to make the appropriate response based upon the brightness of the target. For the second group, the choice was between a square or a diamond shape; although each stimulus was painted black or white, it was its shape, square or diamond, rather than the brightness, that was relevant. In a second phase of the experiment all subjects were trained to make a brightness discrimination; here, all the stimuli were squares and they were painted either black or white. It should be clear that the first group of subjects would be making an intradimensional shift response during Phase 2 (the same stimulus dimension, namely, brightness, was relevant), whereas the second group would be required to make an extradimensional shift discrimination in Phase 2 (the appropriate dimension changed from shape to brightness).

The results showed that the intradimensional subjects, Group 1, took 45.6 trials before learning the discrimination in Phase 2. The extradimensional animals, on the other hand, were much slower, requiring about 61.2 trials before they were able to reach the discrimination criterion. In other words, learning a second discrimination was much easier when the initial problem involved the same dimension than when the initial discrimination involved a stimulus dimension that was different. Similar results have been observed for pigeons (Mackintosh & Little, 1969), rats (Shepp & Eimas, 1964), and monkeys (Shepp & Schrier, 1969).

The finding that intradimensional shifts are learned more easily than extradimensional shifts cannot be explained very well by the Hull-Spence theory or by a hypothesis which claims that general orienting or response strategies are learned during the initial discrimination that later facilitate learning the second problem. Like the overlearning reversal effect, attention to relevant stimulus dimensions seems to be an important factor. The analyzer for the relevant stimulus dimension in the first phase of the study, that is, the attention to the appropriate stimulus dimension, is strengthened during Phase 1. Then, after the shift in Phase 2, the intradimensional subjects already possess a strong attention response to the appropriate dimension; all that is required is for those subjects to learn a new motor reaction. On the other hand, extradimensional subjects must respond to an entirely new dimension after the shift; they must suppress attending to the formerly relevant dimension and learn both the new appropriate dimension as well as the new motor response.

Learning Sets

As discussed above, intradimensional shifts, using the same stimulus dimension, are more easily performed than extradimensional shifts, in which the stimulus dimension is changed. This fact, however, does not mean that discrimination learning using different stimulus dimensions cannot be improved by prior training. In fact, exposure to *many* different discrimination problems produces a dramatic improvement in the ability to learn new problems. This phenomenon is called *learning set;* it refers to the progressive improvement in a subject's ability to solve discrimination problems as a function of repeated exposure to many different problems (see Medin, 1972; Miles, 1965; Reese, 1964).

The classic study on learning sets was done by Harlow (1949) who used rhesus monkeys as subjects. The animals were positioned before an apparatus that consisted of a tray with two food wells; various stimuli like blocks or other objects, could be placed on top of these food wells. The monkey's task was to choose one of the stimulus objects. If it chose correctly, it was allowed to eat the reward that was concealed beneath the object, but if it chose the incorrect stimulus, the objects were withdrawn from the monkey's view, rearranged, and then presented once again for another trial. From 6 to 50 trials were given for each pair of objects. After the monkey had mastered discrimination for one set of objects, a different set was used. This procedure continued until a total of 344 different problems had been presented to the monkey. The results of this experiment are shown in Figure 9–8. Percent correct responses are plotted for the first 6 trials. It is clear that learning on each of the first 8 problems was somewhat slow. After 6 trials with the same stimulus objects, the subjects were correct only about 75 percent of the time. However, as the animal became experienced with more and more problems, the rate of learning improved. For instance, consider the average performance on problems 33–132 (see Figure 9–8). Animals were correct almost 80 percent of the time on the second trial of those problems and they reached over 90 percent correct performance after about 6 trials. Toward the end of the training series, that is, problems 289–344, the animals were nearly perfect on the second trial; percent correct performance was 97 percent. In summary, the acquisition rate on each successive discrimination problem improved over training. This dramatic improvement in the ability to learn discrimination problems has been shown in a variety of species from monkeys to rats to birds, and it occurs even when the training techniques involve the avoidance of unpleasant stimuli, as opposed to the reward procedures described in the experiment above (Stoffer & Zimmerman, 1973).

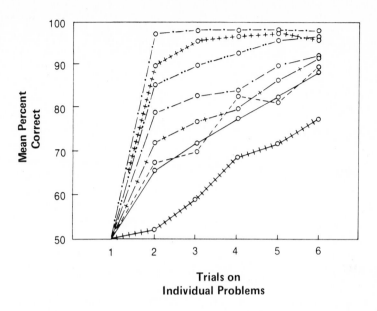

Figure 9-8. Mean percent correct as a function of the trial for groups of discrimination problems. (From Harlow, 1949)

Harlow's work demonstrated an important principle in learning, namely, that the ability to solve problems, the ability to learn itself, may be influenced by previous learning experiences. Too often we assess learning abilities in subjects that have had very limited experience; they have been born and raised in laboratories without having had much chance to learn anything. Animals in their natural environment, of course, are very different; they are constantly faced with problems that they must solve. Their ability to solve any particular problem or discrimination, therefore, will be influenced by their previous encounters with other such problems. Certainly, the investigation of learning sets provides an added dimension to the study of learning, one that is particularly relevant to organisms in their natural environment.

Another issue raised by Harlow's research was the possibility that learning sets could provide an index for general intelligence level. The notion was that the ability to improve in discrimination learning, the ability to utilize previous experience to solve problems, should be related to the evolutionary complexity of the organism. Thus, monkeys would show greater improvement from problem to problem than rats, and humans would show even better learning set performance than monkeys. Although several investigators favored such an approach (for example, Harlow, 1959; Warren, 1965), recent research has not supported this idea. First of all, the methodology used to assess learning set formation can influence the results in important ways. Devine (1970), for example, has shown that two species of monkeys differ in their learning set behavior merely as a function of differences in methodology. More important, perhaps, is the realization that species differ dramatically in terms of the stimuli to which they are sensitive and the responses which they are capable of performing. Rats, for instance, are relatively poor in their learning set formation when visual stimuli are used. But when more "important" stimuli are used, namely, odors, to which rats are extremely sensitive, then their learning set performance is comparable to that reported for primates (Slotnick & Katz, 1974). It would, therefore, be erroneous to conclude that monkeys are more intelligent than rats merely because monkeys can perform a visual learning set better than rats. If the appropriate stimulus dimension is used, to which rats are sensitive, then no difference is found. In short, learning set formation does provide valuable information about the learning capacities of various species, but it does not provide a simple measure of intelligence for comparing different species.

Learning set formation is difficult to explain using conventional theories. Both the attention and the Hull-Spence theories concentrate on intradimensional shift improvements and, therefore, do not handle learning set phenomena very well. The theory developed by Harlow (1959) suggested that learning set formation resulted from a subject's learning what not to do; that is, the subject's tendency to make an incorrect choice gradually was inhibited during training.

Harlow's theory stemmed from his observation that animals committed systematic errors during learning set formation. Some of them perseverated, that is, they continued to choose the incorrect stimulus because of innate or acquired preferences. Others responded to the right or left side regardless of which object had been placed there. It appeared, therefore, that errors were not random; subjects were making consistent errors suggesting that they were following some particular strategy in responding. Later in training, however, many of these error patterns were suppressed; subjects had learned to make the correct choice by suppressing the errors they previously were committing.

A more developed theory was proposed by Levine (1959, 1965). According to this position, animals develop hypotheses about the discrimination problems; they form general strategies or rules that guide their behavior. For example, one strategy might be "respond to the left." If this strategy were successful, then a subject would continue performing in that fashion. However, having entertained such a strategy and discovering that it did not consistently lead to reward, the subject would abandon it and form a different strategy. The strategy that animals seem to follow most closely in learning set formation is: "win-stay with object, lose-shift to other object." In other words, the sophisticated learning set animal will make a choice; if the choice is correct (rewarded), then the animal will continue on the remaining trials to choose that object. However, if its first choice is incorrect, the animal will immediately switch to the alternative object and continue to choose that.

Levine's theory accounts for two important features of learning set behavior. First, errors are systematic; they appear to be the result of a general plan or strategy rather than the result of random variations. Second, the improvement in discrimination learning stems from the eventual confirmation of the appropriate strategy or hypothesis, not from the strengthening of particular habits or choices. The improvement in discrimination couldn't really stem from the strengthening of particular choices; it ought to reflect a change in the animal's overall response strategy. Otherwise, the improvement would not be applicable to all of the problems.

A good deal of research has supported Levine's theory (see Behar & LeBedda, 1974; Schrier, 1974; Schusterman, 1964). Subjects do seem to follow the "win-stay, lose-shift" strategy. However, other factors may also be important. For instance, if the intertrial interval between problems is fairly long, then learning set performance deteriorates (Deets, Harlow, & Bloomquist, 1970; Kamil, Lougee, & Shulman, 1973; Kamil & Mauldin, 1975). The reason is probably that the delay between trials allows the memory trace from the previous trial to decay. If the animals cannot remember which object was rewarded or nonrewarded on the last trial, then they have difficulty choosing the correct object on the current trial. This retention loss, interestingly, develops during learning set formation. That is, naive animals do not have difficulty dealing with a long intertrial interval but experienced subjects do (Kamil & Mauldin, 1975). These results suggest that learning set formation is a conditional discrimination; animals are making their choices based upon the memory of previous trials. If anything, such findings tend to support Levine's theory about learning set formation, and, in fact, they suggest a mechanism or system by which the hypotheses and strategies could operate: animals may develop response strategies as described by Levine and these strategies may operate in terms of memory traces.

Summary

Generalization is the process that causes animals to respond in the same way to novel stimuli that they did to the original training CS. Discrimination, on the other hand, is the reverse process; it is evident when animals respond differently to two stimuli, that is, an S+ and an S−.

The most important factor governing generalization is the degree of similarity between the CS and the generalized stimuli. Studies have shown that responding to these stimuli is a graded function of similarity: the more similar the stimuli, the higher the generalized response. This holds true for both conditioned excitatory and inhibitory responses.

Other factors affect generalization. Higher levels of acquisition or stronger motivation levels cause generalization to diminish. However, increasing the training-test interval causes the opposite effect. Early experience is an important dimension too. Some characteristics of stimuli, like color, can be distinguished at birth by certain animals like birds, but early experience with other characteristics, like noises, does affect later generalization. Another factor is prior discrimination training. If animals are taught to discriminate between two stimuli and then are given a generalization test, generalization is markedly reduced. Studies in this area of research have shown that the peak of responding on such a test shifts from the S+ to stimuli on the side opposite from the S−.

Numerous factors affect discrimination. One is problem difficulty. If animals are trained on an easy problem, then solution of a hard problem later on is easier. The information value of the stimuli is also important. Having experienced the S+ previously helps subjects to discriminate better than having experienced the S−. Discrimination is also easier when the single feature that distinguishes between the two cues is part of the S+ display. A third factor that facilitates discrimination learning is the condition in which two responses lead to different outcomes. Finally, animals are better at discrimination when they are permitted to observe other animals respond in the same situation.

One of the earliest and most prominent theories of discrimination is the Hull-Spence theory. This position claims that differential responding occurs when excitation develops for the S+ cue and inhibition for the S−. An added proposition is that these two tendencies algebraically summate for any given generalized stimulus.

A more recent theory is the attention hypothesis. According to this theory, animals develop attention analyzers for stimulus dimensions that predict reward. These analyzers guide the animals' choice responding. Considerable support has been found for this theory.

The overlearning reversal effect describes the condition in which animals that have received extra practice on their original discrimination problem learn the reverse problem much faster than animals that received ordinary levels of training. The attention theory, but not the Hull-Spence theory, accounts for this effect.

A second phenomenon in discrimination learning research concerns intra- and extradimensional shifts. Animals are able to learn certain discriminations faster after they have solved other similar problems.

Finally, learning set refers to the phenomenon that animals become progressively better at solving problems. The appropriate explanation for this appears to be that animals develop hypotheses or strategies that guide their problem-solving behavior.

10
Motivation

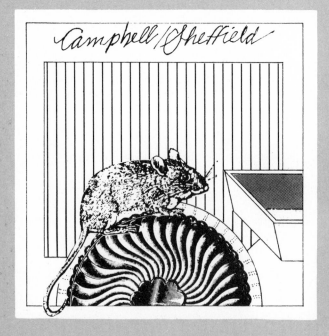

Introduction

Bio-Regulatory Systems

Acquired Motivation

Stimulation Theories

Nonregulatory Motivational Systems

Summary

Introduction

The study of motivation has often been equated with the study of basic learning; certainly the literature is overlapping and many of the concepts are the same. Nevertheless, we have chosen to treat certain motivational variables and theories separately in this text. We have done so partly for the sake of convenience, but also because it is often useful to distinguish between learning and motivational factors.

Part of the reason that motivational and learning variables have been linked together so closely is that motivation concepts often lack a clear focus. Traditionally, motivation was thought to be a temporary determinant of behavior like hunger; this is in contrast to conditioning which was believed to be a relatively permanent determinant of behavior. This distinction is only partly true. Some types of motivational factors, such as deprivation levels, are easily reversed. However, other forms of motivation are not so labile; indeed, many of these forms of motivation are learned and, thus, are relatively permanent. In this chapter, we will discuss both forms of motivation: the biological or regulatory factors, such as food or water deprivation, and the acquired motivation factors.

Let us be more specific about the nature of motivation as compared to learning. Consider this example. A hungry rat is taught to run down an alleyway to a white goal box, not a black one, in order to get food. After repeated exposure to the training regimen, the animal will show signs that it has learned this simple discrimination. We could say that the animal has learned to perform a response that it was not previously capable of performing. Now, imagine that we take this same animal and feed it until it is satiated and then put it back into the discrimination apparatus. More than likely, the animal would not perform the discrimination response; it would sit in the start box, looking indifferent, and perhaps even fall asleep. What is the proper conclusion from this test? Has the animal forgotten the discrimination or the running response altogether? Is the animal frightened to leave the start box? Intuitively, the answer to these questions would seem to be no. The more plausible explanation would be that the animal lacks motivation; it is no longer hungry and, therefore, no longer has the motivation to perform.

This simple example illustrates a basic principle concerning animal behavior: very often (but not always) an animal's performance is a function of two factors. First, an animal performs only those responses that it has learned (or, of course, that are innate). Second, an animal performs a response only when it is motivated to do so. Although there are exceptions to this general principle, as discussed below, it would appear that both learning and motivation contribute in some way to an animal's performance. When either one is lacking, performance of the criterion response is not observed; either the animal has not learned (or it has forgotten), or the animal is not motivated.

The discussion above suggests that motivation is the energy or impetus behind behavior. Whereas learning provides some kind of directive function, that is, it guides the animal in choosing *what* specific response to perform, motivation is the force or energy that sets this response into motion. This is not to deny that some motivational factors may provide direction. For example, it is possible for an animal to make a particular response when it is hungry and a different response when it is thirsty; those drive conditions, in part, determine which behavior will be performed (see Chapter 6). However, impetus or energy is really the more essential concept that underlies a motivational state.

Historical Perspectives

Scholars in the seventeenth and eighteenth centuries believed that human beings alone had reason and intelligence. Lower species, in contrast, lacked free will, intelligence, and foresight; they were governed merely by the mechanical laws of physics and biology. But many animals behaved in ways that *appeared* to be intelligent. In fact, some species performed astonishing feats that defied explanation. This posed a dilemma: if intelligence was presumably possessed by human beings and no other species, how could these intelligentlike behaviors of lower animals be explained without actually appealing to the concept of intelligence?

The answer was instinct. By suggesting that animals had instinct, one could explain their intelligentlike behavior without attributing intelligence itself to the species. Instinct was acting toward a goal without having to learn; instincts were inborn behavior patterns. Moreover, they were forces for action; they impelled or energized behavior. They reflected needs and desires of the animal.

Drive Theory

The problem with instincts was that they were used to explain too many behaviors (for example, see Ayres, 1921). Scientists evoked various instincts, almost arbitrarily, to account for behaviors that were often learned. Investigators, of course, could not manipulate instincts. They could only speculate that instincts were the cause of a certain behavior. Such theorizing was simply too easy; virtually *any* behavior could be attributed to the action of an instinct without fear of contradiction.

What replaced instincts was the concept of drive (see Bolles, 1975, for an excellent discussion of the history of drive theory). Drives were based on biological needs. In that regard, they were mechanistic and tangible; they could be quantified in terms of the extent of the biological need and they could be linked directly to the physiology of the species. It was claimed that behavior was motivated by drives, such as hunger and thirst, that arose from a biological need, that is, the need for food and water. Drives therefore provided a mechanism for the impetus behind behavior, but, because they could be quantified in terms of the physical need that produced them, they were more amenable to investigation than were instincts.

Certain concepts often become popular because they have a particularly persuasive advocate. This certainly was the case for drive theory. The influential psychologist C. L. Hull gave drives a central position in his theory of behavior (Hull, 1943). Although many of the ideas espoused by Hull have been disconfirmed (Hull himself changed his theory radically in a later book), several of his principles are worth reviewing.

First of all, Hull believed that drive was a generalized source of energy or motivation for behavior; he even referred to drive as "D." By generalized, Hull meant that various sources of need combined to contribute to a single drive state. So if an animal were both hungry and thirsty, general drive, D, would be higher than if it were only hungry or thirsty. Similarly, if an animal were both hungry and fearful, then general drive would be greater than if the subject were only hungry or only fearful. Performance, in turn, would reflect the combination of needs. Although it is certainly true that performance is affected by drive level (see later sections of this chapter), the notion that various sources of need combine to produce a *single* drive state unfortunately has been found only under certain limited circumstances (see Bolles & Morlock, 1960).

Another important concept held by Hull was the independence of habit and drive. Behavior was claimed to be a result of the product of habit and drive: performance $= H \times D$. This part of Hull's theory was formulated because, in numerous studies, performance (for example, running speed in a maze) seemed to suggest that the relationship between habit and drive was a multiplicative one. Animals that were trained under either high or low drive showed essentially the same level of performance early in training. This occurred because habit was essentially zero; thus, their performance was also low. As training continued, however, the high and low drive groups diverged, indicating that drive and habit multiplied. If drive were simply an additive component, then high drive groups would be uniformly above low drive groups. Unfortunately, this principle has not been supported by recent research either. Again, while drive affects performance, the drive level does not seem to affect the *rate* of acquisition, that is, learning (see Davenport, 1965).

In conclusion, the specific theory of behavior offered by Hull (1943) appears to be inadequate; his ideas concerning generalized drive and the interaction between drive and habit have not been supported by recent data. Nevertheless, Hull provided an important vantage point for modern theorists. Drive theory is no longer the central concept that it once was, but the idea of drive still plays an important role in our methodology and our theoretical approaches to motivated behavior.

Incentive Theory

Part of the reason for the decline of drive theory was the interest in and support for incentive theory. Quite simply, incentive is a learned source of motivation. Whereas drive was thought to be a central state induced by a biological need, incentive is a motivational state induced by incentive stimuli (reward-related cues). Drive is claimed to push an animal, to goad it into action; incentive is claimed to lure or entice the animal, attract it toward a particular goal. Most importantly, because incentive is a learned source of motivation, the principles of incentive turn out to be the same principles that we use in learning theory itself. We can begin to understand how animals are motivated, how their behavior is energized, in part by studying learning.

Incentive refers to the motivational property of reward objects. Reward not only strengthens the learned reaction, the expectation, but it also motivates performance. Evidence for the incentive concept abounds. One early study was done by Marx and Murphy (1961). Two groups of rats were trained to poke their noses into a small hole in order to obtain food. For one group, a buzzer was sounded before the food was delivered, but not for the other group. In Phase 2 of the study, the animals were trained to run down an alleyway. After a few practice trials, the subjects were given the buzzer while in the start box. The authors found that the running speeds for the animals in the experimental group, the subjects who had received the buzzer-food pairings in Phase 1 of the study, were much higher than those for the animals in the control group for whom the buzzer had not been paired with food. This result showed that behavior could be energized by a Pavlovian CS. The motivational effect was not due to any biological difference between the animals, but merely a difference in their conditioning histories. Furthermore, the difference between groups could not be attributed to differences in learning the alleyway task; both groups were equivalently hungry and both were given the same reward for the running response. The differences between the groups can only be attributed to differences in motivation that were conditioned during Phase 1 of the study (see also Osborne, 1978).

It should be clear from the material discussed above that conditioned motivation, incentive, is entirely consistent with the notion developed in Chapter 5, namely, that Pavlovian CRs are emotional or motivational in nature. During Pavlovian conditioning the organism develops expectations or cognitions about future presentations of the US. But it is also true that the animal's motivational or emotional level is affected. Stimuli that are paired with strong, biologically potent USs are capable of eliciting motivational or emotional states. Incentive is precisely one of these motivational states. The material discussed in Chapter 5 in connection with the interaction studies, therefore, could be viewed as experiments dealing with the interaction of motivations. To study incentive or acquired motivation is to study Pavlovian conditioned emotions.

Bio-Regulatory Systems

Many species in their natural habitat spend much of their time securing food and water. These commodities are essential for all life. It is not surprising, therefore, that considerable attention has been given to the effects of these variables in learning situations. In this section, we shall review many of the findings that relate biological drives, that is bio-regulatory systems, to behavior.

Deprivation and Activity

As discussed above, one of Hull's central concepts was that drive was a general energizer of behavior. All responses, whether startle reactions or instrumental behaviors or simply general activity, would be magnified or energized by any given source of drive. In fact, early experiments showing that motivational manipulations like food deprivation would cause changes in general activity provided a strong impetus for adopting the concept of drive itself (see Bolles, 1975). The typical study on drive and activity involved placing rats in a cage that contained an exercise wheel; their general or "spontaneous" activity was shown to be dramatically increased by food deprivation (Richter, 1922) and by other biological states like estrus (Wang, 1923). Such spontaneous activity had no goal, it was aimless; moreover, it had no reinforcer. For these reasons, it was taken as a relatively pure measure of the energizing effects of drive.

There is no doubt that food deprivation increases the wheel-running activity of rats. The animals spend longer running in the wheel, they do it more frequently, and they run faster when they are deprived than when they are satiated (Jakubczak, 1973). However, even in the early research there were certain problems of interpretation. First, nearly all the studies were done using exercise wheels; drive could hardly be claimed to be a general energizer if the effects of drive were confined to that situation. Second, there seemed to be some indication that the animals' activity was merely in anticipation of food or that they were highly influenced by the time of day. Third, the degree of sensory stimulation that the animals received from the environment seemed to have an important effect on activity.

Considerable progress on this question was made by F. D. Sheffield and B. A. Campbell. Their general agreement was that food deprivation does not, in itself, produce changes in activity; rather, food deprivation changes the animal's *re*activity to environmental stimuli. In one study, Campbell and Sheffield (1953) housed rats in a small rectangular cage that was located in an isolated room. For 10 minutes each day, the fan that ventilated the room was turned off and the lights in the room were turned on. Campbell and Sheffield found that deprivation caused an increase in activity *only* during the stimulus change. That is, virtually all of the increase in activity that is normally associated with food deprivation was performed during the stimulus change (see also Teghtsoonian & Campbell, 1960). Other research by Campbell (1960) and Sheffield and Campbell (1954) found that the increased sensitivity to a stimulus change during food deprivation increased over training sessions. In this latter experiment, animals were given 12 days of training. For 5 minutes each day the fan that ventilated the room and the lights were turned off. Once again, a large stimulus change was observed during this period. In these studies, however, a ration of food was delivered just after the 5-minute stimulus change. What the authors observed was an increase in activity over days. The animals became progressively more active during the 5-minute period, suggesting that some kind of conditioning process, some anticipation of feeding, was taking place.

These experiments challenged previous beliefs by implying that deprivation lowers the animal's threshold or sensitivity to internal and external stimulation. Food deprivation does not simply produce a generalized drive state that indiscriminately activates or energizes all behavior. Rather, it serves to heighten the animal's sensitivity to external stimulation. In addition, the anticipation of food contributes significantly to the activity patterns. The animal's activity may be only a conditioned motor reaction to stimuli that are correlated with food.

Research in the decade after Campbell's work continued to suggest that food deprivation did not produce a generalized drive state (see Baumeister, Hawkins, & Cromwell, 1974). Indeed, many investigators found that the relationship between activity and hunger was more complicated than expected. Campbell and Cicala (1962), for instance, discovered that when animals were placed in a rectangular cage, water deprivation did not induce the same activity patterns that food deprivation did. Although water deprivation may cause activity to increase when the animal is placed in an exercise wheel (Campbell, 1964), it rarely does so when the subject is in a normal rectangular cage. Other investigators also noted that normal cages or mazes did not elicit the same kind of activity that running wheels did (Treichler & Hall, 1962). According to Hull's theory, the nature of the apparatus, of course, should not make such a difference, but research showed that it did. In summary, various measures of activity, including wheel running, locomotion in the home cage, activity in small balanced cages, circular fields, etc., all seem to yield somewhat different results (Tapp, Zimmerman, & D'Encarnacao, 1968). This made it difficult to support any notion of drive as an indiscriminate energizer of all behavior.

In addition to the factors cited above, various investigators showed that changes in activity as a function of deprivation were not uniform across all species of animals. Thirst, for example, did not provoke any increase in activity for chicks, guinea pigs, hamsters, or rabbits. Rabbits showed no increase in activity for food deprivation in a rectangular cage, but small chicks did (Campbell, Smith, & Misanin, 1966b). Again, the general conclusion is that deprivation does not lead to a generalized drive state that indiscriminately activates all behavior.

Finally, studies by Bolles and his associates have helped to clarify the relationship between food deprivation and activity. In one study (Bolles & Stokes, 1965), rats were fed at intervals of either 19 or 29 hours. These times are not synchronous with the normal 24-hour cycle. The authors found that the rats were unable to anticipate the feeding times by becoming active. The anticipation of food that was observed in previous experiments apparently was not controlled by internal or external cues, but rather by a 24-hour "clock." Bolles and his colleagues went further to show that a second important factor was the level of illumination (Bolles, 1968, 1970b). Darkness enhanced activity (which is caused by anticipation of feeding and the 24-hour clock), whereas the presence of light inhibited wheel running.

In summary, food deprivation, or any other deficit (such as lack of water) in a biological system, does not induce a general drive that uniformly energizes all behaviors. The factors that seem most important are anticipation of feeding, the normal 24-hour cycle, and the illumination level. In addition, activity levels show a complex relationship between the type of species and the apparatus that is used as well as the type of biological deficit.

Wheel-running activity and food deprivation, of course, do show the predicted relationship. What accounts for this relationship? We still aren't sure, but a plausible theory is that food deprivation alters the response hierarchy of the animal; that is, hunger has a selective function on the response tendencies of the animal, not a general energizing function. First, Woods and Bolles (1965) found that animals that were reared with food scattered all over the floor of the cage increased their nonobject exploration and decreased their object exploration when they were later food deprived. In contrast, animals that were raised with food placed in a fixed location in their cage decreased their nonobject exploration but increased object exploration. This suggests that previous experience with food sources interacts with hunger to produce a change in the *type* of behavior that is executed during deprivation.

In a more detailed study, Wong (1979) examined various types of activity patterns during deprivation. Each test chamber had six manipulanda (levers, strings, and so forth) as well as various opportunities for exploring and locomotion. The response to deprivation varied. Manipulation and exploration increased with deprivation, and immobility decreased. Males spent more time grooming than females did but, unlike the females, their scores decreased with deprivation. And as shown in Figure 10–1, there were large increases in sand digging with food deprivation but much smaller increases in nosing the food cup. All these results suggest that drive *selects* the response, it does not energize all of them. That is, drive lowers the threshold for behaviors that are ecologically related to food seeking (see also Shettleworth, 1975, and Chapter 6).

Deprivation and Instrumental Learning

One of the most commonly observed phenomena is the relationship between instrumental performance and deprivation level. Many studies have shown that the stronger the motivation, the better the performance in an alleyway (Pavlik & Reynolds, 1963; Zaretsky, 1965) or a lever box (Kimble, 1951; Marwine & Collier, 1971). The relationship holds even when the reinforcer is an unconventional one. For example, animals will press a lever at a faster rate as a function of food deprivation when the reward is merely the smell of the food (Allen, Stein, & Long, 1972).

The length of food deprivation also has a strong effect on resistance to extinction. The stronger the motivation during extinction, the greater the resistance to extinction. This is not surprising, but what is more interesting is the finding that high drive during acquisition affects later resistance to extinction (Barry, 1958; Campbell & Kraeling, 1954; Theios, 1963). In each of these studies, all animals were tested during extinction under the same

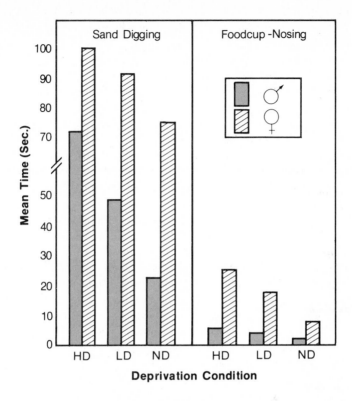

Figure 10-1. Mean time in seconds performing either the sand-digging or foodcup-nosing behavior for male and female rats under either no drive (ND), low drive (LD), or high drive (HD) conditions. From Wong, 1979)

deprivation level. But some had experienced different levels of deprivation during acquisition; this affected their extinction performance. In particular, stronger drive during acquisition tended to "carry over" and increase extinction performance beyond the level of subjects whose drive was constant throughout. Such results suggest that drive level affects learning: the stronger the motivational level during acquisition, the stronger the response expectancy and performance. We should quickly point out that this persisting effect of drive level on extinction behavior has not always been found (Leach, 1971). Furthermore, Capaldi (1972) has shown that the effect may depend upon a variety of conditions including the reinforcement schedule during acquisition. If motivational level does affect response learning, as opposed to merely affecting performance level, then it probably does so by providing a very salient set of internal cues that become part of the stimulus complex (see Logan, 1956).

Finally, let us consider contrast effects as a function of shifts in deprivation level. Recall from Chapter 4 that changes in reward magnitude (incentive) and delay cause overshooting to occur following the shifts: animals that are shifted to a lower level of reward perform more slowly than those maintained at that lower level all along; similarly, animals shifted to a higher level of reward overshoot the level of performance of the animals maintained at that high level throughout. The same sort of phenomenon is observed following shifts in deprivation level (Capaldi, 1971; 1973; Mollenauer, 1971). In the Mollenauer experiment, animals were trained to run down an alleyway for food. The deprivation levels were either .5 to 1 hour or 22 to 22.5 hours. After many trials were given, some of the animals were shifted to the alternative deprivation level. Mollenauer observed contrast. As shown in Figure 10–2, the speed of running for the animals that were shifted to a higher deprivation level exceeded the speeds of those animals that were maintained at that deprivation level all along. Likewise, animals shifted to a lower deprivation level, from 22.5 hours to 1 hour, overshot the performance level of animals in the constant 1-hour group. Shifts in deprivation level therefore appear to function like shifts in reward magnitude. Mollenauer's experiment highlights not only the effect of these deprivation shifts, but also the fact that in order for contrast to occur, extended training must be given (see also Capaldi & Hovancik, 1973).

Irrelevant Drives

One of the important predictions from Hull's theory is that different drive sources combine to produce a single generalized drive state. If a hungry animal were responding in order to receive food, the addition of thirst, an irrelevant drive, should enhance performance because thirst and hunger should combine as one source of generalized drive. As we discussed previously, the evidence for this idea is quite weak (for example, Young & Black, 1977; but see Millenson & de Villiers, 1972). But this issue is more complicated. Some results have been found showing that irrelevant drives do influence performance. In one study (Capaldi, Hovancik, & Davidson, 1979) hungry animals were placed in an alleyway and allowed to run to the goal box for reward. For some animals, the reward was always food. Other animals received food on some trials and water on other trials. Finally, some animals received food on some occasions but no reward on others. In Phase 2 of the study all the subjects received the food/no reward treatment. The authors found that performance for animals in the group that had received water on alternate trials during Phase 1 was disrupted. This confirms that they had learned about the "irrelevant" water reward during Phase 1; they had come to expect water on alternate trials and when, in Phase 2, they experienced no reward in the goal box, their performance was disrupted. Thus, animals appear to learn about the rewards for irrelevant drive states.

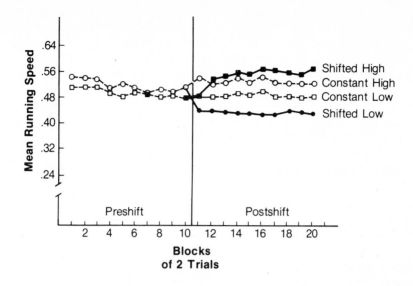

Figure 10-2.   Mean running speed as a function of blocks of two trials during the preshift and postshift phases of the study for groups that were high or low deprived, or were shifted at trial 75. (From Mollenauer, 1971)

Part of the explanation for the findings cited above is that hunger and thirst are never isolated. When an animal is food deprived, it voluntarily reduces its drinking so that it ends up being thirsty as well. Similarly, when an animal is deprived of water, it voluntarily cuts down on its food, thus making itself both hungry and thirsty (McFarland, 1964; Verplanck & Hayes, 1953). Such an interaction can have interesting consequences. For instance, Capaldi, Hovancik, and Lamb (1975) gave subjects one trial per day in a straight alleyway. The reward was food and the motivation was either hunger or both hunger and thirst. They discovered, contrary to the prediction of Hull's theory, that the addition of the thirst drive reduced performance levels because of a reduction in the value of the food reward. It is well known that rats, among other animals, require water in order to eat; a given amount of food simply isn't as reinforcing if the animal is extra thirsty since the animal needs water to eat the dry food (see also Hsiao & Trankina, 1969).

Drive Measurements

As we discussed in Chapter 1, we use performance to infer learning; we cannot observe learning directly, but only indirectly through performance. The same argument must be made in the case of motivation: we cannot observe motivation directly (except perhaps in the sense of particular biological changes that accompany deprived states); we must assess the motivation by means of the animal's performance. The challenge of measuring drive in a precise manner, therefore, has stimulated many scientists. One early theorist was Warden (1931). In his experiments, animals were allowed to cross an electrified grid in order to receive a goal object. The more readily the animals crossed the grid, the more motivated they were claimed to be. For example, if an animal were placed on the start side while it was hungry, its score, its latency to cross the grid, could be compared to another animal that was placed on the start side while it was thirsty. Furthermore, various levels of hunger and thirst could be assessed. It was even hoped that eventually the comparative strengths of many different sources of drive could be computed. For instance, it might be found that 12 hours of food deprivation were needed to produce the same number of grid crossings that 24 hours of water deprivation produced. Similarly, 24 hours of water deprivation might be equivalent to, say, depriving a young rat pup of access to its nursing mother for 1 hour. Warden's work did provide a good starting point for investigating the comparative strength of different sources of motivation, but unfortunately modern critics believe that this early research is of limited value (see Bolles, 1975).

Quite a different approach to the problem of measuring motivational states has been taken recently by McFarland and his associates (for example, McFarland & Lloyd, 1973). The purpose of the research program has been to measure the momentary state of the animal, to determine whether it is primarily hungry or thirsty at any given point in time, as well as to provide an elaborate theoretical framework for understanding the interaction of motivational states in general (see McFarland & Sibly, 1975). The technique assumes that an animal that is both hungry and thirsty is predominately hungry *or* thirsty at any given point in time. At one time, the animal might be primarily hungry; hunger is dominant. At some later time, presumably because the animal has eaten food, the animal might be primarily thirsty; thirst is said to be dominant.

The way to assess when an animal is primarily hungry or thirsty was shown by Sibly and McCleery (1976). Doves were put in a box containing two plastic disks. Pecking one disk provided food, whereas pecking the other disk resulted in water reward. It was observed that the animal, when first placed in this situation, began to peck at whichever key provided the reward that satisfied the "strongest" drive. For example, the dove would begin pecking at the food disk and continue to do so until it had achieved a number of food rewards. However, the animal would eventually switch to the other type of reward at which point the authors would interrupt it by turning out the house lights for one minute. The purpose of the interruption was to ensure that the switch was not merely a temporary phenomenon.

If, after the interruption period, the animal resumed the activity it was doing previously, then it was claimed that it was still in the same motivational state as before. However, if the subject continued to respond to the key to which it had *changed,* then the authors suggested that its motivational state had switched. In other words, merely because the animal changed from one type of reward to the other did not necessarily mean that its dominant motivational state had changed. Only when the animal switched and *remained* with its new choice for at least more than one key peck could it be said that the dominance had switched from one motivational state to the other. Of course, after responding to this new key for a period of time, the animal may switch back to the original reward. In fact, animals show a predictable switching from one reward to the other and back again. But by using the interruption technique, it is possible to know precisely what motivational state the animal is in at any given moment. This technique for measuring motivational states does not explain in any sense how those states are operating, but it does provide a framework for measuring and interpreting those states. In addition, it represents a method by which the action of other factors, such as response difficulty (Larkin & McFarland, 1978) and reward magnitude (Sibly, 1975), can be assessed.

Acquired Motivation

The section above considered the regulatory systems, such as hunger and thirst, and how they affect various behaviors. In this section we want to consider acquired motivation (see D'Amato, 1974). Recall our discussion in Chapters 3 and 5 in which we claimed that Pavlovian conditioning involves behavioral reactions, expectancies, and emotion or motivation. We shall focus on this third idea, that motivational states can be conditioned; these states, in turn, serve to energize and select behavior.

Fear

It has long been known that fear can be learned. One of the earliest and most important demonstrations of this was the work of N. E. Miller (1948, 1951). In the earlier study, rats were placed in the black side of a two-compartment box and then were shocked. This, of course, made the animal run to the other side of the box, which was painted white, to escape shock. After this procedure was repeated a number of times, the animal was placed in the black side again but no shock was administered. Miller found that the animal ran to the white side without even being shocked. He also showed that the rat would learn to press a lever near the door to gain access to the white side. The black color, the aversive CS, had elicited fear, which, in turn, motivated the acquisition of a new behavior (the lever press).

Experiments of this sort provided strong support for the two-factor theory of avoidance learning discussed in Chapter 5. The important point for our purposes here is that fear, a motivational state, may be learned. Bear in mind, though, that certain other fears may be innate (see, for example, Hebb, 1946).

Conditioned Appetitive States

Although we have emphasized here and in Chapter 5 that appetitive motivation can be conditioned, that a CS that predicts food will elicit hope, the literature historically has failed to show that the subject will *act* motivated during the CS. In other words, although the interaction studies reviewed in Chapter 5 suggest that appetitive motivation is, indeed, conditioned, other evidence, which has examined this issue more directly, has not been sympathetic to the idea of conditioned appetitive drive states (see Cravens & Renner, 1970; D'Amato, 1974). What must be shown is that a Pavlovian CS will elicit a state of, say, hunger such that the biologically satiated animal will actually eat food because of the incentive cue, not because it is "biologically hungry."

The general technique used in the early research was to deprive an animal of food while it was in the presence of a distinctive cue. Later, when the animal was fully satiated, the CS was presented. It was believed that the CS would induce hunger because it had been associated with that state previously; if it did, then the animal would consume more food than it otherwise would consume. Although some investigators reported positive results (for example, Wright, 1965), the overwhelming evidence did not support the notion of conditioned hunger (for example, Cravens & Renner, 1969; Wike, Cour, & Mellgren, 1967). The conclusion, therefore, was that while external cues could become associated with rewards and thereby elicit incentive motivation (Zamble, 1974), external cues that were associated with deprivational states did not become conditioned motivators; they did not induce a conditioned drive state.

Part of the problem was that deprivation is slow to appear in an animal. This is in sharp contrast to aversive stimuli like shock which are very quick acting. It may be that conditioned hunger is difficult to show because the US, the state of hunger produced by food deprivation, has such a slow onset. In order to overcome this difficulty, investigators began using biological agents which would rapidly induce the state of hunger. For example, Balagura (1968) housed rats in a cage containing a lever. The subjects could press the lever at any time in order to receive a food pellet. During training, the animals were injected with insulin and then placed in the lever cage for two hours. The CS in this instance was the injection procedure itself, that is, the handling and insertion of the needle. The US, of course, was the insulin. It is known that insulin causes a change in the blood sugar content of the animal (UR) and it induces eating. Alternatively, this means that insulin rapidly induces hunger. Following training, Balagura gave the animals an injection of normal saline. This substance, as discussed in Chapter 6, is biologically inactive; it will not produce any change in the blood sugar. And yet the animals that had received injections of insulin pressed much more on the bar press tests after they received the saline injections. In other words, the CS, the injection procedure, induced a state of hunger based upon the previous insulin injections. Even though the animals on the test received saline, and thus were "physiologically" normal, nevertheless the CS induced hunger in them such that their bar pressing, and food consumption, was maintained at a high level.

Other investigators had trouble confirming the notion that hunger could be conditioned even with a rapid-acting drug like insulin. Siegel and Nettleton (1970), for example, conducted an experiment similar to Balagura's. The difference was that some animals were not permitted to press the lever for food while they were under the influence of insulin. That is, they were given an insulin injection, and placed into the lever cage, but the lever was withdrawn during that period of time. On the test, when all animals were injected with saline and placed in the lever cage with the lever now present, only the animals that had received insulin *and* had been given access to the lever previously showed the conditioned effect. Animals that were injected with insulin but did not have access to the lever during the training phase did not show a conditioned hunger effect on the test; they did not press the lever to get food even though they knew perfectly well how to obtain food by lever pressing. This finding suggests that Balagura's experiment was in error. In his study, the lever pressing by the animals that had access to the lever all along had simply become an acquired instrumental response. This seems plausible since insulin naturally lowers glucose levels which can be relieved by feeding (that is, by lever pressing).

There is one other methodological consideration that may have an important bearing on the issue of conditioned hunger. Recall that taste aversion conditioning is most potent in rats when a flavor or odor is used as a CS. When external cues are used, such as lights or tones, conditioning is much weaker (see Chapter 6). In addition, slow-acting gastric distresses, like vitamin deficiencies, can easily become associated with the taste of a diet. If we extrapolate from these results, we might conclude that internal appetitive states, in this case hunger, would be more easily associated with internal cues such as taste. Indeed, several studies have shown that a state of satiety can be conditioned. Booth (1972) paired an odor or taste with an injection of a calorically dense substance. The substance caused the animal to be satiated, that is, to lack hunger. Later, the CS was found to inhibit feeding behavior, presumably because it induced a conditioned state of satiety in the rat. Given this result, it surely would appear that the use of odors or tastes as CSs would provide an effective methodology for the conditioning of hunger states too.

This strategy was used in several studies by Mineka (1975). In one experiment, animals were given a distinctive taste prior to experiencing intense hunger; a different taste was paired with mild hunger. During the test, while the animals were moderately hungry, they were presented with either the "strong hunger" taste or the "weak hunger" taste and were given access to food. If the taste induced hunger, then the animals should show a quicker reaction to eat food after the "high hunger" taste than after the "low hunger" taste. Similarly, they should eat more food when given the "high hunger" taste than the "low hunger" taste. Evidence *was* shown for a conditioned hunger state; the animals behaved according to the predictions. It would appear therefore that conditioned hunger is possible and it is demonstrated most readily when an internal cue such as taste rather than an external cue is used. Unfortunately, additional studies by Mineka (1975) failed to show a very strong conditioned hunger effect and her final conclusion was that conditioned hunger is, at best, elusive. Certainly the use of taste cues as CSs is more appropriate than external stimuli, but the phenomenon is neither reliable nor strong.

Before we comment further on the theory about why conditioned appetitive states are not as strong as conditioned aversive states like fear, let us first consider the efforts to condition thirst. One study by Seligman, Ives, Ames, and Mineka (1970) used a technique in which thirst could be rapidly induced by the injection of hypertonic salt plus procaine (an anesthetic which eases the pain of the salt injection). The CS in this study was 1 hour of deprivation plus placement in a white box. Note that the 1-hour deprivation should also produce some internal cues that, in turn, would function as CSs. Strong effects were observed. Virtually all the animals showed heightened drinking later in the presence of the CS. The authors concluded that conditioned drinking had, indeed, occurred.

Subsequent research, however, cast doubt on that conclusion. First of all, no conditioned drinking was observed to a hypertonic salt solution by itself; the salt really is what should have induced the thirst to begin with. It appeared that the procaine was more responsible for the conditioned drinking observed previously (Seligman, Mineka, & Fillit, 1971). Even more important were experiments demonstrating that these thirst-inducing agents produced standard taste aversions in rats (Mineka, Seligman, Hetrick, & Zuelzer, 1972). It was not the CS, therefore, that was inducing thirst so much as it was an ordinary aversion, the reaction to which was drinking (because drinking relieved the illness). Other studies also cast doubt on this phenomenon. For example, Weisinger and Woods (1972) used a fast-acting thirst inducer (formalin) as well as a more naturalistic CS (an odor). Conditioned drinking was observed but a later study in this series suggested that it was due merely to pseudoconditioning.

In summary, Pavlovian conditioning certainly can produce motivational states; aversive states like fear can be conditioned routinely. Conditioned appetitive states, however, such as hunger or thirst, are difficult to demonstrate reliably. The effects are marginal even when fast-acting drive-inducing agents and naturalistic stimuli are used. The "problem" lies not in demonstrating the power of appetitive CSs; the interaction studies discussed in Chapter 5 confirm that they are indeed powerful. The real problem has been in showing that appetitive CSs have immediate and *direct* effects upon the food or water consumption of the animal. Only if they do can we be sure that conditional hunger or thirst, and not some other emotional state, is present.

The reason for this discrepancy is not clear but evolutionary considerations suggest an answer. Consider an animal in its natural environment. The food that it consumes certainly can serve as a signal for calories or poisons or even vitamin deficiencies. But it is hard to imagine how the ingestion of food could ever become a signal for a decrease in calories or the depletion of energy. Animals naturally feel some degree of satiety after a meal, or they might feel ill because they ingested a poison. If their diet is very constricted and limited to one food item, then they could even develop an aversion for that vitamin-deficient diet. But it makes no sense for the taste of food to be a signal for increased hunger. Food ingestion results in *less* hunger, not greater hunger. In short, throughout evolutionary history, food has been a signal for satiety, or in some cases, illness. Since it cannot be a signal for hunger, it is possible that the conditioning mechanisms are lacking.

Conditioned Physiological Reactions

Despite the very modest demonstrations of conditioned hunger and thirst, and despite the theoretical reasons why those conditioned drive states should be difficult to demonstrate, it still would seem that hunger should be readily conditionable. The reason is that Pavlovian research has shown conditioned effects on blood sugar, thus providing a possible *mechanism* for acquired hunger states. Various experiments have demonstrated conditioned changes in the levels of blood sugar of rats (Hutton, Woods, & Makous, 1970; Woods, Makous, & Hutton, 1969). In the study by Hutton et al., a change in the blood sugar of rats was conditioned by pairing a CS with insulin. Specifically, the authors took a blood sample, injected the animal with insulin, and then placed it in a holding cage that was permeated with the odor of menthol. This procedure constituted the CS. After a 20-minute waiting period, a second blood sample was taken for the purpose of measuring any change in blood sugar that had occurred while the rat was in the cage. Two groups were used in this study. The experimental subjects were injected on 5 conditioning trials with insulin; on other occasions throughout the experiment, these animals were given a placebo injection (a biologically inactive drug). The control subjects received the same treatment (that is, the same number of placebo and insulin injections), except that their 5 insulin injections did not follow the CS presentation. Thus, for the experimental animals, the CS was paired 5 times with insulin, but the control subjects received a placebo on those 5 CS trials and insulin was administered at other times. After the training procedure was completed, all animals were tested for a conditioned blood sugar reaction to the CS alone.

The results of the experiment are illustrated in Figure 10–3. The 5 CS trials are shown on the left portion of the figure. It is clear that blood sugar was noticeably reduced in the experimental subjects but not in the control animals. This is due to the biological action of insulin. The test results are shown on the right portion of Figure 10–3. During this phase all subjects received the CS plus a placebo injection. The control subjects for whom the

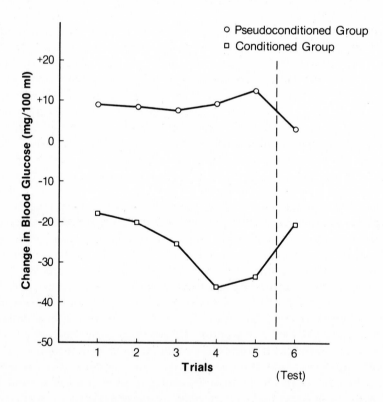

Figure 10-3. Mean change in blood glucose level as a function of training and testing for the conditioned and control groups. (From Hutton, Woods, & Makous, 1970)

CS had never been paired with insulin showed no change in blood sugar. The experimental animals in contrast showed a decrease even though the injection on that test trial had been a placebo drug. This physiological change was brought about by the CS alone; it was a conditioned blood sugar reaction, not merely a reflex to insulin.

The control of physiological reactions like blood sugar level is an enormously complex process. Research has indicated that if the proper drugs are used, procedures nearly identical to those just described can result in conditioned hyperglycemia, elevation rather than reduction of blood sugar (Siegel, 1972a). Moreover, the changes that are conditioned are not necessarily dependent upon actual blood sugar changes. Woods (1976) has shown that merely receiving insulin will lead to a conditioned hypoglycemia in the future. In that experiment, the blood sugar changes that normally are produced by insulin were eliminated by adding the appropriate amount of glucose to the insulin injection. Conditioned effects were still found though.

It is interesting to speculate that these conditioned blood sugar changes might be a mechanism for conditioned motivational states. Conditioned hypoglycemia, in one sense, is a conditioned hunger state since low blood sugar is associated with starvation. Conditioned hyperglycemia, on the other hand, is associated with satiety. If future research clarifies how this type of mechanism produces motivated behavior (an increase in food or water consumption), then the area of conditioned appetitive states will have been clarified a good deal. We do know that these conditioned blood sugar reactions can operate in the natural environment to control food intake. For example, the ingestion of a sweet-tasting substance is invariably followed by an elevation of blood sugar. Sweet tastes come from sugars which cause the blood sugar to rise and the natural defense mechanism against an elevation in blood sugar is the secretion of insulin. Therefore, if an animal ingested a sweet-tasting substance, the body would normally react by secreting insulin to counteract the elevation in blood sugar. This reaction occurs even when the sweet-tasting substance is nonnutritive, that is, when it has no biological effects on the body. As shown by Deutsch (1974), saccharin will induce hypoglycemia in rats; in addition, a novel flavor paired with a glucose injection also will elicit hypoglycemia. It is likely that such a mechanism helps to control food intake. When an animal experiences the sweet-tasting substance, it will inhibit further food intake. In summary, the research on conditioned physiological changes extends the normal Pavlovian technique to an area of considerable interest. It may be that these physiological reactions serve as mechanisms for diet regulation. As such they may represent some of the underlying changes that accompany motivated states.

Stimulation Theories

Experimental psychology has had a variety of stimulation, or activation, theories of motivation. The one that we shall discuss in detail here is the opponent process theory developed by Solomon and Corbit (1974). The overall goal of the opponent process theory is to account for the dynamics of motivation, that is, the various changes due to sensory stimulation. In doing so, it views the subject as a dynamic element in the motivational process.

Opponent Process Theory

The opponent process theory of motivation is a stimulation theory; that is, it provides a theoretical account of the motivational effects of strong stimulation on an organism. The theory is especially suited to explain how sudden onset and offset of stimulation affect the motivational state of the organism. Slow-acting events, like the slow onset of hunger, or even conditions that are slow to terminate, like low body temperature after severe cold, are less appropriate to this theory. But in terms of rapidly acting stimuli, for example, the sudden appearance of a predator or the sudden presentation and withdrawal of a valued object, the theory is quite adequate.

Our discussion of the opponent process theory would be facilitated by citing an example which the theory is designed to explain. Church, Lolordo, Overmier, Solomon, and Turner, (1966) reported the results of an experiment dealing with the effects of shock on heart rate. A dog, wired to a heart rate monitor, was placed in an apparatus and given shock. On the first few occasions, there was a very large cardiac acceleration; this was accompanied by signs of stress such as vocalization and severe struggling as well as pupil dilation and defecation. At shock termination, the heart rate suddenly decreased dramatically; in fact, it "rebounded" well below the normal baseline of the animal. Only then did it gradually recover to baseline. After repeated stimulation, the pattern changed somewhat; the animal still showed heart rate acceleration and stress during shock but the levels were much lower. However, the rebound effect was still present; in fact, the heart rate deceleration seemed to be a stronger and more lasting reaction to shock as training continued. Despite the changes in intensity of the acceleration and deceleration, the general pattern was the same: an initial period of heart rate increase followed by a period of decline. These results are shown in Figure 10–4.

The opponent process theory of motivation explains these data quite nicely. First, strong stimuli induce not one motivational state, but two. In particular, cues initiate a primary motivational state, labeled State A, and at the same time, a second motivational state, labeled State B. State B is an

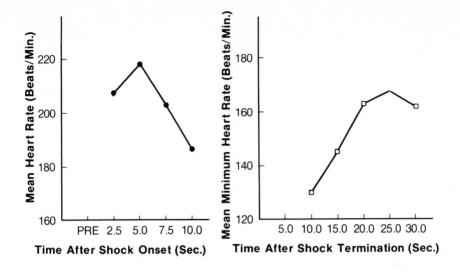

Figure 10-4. Mean heart rate (beats per minute) as a function of time after shock onset (left) or shock offset (right). (From Church, LoLordo, Overmier, Solomon, & Turner, 1966)

opposing state, hedonically opposite to that of State A. So, for example, if an animal is shocked, the aversive State A is first apparent, but, at the same time, State B, the "pleasant" state, is also developing. In short, the opponent process theory claims that two emotional or motivational states are induced by strong stimuli. The first, State A, is opposed by the second, State B. It is important to note that it makes no difference whatsoever whether the strong stimulus is an aversive cue, such as shock, or an appetitive cue, such as food. State A will be aversive (and B will be pleasurable) if the stimulus is aversive; likewise, State A will be pleasurable (and B will be aversive) if the stimulus is pleasurable.

Let us be more specific about these motivational states. State A has a very short onset time and its maximum intensity is felt almost immediately. Furthermore, State A does not habituate very readily. It is a UR to a strong US and, by definition, is a persistent reaction. State B, in contrast, is a "slave" process; it is activated by the onset of State A. It is not a "rebound" in the normal sense because it does not occur as a result of the *termination* of State A, but rather it occurs as a result of the *onset* of State A. State B, unlike State A, is slow acting; it takes a much longer time to reach its peak and it subsides only slowly.

Two additional points are crucial to the theory. First, the overall emotional or motivational consequence is a function of the summation of

States A and B. That is, the motivation felt by the subject is not the full intensity of State A, but rather the motivation caused by State A minus that caused by State B. Consider Figure 10–5. The marker at the bottom of the figure indicates when the stimulus event is turned on and off. At the top of Figure 10–5 we have represented the two underlying opponent processes, States A and B. Baseline is at point 0. When the stimulation is turned on, State A immediately rises to its maximum, but State B, as discussed above, only slowly develops its strength. Now consider the middle portion of Figure 10–5. This represents the net motivational response, that is, the summation of States A and B. If we add together the opposing processes at each point in time, the curve shown there is the result. Specifically, the animal experiences a sudden onset of State A followed by a slight cessation in the strength of this state; this corresponds to the development of State B. When the stimulation is turned off, State A quickly subsides; but because State B only slowly declines, there is a rebound effect. In other words, following stimulus offset, State A is rapidly eliminated leaving only State B operating. The animal experiences this rebound and then slowly returns to baseline. This is precisely the reaction observed by Church et al. (1966) in connection with their heart rate experiment. When shock was presented, the aversive nature of the shock was manifest by the acceleration of the heart rate. Once the shock was terminated, however, the rebound effect was evident; State B caused a deceleration of the heart rate below baseline which then slowly recovered.

The second additional point, illustrated on the right portion of Figure 10–5, is that the State B is strengthened with repeated experiences of the stimulation (although State A is *not* strengthened with repeated trials). To be specific, State B gets stronger, it lasts longer, and has somewhat faster onset and offset times as a function of repeated exposure to the stimulus. Consider the consequences of these results on the right portion of Figure 10–5. After many stimulations, with State B now being much stronger, the net effect shown in the middle of Figure 10–5 is a much smaller reaction to the stimulation onset but a much larger rebound effect after the stimulation has been turned off. What this would mean in terms of the experiment on heart rate in dogs would be that after many shocks, the animal would not demonstrate the massive increase in heart rate during shock but would show a much larger rebound effect once shock was turned off. In terms of the language of motivation, the animal would "feel" less aversion during shock than it did early in training, but much greater "relief" following the offset of shock than it did early in training.

Let us summarize the opponent process theory of motivation. Intense stimulation induces two motivational states, A and B. State A is the primary state as it reflects the direct consequences of the stimulation (either aversion or pleasure). State B is a slave process; it also begins with the onset of the stimulation but it is a process that opposes State A; it is hedonically opposite

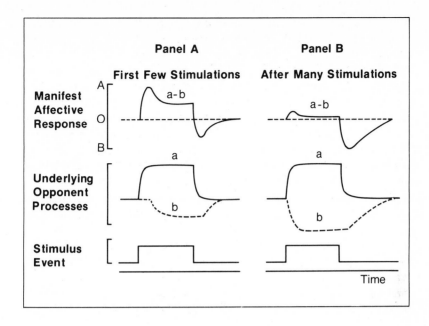

Figure 10-5. Schematic diagram showing the operation of the opponent process theory of motivation. (From Solomon & Corbit, 1974)

to it. The net effect, that is, the resulting motivational state of the animal, is the summation of these two states. But because State B has a slower onset time, has a slower offset time, and increases in strength and duration with repeated stimulation, a rebound effect is observed when stimulation is suddenly terminated. And this rebound effect increases with continued experience.

Illustrations of the Opponent Process Theory

The opponent process theory has provided an interesting perspective on a variety of behavioral phenomena including mood shifts in human beings and, most notably, drug addictions (Solomon, 1977). In the animal learning and motivation literature, many phenomena have been reinterpreted in light of Solomon's theory. For example, it is known that animals that are given shocks while pressing a lever for food will increase their response rate immediately following the offset of shock (Weiss & Strongman, 1969). The shock offset induces a burst of responding. Traditionally, it was believed that this phenomenon resulted because the inhibition due to shock was relieved. The animal's lever pressing was suppressed by the shock (and certainly by any CS that would accompany the shock such as in a standard CER experiment), but the offset of shock was an event that predicted a relatively safe period of time. At no other point during the session would the subject be as "sure" that the next period of time would be shock free. The burst in responding, therefore, suggested that the animal "wanted" to respond maximally during this safe period before it would have to concern itself with future shocks.

The account provided above is a plausible one. However, LaBarbera and Caul (1976a, 1976b) have suggested a different explanation based upon the opponent process theory. In one experiment, the CS-shock pairing was preceded by shock; that is, animals received one shock, then a CS, then a second shock. The control subjects received the usual CER training; that is, CS-shock pairings. Lever press responding was predictably suppressed during the CS for both groups. But lever pressing was more severely suppressed in the control animals that did not receive the first shock. In contrast, the experimental animals, those that received shock-CS-shock, pressed the lever more during the CS. On the surface at least, this difference makes no sense because both groups should anticipate the shock following the CS to an equal degree. However, the opponent process theory offers a good explanation of this phenomenon. The first shock, the one given to the experimental subjects prior to the CS, induced an aversive state and then a pleasurable "rebound." The rebound, in fact, was occurring *during* the CS presentation, thus counteracting the conditioned fear that was aroused by the CS. The control animals had only the CS-induced fear; they were not experiencing a rebound (State B) during the CS and, therefore, their suppression during the CS was not reduced. In summary, the postshock burst observed frequently in CER experiments appears to reflect a rebound phenomenon, that is, the operation of State B, rather than the anticipation of future shocks.

Another phenomenon in animal learning that has been explained in terms of the opponent process theory of motivation is imprinting (Hoffman & Solomon, 1974; Starr, 1978). First, it is well known that young organisms, such as ducklings, emit high-pitched distress calls when they are in a novel or aversive situation (see Chapter 6). The suppression of these distress calls by the presentation of an imprinting object is taken as evidence that the object is reinforcing. Moreover, distress calling increases when the duckling is separated from the imprinting object. There is, however, a paradox here: the distress calling following the withdrawal of the imprinting object obviously cannot be due to novelty or aversion. For one thing, the separation-induced distress develops with repeated exposures to the imprinting stimulus (Hoffman & Solomon, 1974). Repeated exposures, by definition, mean that the stimulus is familiar, not novel. Initially, distress calling following the removal of the imprinting object is moderate but later it becomes much stronger. The cause of the distress, therefore, could not be the novelty of having the stimulus removed. Second, if a duckling that is not giving any distress calls is given a single extended exposure to the imprinting object, distress calling occurs upon the withdrawal of the object. Again, the distress does not seem to be induced by fear or novelty as is usually the case.

The opponent process theory of motivation has provided an explanation for this phenomenon. According to the theory, the imprinting object induces both State A (pleasure) as well as the opponent motivational process, State B (distress). Upon the withdrawal of the stimulus, State A immediately subsides to baseline, but State B only slowly recovers. This means that the subject has a residue of motivation, the opponent State B or distress, following withdrawal of the imprinting object. Also in accord with the opponent process theory, repeated stimulations lead to a magnification of State B; this is shown by the more prolonged and intense distress calling (State B or rebound) following object removal (Starr, 1978). The use of the opponent process theory to explain some imprinting phenomena confirms that the initial motivational process, State A, can be pleasurable or positive as well as aversive or negative.

Nonregulatory Motivational Systems

It is clear that not all motivation stems from the bio-regulatory systems; organisms also seem motivated by the action of nonregulatory systems (see Eisenberger, 1972). An example of a nonregulatory motivational system is the exploration drive. It is thought that learning is influenced by this factor to the extent that animals are curious and have a need for exploration. In addition, various factors which affect exploration tendencies conceivably modify the incentive value of reward objects. Animals might respond to particular situations involving reward not only in terms of their biological needs—hunger and incentive—but also in terms of their exploration tendencies.

Exploration

The literature on exploration, curiosity, and so forth is extensive. In terms of learning theory, research has clearly shown that exploration, or more precisely the opportunity to explore, is highly reinforcing to animals. For example, rats will learn to make the correct choice in order to gain access to a more complex maze (Montgomery, 1954). Similarly, monkeys will learn a complex discrimination response in order to gain an opportunity to explore visually the environment outside their cage (Butler, 1953). These effects are not simply momentary; exploratory tendencies can develop at an early age and have a strong influence later on. Sahakian, Robbins, and Iverson (1977) compared subjects that were reared in isolation with those that were not. The authors discovered that exploratory behavior was dramatically affected by the isolation experience. Isolated animals were more active and had a higher frequency of contact with a novel object when tested later on. In addition, these animals showed a stronger preference for a novel environment as compared to a familiar environment.

One interesting aspect in this literature was the finding that exploration is responsive to sensory deprivation. A series of experiments by Fowler (1967) showed that brief deprivation of sensory experience heightened an animal's exploratory tendencies. Animals that were deprived of patterns in their environment learned to choose the correct arm in a T maze in order to be rewarded by such patterns. The more they were deprived, the more the patterns were reinforcing. In many ways, these results suggest that exploration or curiosity operates like the physiological drives of hunger and thirst. Periods of deprivation lead to heightened drive states (heightened exploratory tendencies) in the same way that food deprivation leads to increased hunger. Similarly, animals that are presented with novel objects show a satiation effect; their curiosity and exploration declines with such experience just as hunger declines after getting food.

It should not be forgotten that novel objects or environments can induce fear (neophobia) as well as curiosity. Blanchard, Kelley, and Blanchard (1974) placed rats in a novel environment for only a few minutes. The next day, the animal's familiar home cage was attached to this environment and the latency to cross from the home cage into the environment was measured. Compared to control subjects, the animals that were previously exposed to the environment showed a reduced latency to enter it. The experience had caused the fear to diminish. In a second study, the authors placed animals in a novel environment and permitted them to cross an electrified grid in order to escape this environment. They found that the rats did, in fact, escape; again, the novel environment seems to have elicited fear. It would therefore appear that animals strive for some kind of optimal stimulation level (Fiske & Maddi, 1961). Changes to both a more complex and a less complex environment seem to be reinforcing (Taylor, 1974). When a subject is deprived of stimulation, change is reinforcing; however, too much change can be fear provoking.

Spontaneous Alternation

The relationship between exploration and learning is nicely exemplified in the literature on spontaneous alternation. If a rat is placed in a T maze and allowed to choose between the arms (without receiving reward for either), the animal invariably alternates its choices, first choosing one arm and then the other arm on the next trial. Alternation behavior is observed even when a long delay, an hour or more, is given between the first and second trial (Still, 1966) or when animals have a choice between a familiar and novel flavor (Holman, 1973). The phenomenon of spontaneous alternation has intrigued a number of psychologists because it is not clear why the animals do not choose in an essentially random fashion; after all, there is no obvious reward and, therefore, nothing in particular should influence their choices.

The early literature, summarized by Dember and Fowler (1958), suggested that the spontaneous alternation behavior was due to two factors: avoidance of the choice just made, presumably because the animals were "bored" with that choice, and an active approach to the novel choice, presumably because the animals were curious about this new choice. These two factors are precisely those discussed above; exposure to one set of stimuli can lead to a sensory satiation effect and enhance curiosity.

More recent studies have attempted to identify the stimulus that is responsible for the alternation behavior, that is, the cue that the animals use to remember what choice they made previously. The most thorough study was by Douglas (1966). He attempted to evaluate the various sources of information, including odors, visual and orientation cues, etc. He found that there were two sources of stimulation that aided the animal in making its spontaneous alternation response. First, the animals had a relatively weak tendency to avoid their own odor from previous trials. Second, rats had a much stronger inclination to use special cues coming from outside the apparatus, for example, lights and other visual features of the room. Although other investigators have not replicated all of Douglas' findings (Bronstein, Dworkin, Bilder, & Wolkoff, 1974), the orientation of the maze within the room does seem to be a critical factor. For instance Douglas, Mitchell, and Del Valle (1974) varied the angle of T maze at the choice point. For some animals, the two choice arms of the maze were essentially straight continuations of the approach alleyway. For others, the angle of displacement was as great as 330 degrees. The authors found that unless the angle of the choice alleyways was 45 degrees or greater, little spontaneous alternation was observed; animals chose in essentially a random fashion. However, as the angle of the two choice stems increased beyond 45 degrees, the percent spontaneous alternation behavior increased as well. A more divergent angle, of course, means that the two alleyways are more discriminably different in terms of the spatial cues coming from the room. It is easy to distinguish between the two alleyways when, in terms of the lights in the room, they are diverging noticeably.

Let us now consider another example of alternation behavior as reported in a study by Olton and Samuelson (1976). Their experiment is especially interesting because it illustrates some of the behaviors that might be important to an animal in its natural environment. Olton and Samuelson constructed a special maze consisting of a central start box and eight arms radiating out from the center like spokes on a wheel. Rats were placed in the start box and were allowed to choose any of the eight arms. Food reward was situated at the end of each arm. After one response had been made, the subjects were put into the start box again and given the opportunity to make another choice. The results indicated that animals chose, on the average, about seven different arms within their first eight choices. In other words, animals did not persist in choosing the first arm that was rewarded. They seemed to "ignore" or avoid returning to those locations until all arms had been visited at least once. Additional experiments showed that this alternation behavior was accomplished through a memory process: The rats remembered which maze arm they had visited in the past and, on that basis, chose a new arm; choices were not made on the basis of odor cues or other environmental stimuli.

Much of the work in learning focuses on the repetition of rewarded behaviors. After all, rewarded behaviors are successful and, therefore, from an evolutionary perspective, they should be repeated if the animal is to behave adaptively. However, the work by Olton and Samuelson on alternation, as well as the general work on exploration, provides an interesting and important perspective on motivated behaviors. In particular, it suggests an important principle that is operating in complex situations: animals must explore their environment if they are to gain maximum information about the available resources; they cannot simply rely on revisiting those few locations where they have experienced reward in the past. If an animal performed only those acts that previously led to reward, then it would fail to learn about other potential resources in the area. In short, an animal must sample its environment, even though such sampling behavior may detract it temporarily from a rewarded behavior.

The notion that exploration is essential for adaptive behavior complements our understanding of motivated behavior by allowing us to differentiate between three essential types of systems. (1) Bio-regulatory systems activate the animal whenever temporary physiological deficits are experienced. (2) Acquired motivation also energizes the animal but this is based on previous learning. (3) Nonregulatory systems such as exploration, on the other hand, provoke the animal to learn about its environment *prior* to any danger; they allow the animal to plan for the future when the present reward contingencies may no longer be in effect.

Summary

Drive has been a useful concept in learning research because it is thought to combine with learning mechanisms to influence performance. Whereas learning primarily gives focus or direction to responding, drive primarily gives energy or impetus. This at least was the essence of Hull's influential theory. More recently, however, incentive motivation has been viewed as an important factor in performance. Incentive is acquired motivation; it stems from the goal object. Incentive energizes behavior by enticing the subject toward the goal.

Behavior is activated through the action of bio-regulatory systems. For instance, general activity may increase with longer food deprivation periods. This relationship, however, is complex and research has tended to show that only certain behaviors are energized by deprivation. A second area of research has shown that deprivation affects instrumental learning. Performance levels increase as a function of hunger and appropriate contrast effects are observed when these levels are changed. Finally, research has shown that animals can learn about and utilize irrelevant drives and that drive states can be quantified using specialized techniques.

Acquired motivation is generated during Pavlovian conditioning (see also Chapter 5). While fear is a particularly vivid example of acquired drive, appetitive states have also been studied. The picture is complex though. Little success has been achieved in demonstrating that Pavlovian CSs will actually induce motivated behavior such as eating. This is true even when internal CSs are paired with fast-acting, hunger-inducing drugs. However, there is considerable evidence showing that physiological changes that underlie motivated states, for example, changes in blood sugar, etc., are readily conditioned. These changes should provide mechanisms for conditioned appetitive states.

The opponent process theory describes the dynamics of motivational or mood changes. According to this theory, strong stimuli induce State A which may be either pleasurable or aversive. State A increases quickly at the stimulus onset and decreases rapidly after the offset. At the same time, State B is initiated. This motivational state is hedonically opposite to State A; it has a slow rise and fall time. Finally, the momentary emotional or motivational state of the animal is the algebraic sum of States A and B. Thus, when the stimulus suddenly goes off, State A quickly disappears leaving traces of State B; this corresponds to an emotional rebound effect. Such rebound effects are observed in many situations including aversive conditioning studies and imprinting.

Finally, from studies of nonregulatory systems it appears that novel stimuli are reinforcing and that the reinforcement effect increases with sensory deprivation. Moreover, animals alternate between various responses when they are given the choice. This behavior is related to curiosity and is accomplished largely through the use of memory and cues such as room lighting from outside the apparatus.

11

Punishment
and Learning Disorders

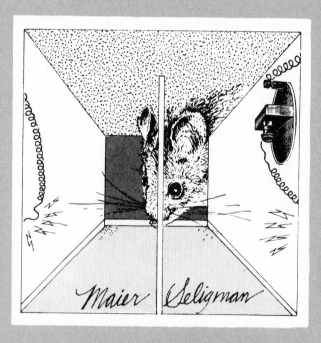

Maier Seligman

Introduction

Avoidance Extinction

Response Blocking

Punishment of Avoidance Extinction

Learned Helplessness

Summary

Introduction

Throughout this text we have stressed the idea that animals learn stimulus and response expectancies and that these expectancies interact with the species-typical behavior potentials to produce adaptive behavior. In addition, we have emphasized the notion that motivational or emotional states, both those that are biologically based as well as those acquired through Pavlovian conditioning experiences, energize behavior. It is, however, fair to ask whether learned behavior is always adaptive. Isn't it possible that an animal could learn the *wrong* response, that is, one that is not in its own best interest? For example, what if the subject attributed a recent illness to the "wrong" food? What if the animal had become sick but the food had been only temporarily unsafe? Normally the food is a perfectly healthy substance. The animal would mistakenly avoid that food in the future and thereby fail to utilize an important food resource. Over the long run, such a reaction would likely diminish, but the learning of inappropriate behaviors nevertheless is a risk. There is nothing inherent in the laws of stimulus and response learning to ensure that each and every response that is learned will be adaptive.

Consider another problem, namely, the possibility of having a more general learning disorder. Such a disorder could stem from a variety of sources. If, for instance, an animal were so persistent in one of its behaviors that it could not come into contact with and learn about new stimuli or responses, then we would say that it was behaving maladaptively. What about emotional excesses? These too could prevent an organism from learning or performing appropriately. For instance, an animal could be "frozen" with fear. Wouldn't it be more adaptive for it to experience less emotion and therefore greater mobility? Conversely, an animal could be in a "frenzy." Wouldn't it be more sensible for it to calm down so that it could cope effectively? The answer to such questions is not simple; as we discussed in Chapter 6, freezing or mobility both are appropriate avoidance reactions depending upon the species, the situation, the type of stimuli, and so forth. What seems rational to us may not be the SSDR that has developed for other species. But the problem still remains. There is always the possibility that *inappropriate* reactions could be learned to stimuli or that the motivating forces that impel behavior could be excessive. Our general purpose in this chapter, therefore, is to consider some of these problems, especially some important learning disorders and therapeutic techniques. Although our knowledge still is very incomplete, substantial progress has been made in recent years.

Let us proceed, however, with one very important caution in mind. The learning disorders and the various effects of punishment that we shall discuss in the sections below have all been produced under rigid laboratory conditions. There is no guarantee that they will operate in an equivalent fashion in the real world. Laboratory conditions are, by definition, highly constrained situations. The animals are not permitted to respond as they would in nature, nor are naturalistic stimuli very often used in the laboratory. For this reason, we have no knowledge about whether these learning disorders develop "naturally" in the animal's own environment. The laboratory experiments certainly tell us that these learning disorders *can* develop under certain circumstances, but the conditions that are typical of laboratory situations may, in fact, be crucial to their development.

Avoidance Extinction

As we discussed previously, avoidance responding is a normal means for coping with stress. The animal is able to avoid future punishment by attending to various warning signals. But consider the problem that the animal faces during normal avoidance extinction. When the US is never scheduled to be presented, how does the subject, in fact, know that the US will never be presented; how can it ever learn that extinction is in effect? The dilemma is as follows: during acquisition, the animal is responding to the warning signal because it expects that if it doesn't, the US will follow shortly. During extinction, however, when the animal needn't respond at all, continued "avoidance" reactions merely prevent the animal from ever seeing that the US is no longer scheduled to be presented. The animal has no way of knowing that extinction has started; its continued rapid responding, based on the belief that it would be shocked if it didn't respond rapidly, prevents it from ever experiencing the nonoccurrence of the US. From the subject's viewpoint, the US is omitted in either case: it was avoided if the acquisition conditions were in effect and the shock was "avoided" if extinction conditions were in effect because the shock was never scheduled to be presented at all. It would seem that the animal must experience the fact that no US is scheduled to occur before it can change its behavior to the warning signal.

The implication is that animals ought to find it difficult to extinguish their avoidance reactions. They would continue to respond because they would fail to learn that the noxious US is never going to be presented. In reality, the findings on this matter are not very clear and the theoretical issues are also complex. For one thing, we know that avoidance conditioning operates in terms of the animal's SSDRs (see Chapter 6). Strong SSDRs presumably would take longer to extinguish than weak SSDRs, thus confirming the prediction stated above. But little research has been done to confirm these and other points directly. In fact, some investigators have suggested that avoidance extinction takes place very rapidly; it would appear that such would be the case especially if the avoidance reaction was a poor SSDR.

Our original notion, however, that avoidance extinction involves a unique problem, one that may lead to unusually persistent responding, was first considered in an experiment by Solomon, Kamin, and Wynne (1953). The authors trained dogs to jump from one compartment to another and then back again on the next trial (a standard shuttlebox avoidance experiment). Once the response was firmly established, they attempted to extinguish it by eliminating all US presentations. Solomon et al. found that most subjects continued to respond during extinction for hundreds of trials even though no shock was ever presented. In other words, the avoidance reaction was unusually persistent. Indeed, it could be claimed that the animals were behaving in a seriously maladaptive fashion.

In a later paper, Solomon and Wynne (1954) developed a theory about why avoidance responding had been unusually persistent. They claimed that the fear or anxiety upon which the avoidance reaction was based was "conserved." According to this notion, fear motivates avoidance behavior (Mowrer's two-factor theory, see Chapter 5); since the subject continues to respond quickly during extinction, that is, prior to the time when the US would have occurred if the acquisition contingencies were still in effect, the Pavlovian fear never extinguishes. Fear, after all, can extinguish only if the subject comes to learn that no US is scheduled. But because the subject never discovers that no US is scheduled, the Pavlovian fear goes unabated.

Since fear is not extinguished, the avoidance response continues to be motivated and performed.

It should be obvious that such a reaction is maladaptive because it does not reflect the real contingencies that are operating in the environment. The animals are wasting their time and energy avoiding a noxious stimulus that never will appear anyway. Realizing this, Solomon, Kamin, and Wynne developed "therapy" techniques to promote fear extinction and eventual avoidance extinction. Two such therapies were used in the original study (1953). The first and most effective technique involved physically blocking the animal's response. The animal was forced to experience the nonoccurrence of the shock without being physically able to respond. Such a procedure led to fear extinction and eventual cessation of responding. As a second therapy, Solomon et al. punished the dogs for making the inappropriate avoidance reaction. Again, the argument was that such a procedure should force the subjects to abandon their avoidance reaction and, in turn, learn that no US is scheduled to be presented. We shall consider both of these treatments in the following two sections.

Response Blocking

As noted above, one of the techniques used by Solomon and his associates to hasten extinction was to block the response. Presumably, this allows the animal to learn the new conditions prevalent during extinction. Whether or not avoidance extinction takes place more slowly than expected, blocking the response should have a pronounced effect on behavior.

Response prevention, or, as it also is called, *flooding,* has been studied extensively in recent years (see Baum, 1970; Smith, Dickson, & Sheppard, 1973). The basic procedure is as follows. Rats are trained to make a typical shock avoidance reaction, such as shuttling back and forth at the sound of a tone in order to avoid electric shock. After a criterion of learning is reached, animals in one group are placed in their home cage until extinction. Animals in the experimental group, however, are given flooding or response prevention. That is, these animals are placed in the apparatus and are given the CS, but a glass or wooden barrier is now positioned so that they cannot run from one side of the box to the other. They must merely remain in the presence of the CS (they are "flooded" with the CS) without being able to execute the avoidance reaction that they were taught previously. In the third phase of the study, both groups are given typical avoidance extinction. Here, the CS comes on as usual and the animals are able to make the proper reaction, although no shock is ever delivered. The important result is that the flooding procedure, experiencing the CS but without being able to execute the CR, hastens the response extinction;

animals that were flooded previously stop performing long before the control subjects that did not receive the flooding procedure.

Factors Influencing Avoidance Extinction

Research has shown that several variables influence the rate of avoidance extinction after a flooding session. One of the most important is the degree of control exercised by the subject during flooding. Katzev and Berman (1974) examined the difference between being exposed to the CS and actually having control over its offset. Three groups of rats were trained to make an avoidance reaction in the first phase of the experiment. In Phase 2, during the flooding session, animals in one group (response-contingent group) could control the CS offset by making the avoidance reaction. Rats in another group (the noncontingent subjects) could not; they merely received the same duration of CS exposure that the former group experienced but had no control over its onset or offset. Subjects in a third control group were not given the flooding treatment. In the third phase of the study, all subjects were given normal avoidance extinction.

As shown in Figure 11–1, the two groups that experienced flooding showed a decrease in their resistance to extinction relative to the control group. The number of responses executed during extinction was much lower for those two groups. However, the more interesting finding was that the group that lacked control over the CS (noncontingent group) showed the largest reduction in resistance to extinction. That is, extinction took place most rapidly when the animals received the CS during flooding but couldn't control it. Merely being exposed to the CS without shock also facilitated avoidance extinction as shown by those subjects who could control the CS offset—the response-contingent group.

A second manipulation that affects avoidance extinction is the duration of the CS or the number of CS trials given during the flooding stage. Bersh and Keltz (1971) trained rats to make a shuttlebox avoidance reaction to a criterion of three consecutive avoidance responses. The subjects then received 5 flooding sessions during which the CS was presented for 5, 15, 120, or 300 seconds. Relative to a control group that did not receive flooding, resistance to extinction was reduced in all the experimental groups, and the effect was greatest for the 120- and 300-second groups. These data suggest that the longer the subject experiences the CS during the flooding stage, the quicker the subsequent avoidance extinction.

The relationship between the number of flooding trials and the duration of the CS is a complicated one, however. One experiment (Schiff, Smith, & Prochaska, 1972) suggested that it makes no difference whether the animals are given many trials with short duration CSs or fewer trials with longer lasting CSs. In their experiment, rats were trained to run down an alleyway to avoid shock. During the flooding phase, separate groups of

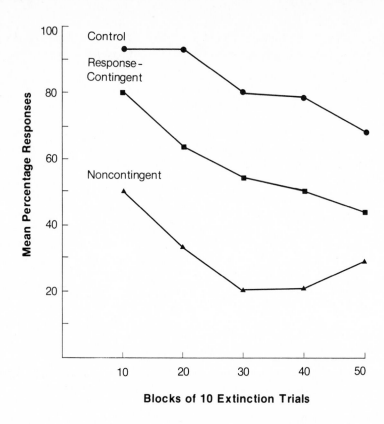

Control

Response-
Contingent

Noncontingent

Figure 11-1.   Mean percentage of responses that qualified as avoidance reactions during blocks of extinction trials for the control (no flooding), response-contingent (control of CS offset during flooding), and noncontingent (no control of CS) groups. (From Katzev & Berman, 1974)

subjects received either 1, 5, or 12 response-blocking trials each lasting either 0, 5, 10, 50, or 120 seconds. As usual, in Phase 3, all subjects were extinguished. The extinction data indicated that both the number *and* the duration of the flooding trials were effective. Shorter durations could be offset by a greater number of trials, while fewer trials could be offset by longer duration trials.

Although Shiff et al. showed that increased duration and number of flooding trials were equally effective in reducing later resistance to extinction, there was some indication in their data that the number of trials was more critical than the duration. In fact, other investigators (Baum & Myran, 1971; Franchina, Agee, & Hauser, 1974) have shown that the presentation of several flooding trials is more effective than a single, long lasting trial. In addition, Tortora and Denny (1973) have suggested that longer lasting CS presentations led to faster extinction but only when low to moderate shock intensities were used. At the very least, it appears that the more experience a subject has with the CS during flooding, the faster extinction takes place later on.

Shock intensity, of course, also has an important impact on the efficacy of flooding. Baum (1969a) trained rats to avoid shocks of .5, 1.3, or 2.0 milliamps. The subsequent flooding session lasted 3 minutes. Such a session was more effective for the lower shock groups than for the high shock groups. In other words, the ability for flooding to promote extinction was inversely related to prior shock intensity.

There are numerous other factors that influence the effectiveness of the flooding procedure. Simply changing the activity level of the subject seems to have a strong effect. For example, Lenderhendler and Baum (1970) found that gently prodding the animals with a wooden paddle, so that they became more active in the cage during the flooding session, increased the effectiveness of flooding. Similarly, Baum and Myran (1971) showed that confining the animals, so that their activity was unusually restricted, decreased the efficacy of the flooding procedure. Second, it has been found that flooding is more effective if another nonfearful subject is placed in the apparatus with the test animal (Baum, 1970). No adequate explanation is provided for this social facilitation phenomenon although certain emotional states are known to be transmitted via odors. Finally, Mineka (1976) has shown that the flooding of one avoidance reaction can reduce the resistance to extinction for a second, different type of avoidance response. In Mineka's experiment, animals were trained to make both a shuttlebox avoidance reaction and a jump-up response to avoid shock. Later, experimental subjects were flooded in the jump-up apparatus. This procedure had the effect of promoting the extinction of the shuttlebox response relative to a control group that was not flooded. Thus the flooding of one response can have important implications for the extinction of a second behavior. How the transfer is accomplished is not entirely clear at this time.

Theories of Flooding

Flooding is, quite clearly, a form of reality testing. The animal's potential for later performance during extinction is changed through forced exposure to the CS without being permitted to perform the response. But what accounts for this change, the action of flooding upon later avoidance extinction? Several theories have been proposed although none at this time seems to be entirely appropriate (see Mineka, 1979). The most pervasive notion is based upon Mowrer's two-factor theory of avoidance learning (see Chapter 5). Recall that this theory hypothesizes that avoidance learning is motivated by fear. Therefore, it would be fear that is extinguished during the flooding session. That is, since the CS is presented without the US, the Pavlovian fear, which underlies the avoidance reaction, would diminish. Later, during extinction, the avoidance reaction itself would decline more quickly because the fear, which presumably motivates the reaction, is lower.

The fear theory enjoys a good deal of support. It has been shown that animals are, indeed, less fearful after experiencing a flooding session (Bersh & Paynter, 1972; Linton, Riccio, Rohrbaugh, & Page, 1970; Wilson, 1973). Other evidence exists too. Extinction should take place more rapidly with flooding sessions of longer duration because the Pavlovian fear is lower after such long sessions than it is after shorter duration sessions. As discussed earlier, this result is, indeed, found. Third, there is an inverse relationship between shock intensity and the effectiveness of flooding. The greater the shock, the more intense the fear; the more intense the fear, the less effective the flooding session will be. Finally, it has been demonstrated (Shipley, Mock, & Levis, 1971) that fear reduction is a direct function of the CS duration during flooding. The longer the duration, the quicker the extinction and the less the Pavlovian fear. In summary, there is strong evidence to suggest that avoidance extinction is facilitated by flooding because flooding helps to reduce fear, which motivates the avoidance reaction.

Other investigators, however, have been critical of the fear theory, at least as the sole explanation for the flooding phenomenon (see Mineka, 1979). One of the most notable problems is the finding that fear remains even after flooding has been administered. In an experiment by Coulter, Riccio, and Page (1969), response blocking decreased later resistance to extinction. However, considerable fear was still present at the end of extinction, suggesting that extinction was not necessarily facilitated because of a reduction in fear. After all, if extinction in fact depended upon the lack of fear, how could avoidance extinction take place with fear still being quite substantial? In another study, Linton et al. (1970) found not only that residual fear was still present after flooding, but also that the previously blocked subjects were *more* fearful than the animals that were not flooded at all. In short, flooding does seem to facilitate avoidance extinction, but it certainly does not eradicate all Pavlovian fear.

The alternative explanation offered by Coulter et al. (1969) is termed the competing response theory. According to this position, subjects develop freezing responses that compete with the avoidance response. In other words, the decrease in resistance to extinction of the avoidance behavior is not due to a loss of motivating fear. Rather, it is due to the fact that freezing reactions, acquired during the flooding stage, compete or interfere with the execution of the avoidance reaction during the extinction phase. Subjects cannot perform their avoidance response and, at the same time, freeze. What *appears* to be a loss of fear, in reality, according to this theory, is merely the execution of a freezing reaction. In any event, the finding that CS exposure affects extinction but does not eliminate fear is a compelling argument against the two-factor theory (see also Mineka & Gino, 1979b).

Other investigators have suggested a different type of competing response theory. Baum (1970) observed that animals engaged in locomotor behavior during the flooding stage. Quite simply, they tended to walk around the cage rather than freeze. Baum claimed that this general activity, which increased over the flooding session, was reinforced by the nonoccurrence of shock and was later performed during avoidance extinction. In other words, gross locomotor activity was conditioned during the flooding stage and it was this activity that later competed with the avoidance reaction.

Neither form of the competing response theory seems to be completely adequate. Most notable is the finding (Bersh & Miller, 1975; Marrazo, Riccio, & Riley, 1974) that response blocking *plus* shock also leads to quicker extinction of the avoidance reaction. In these experiments, animals were trained to make an avoidance reaction and then were given a flooding session. For some of the subjects, flooding was conducted in the normal fashion, whereas for others the presentation of the CS was accompanied by shock. The result was quicker extinction later on for both groups (relative to control subjects that were not flooded at all). Merely blocking the response was critical. Such a finding is dramatic because it contradicts both the fear theory (fear increased rather than decreased due to the presentation of shock during flooding) as well as the two forms of the competing response theory (it is hard to imagine that animals were freezing or were engaged in general activity while being shocked).

It would appear that extinction following flooding is a function of the interaction between fear reduction *and* competing responses. The subject's SSDRs may change substantially during the flooding session too (Crawford, 1977; but see Mineka & Gino, 1979a). Although no single theory is capable of explaining the flooding phenomenon at the present time, it is hoped that future research will clarify these mechanisms. The very fact that flooding works to reduce avoidance extinction responding is significant, however. To the extent that avoidance responding can be maladaptive, that is, it is executed even when no noxious US is scheduled to be presented, the phenomenon represents a learning disorder of considerable interest and importance.

Punishment of Avoidance Extinction

Recall that the principal learning disability that we are discussing is the inappropriate persistence of an avoidance response (making such a response when no shock is ever scheduled to occur). Recall also that in the original studies by Solomon and his associates two types of therapies were attempted in order to facilitate avoidance extinction. The first, which we have just discussed, was response blocking. The second technique was punishment. Specifically, during extinction, when the animals were inappropriately performing the avoidance response, punishment was given in the goal box. It is well known that the major effect of punishment is suppression of behavior (see Chapters 2 and 4). Therefore, in their experiment, Solomon and his colleagues speculated that punishment should suppress the continued responding.

Although punishment should, in theory, suppress the avoidance reaction during extinction, many experiments have demonstrated just the opposite: paradoxically, a subject that is punished during avoidance or escape extinction may persist even longer, at higher response speeds, than subjects that are not punished at all. This phenomenon, called the vicious circle or self-punitive phenomenon, is most clearly maladaptive. By continuing to respond in this situation, the animal merely receives more shock. In fact, the more the animal responds, the more punishment it will receive (see Brown, 1969).

An example of the vicious circle phenomenon was shown by Brown, Martin, and Morrow (1964). Those authors trained rats to escape shock in an alleyway. The 20 acquisition trials were followed by 60 extinction trials. On these extinction trials, the animals were placed in the start box and the door into the alleyway was raised, but shock was never delivered to the floor. Some subjects, those in the No-Shock group, were never punished during extinction; their escape response was merely extinguished in the normal fashion. Animals in a second group, the Short-Shock group, were given a shock in the last section of the alleyway, that is, just before the subjects reached the goal box. Subjects in a third group, the Long-Shock group, experienced shock throughout the entire 6 feet of the alleyway, that is, from the point immediately outside the start box door to the entrance of the goal box.

The results of the study are shown in Figure 11–2. Speed of running was dramatically affected by the punishment conditions during extinction. The No-Shock group (the regular extinction group) gradually slowed down over the extinction sessions. In contrast, the groups that were punished during extinction ran faster throughout the extinction period. To be specific, subjects in the Long-Shock group continued to run almost as fast as they did during acquisition; the shock maintained responding throughout extinction even though those animals could sit in the start box and not receive shock. Animals in the Short-Shock group were intermediate. Punishment facilitated their response speed although not as much as the other punished group. In summary, the groups that were punished for leaving the nonelectrified start box continued to run fast during extinction, especially the Long-Shock subjects that were punished throughout the entire alleyway. The controls, in contrast, extinguished normally. The results suggest that punishment of an aversively motivated response during extinction produces a form of self-punitive or masochistic behavior. Rather than suppressing the behavior as it commonly does, punishment seems to enhance the behavior. This vicious circle, or self-punitive phenomenon, has been shown to occur in a variety of species and for a variety of different types of behaviors.

Theory of Self-Punitive Behavior

The earliest and most pervasive theory of the vicious circle phenomenon involves the concept of fear. The theory claims that punishment during extinction merely maintains or even increases the level of fear. This, in turn, leads to continued avoidance behavior because, as the two-factor theory suggests, the avoidance response is based upon fear motivation. It is easy to see how this phenomenon is, indeed, a vicious circle: Fear motivates the avoidance reaction; the behavior is punished during extinction, which leads

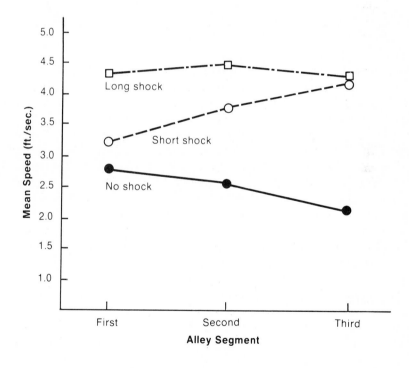

Figure 11-2. Mean running speed in the three segments of the alleyway during extinction for the groups that received no shock, long, or short shocks. (From Brown, Martin, & Morrow, 1964)

to greater fear; greater fear leads to continued avoidance behavior, and so on.

There are some impressive studies that support the fear theory of the vicious circle. For example, Klare (1974) measured freezing prior to each trial. It was argued that such freezing behaviors reflect fear. Indeed, Klare showed that nonshocked animals tend not to freeze in the start box, whereas animals that were punished during extinction show a substantial amount of freezing on the next trial.

If punishment during extinction enhances fear, then manipulation designed to reduce this excessive level of fear should facilitate the avoidance extinction. This has been shown by Delprato and Carosio (1976). During extinction, when some of the subjects were being punished, all of the animals got a tone that had previously been associated with the goal box. According to the notions developed in Chapters 3 and 5, the tone was a safety signal because it was correlated with a period of no shock. The results of the study showed that the safety signal, the tone, facilitated extinction in the punished subjects. According to the fear theory it did so because it reduced the excessive levels of fear.

A third result that supports the fear theory was shown by Bender (1969). He demonstrated that the vicious circle may occur when a fear signal, one that has been associated with shock in the past, is presented to the subjects during extinction. That is, animals that were punished, not with shock but with a signal that had been associated with shock, also show persistent responding during extinction. Again, manipulations that affect the level of fear tend to affect the rate of avoidance extinction.

Several scientists have challenged the fear theory however. They claim that the punished subjects are merely failing to discriminate between acquisition and extinction. That is, during extinction subjects normally come to learn that shock is not scheduled to occur. If, however, shock is continued, even though it is administered only in the alleyway, then this new learning becomes more difficult. The animals are confused; punishment during extinction makes it hard for them to learn that a new contingency is in effect (see Dreyer & Renner, 1971).

The confusion theory has received some support. For example, Campbell, Smith, and Misanin (1966) conducted an experiment similar to that of Brown et al. (1964). They used three groups of rats: a regular extinction group, a group that was punished only in the first portion of the alleyway near the start box, and a group that received shock only in the *last* portion of the alleyway closest to the goal box. The fear interpretation presumably would claim that both groups of punished subjects should show the same degree of self-punitive behavior. After all, the amount of shock that each was receiving during punished extinction was identical. The confusion theory, on the other hand, would predict a different result. It would claim that the subjects that received punishment early in the alleyway would persist in their responding much longer because that is the location where they had received shock during acquisition. In contrast, the performance of the rats that were shocked just outside the goal box should be suppressed; punishment at that location is less like the conditions that prevailed during acquisition than punishment early in the alleyway. The results confirmed this latter hypothesis: subjects that were punished near the start box showed the strongest self-punitive behavior, but the animals that were shocked just before entering the goal box showed suppression of avoidance responding.

For a number of years, evidence was gathered that supported *both* theories. For instance, some investigators found that the degree of self-punitive behavior was decreased when various physical characteristics of the alleyway or goal box were changed (Hom & Babb, 1975). This presumably should happen if the animals are confused; that is, a change in the environment between acquisition and punished extinction should help the animals to perceive the new conditions and therefore should promote extinction. Other investigators, however, did not find this result (for example, J. S. Brown, 1970; Brown, Beier, & Lewis, 1971). They found that changes did not promote extinction. Moreover, Melvin and Martin (1966) showed that a different type of punishing US could be used during extinction. Animals that had been trained to avoid shock and then were punished during extinction with a loud buzzer showed the vicious circle phenomenon too. According to Melvin and Martin, confusion could hardly have been present with such a dramatic change in the US.

The controversy concerning which theory is more correct appears to have been resolved in an insightful paper by Dreyer and Renner (1971). They argue that the fear and confusion hypotheses actually are compatible. Punishment should certainly enhance fear; this is in accord with the principles outlined for Pavlovian conditioning in Chapters 3 and 5. On the other hand, continued responding during extinction may also be due to confusion. The animals may have heightened fear *and* be uncertain about the current contingencies. To maintain that confusion is not a contributing factor would be to insist that the subjects can discriminate between acquisition and the punished extinction conditions. There is certainly no independent evidence to suggest that subjects do in fact make this discrimination. Indeed, the procedures that are most effective in producing self-punitive behavior are precisely those that would retard such a discrimination (Brown, 1969). In conclusion, the self-punitive phenomenon does not appear to be a model of masochistic behavior, but it does illustrate that, under certain circumstances, punishment may produce a highly maladaptive behavior if precautions are not taken to clarify the response-punishment contingencies (see O'Neil, Skeen, & Ryan, 1970).

Stress and Stimulus Predictability

The notion developed just above, that punishment during times of uncertainty can lead to learning disorders, has been considered from a different viewpoint. In particular, scientists have shown that the procedure may result in physical pathologies as well. In some cases the ulcers are due to prolonged periods of shock followed by relief (Desiderato, Mackinnon, & Hisson, 1974), being tightly restrained (Mikhail, 1973), or conflicts between food acquisition and unavoidable shock (Moot, Cebullà, & Crabtree, 1970). In other cases, though, ulcers may be produced when the subjects are punished while in a state of uncertainty. For instance, unsignaled shock can have dramatic effects upon an animal's behavior and physiology. Subjects normally prefer signaled shock to unsignaled shock (Badia, Harsh, & Abbott, 1979). Moreover, unsignaled shock tends to be more aversive than signaled shock; animals routinely will show greater suppression of ongoing behavior when given unsignaled shock than when given signaled shock (Hymowitz, 1979). And most importantly, subjects that cannot choose signaled versus unsignaled shock, or those that receive unsignaled shock, may develop ulcers (Caul, Buchanan, & Hays, 1972; Gliner, 1972). In summary, punishment, when animals are uncertain about the delivery of the punishment, can cause not only learning disorders but physiological disorders as well.

Punishment and Aggression

We noted above that punishment under some conditions can induce inappropriate reactions, continued responding in the vicious circle phenomenon, and even physical pathologies. Punishment may also induce behaviors that are socially undesirable or inappropriate. Specifically, animals may become unusually aggressive when they are punished. Ulrich and Azrin (1962) observed that shock elicited a reflexivelike fighting in rats. The aggressiveness of the animals increased with more frequent or more intense shocks or when the subjects were confined to a smaller apparatus. Although fighting was closely dependent upon the actual shock presentation in this study, other investigators (for example, Baenninger & Ulm, 1969) have shown that a rat will kill a mouse even when the mouse is placed in the cage *after* shock has been terminated. Punishment-induced aggression, therefore, can persist long after the shock has ended.

Other studies have shown that shock-induced aggression occurs in a wide variety of species including monkeys (Azrin, Hutchinson, & Hake, 1963). In this study, the fighting between these primates was more vicious, less stereotyped, and longer lasting than the aggression between rodents. Monkeys have been shown to attack and bite even inanimate objects such as a stuffed doll or a tennis ball (Azrin, Hutchinson, & Sallery, 1974). These sobering results suggest that punishment may, indeed, have highly undesirable effects. Although considerable work is needed to clarify the relationship between punishment and aggression, it appears that when aversive procedures are used during states of ambiguity or stress, inappropriate reactions may occur.

Learned Helplessness

Recall from Chapters 2 and 4 that the correlation between a response and an outcome is the basis for response learning. A response expectancy develops because the response-outcome correlation is high. The subject learns to expect the outcome on the basis of its behavior; its behavior, in a sense, controls the outcome. But what happens to an animal when the correlation between the response and the outcome is zero? To answer this question, consider the analogous question for Pavlovian conditioning: What happens when a stimulus and the US are not correlated? According to the notions developed in Chapters 2 and 3, the random presentation of a CS and US leads to no conditioning. But something else also happens. Specifically, it has been shown (Baker & Mackintosh, 1979) that the random presentation of a CS and US also leads to retarded excitatory conditioning in the future when that CS is then paried with a US. The random relationship between a stimulus and a US, therefore, not only fails to produce excitatory or inhibitory conditioning initially, but it also renders that CS weaker in the future; conditioning is retarded. A possible reason for this decrement in learning is that the animal comes to believe that the CS is irrelevant to the US. In the past, the CS never predicted the US; for that matter, it never consistently predicted no US either. Therefore, the retardation of later learning is due to a "learned irrelevance." The animal has difficulty learning that the CS predicts a US if, previously, it learned that the CS was irrelevant to the US.

Now turn back to response learning. What happens when a response and a reward are independent of each other? The answer is the same as that given for the stimulus learning case: not only does random response-reward presentation fail to produce conditioning (because the response-outcome correlation is zero), but the random relationship between a response and reward also can generate learned irrelevance. In this case, however, the subject is learning that its response, not a stimulus, is irrelevant to the reward presentation. Being unable to control reward through responding, that is, having a response-reward correlation of zero, leads to the expectancy that the response is irrelevant to the reward.

We are claiming, in essence, that stimuli and responses operate in a parallel fashion. As discussed in Chapter 2, conditioning is the development of an expectancy, based upon either a stimulus or a response. These expectancies reflect a correlation between the element (the stimulus or response) and the outcome (a US or no US). When the element and outcome are unrelated or randomly presented, no conditioning is observed. Furthermore, as discussed just above, when they are unrelated, it is more difficult in the future to use either element in a normal conditioning situation.

The result, that future response learning is retarded if the response was unrelated to reward in the past, is called learned helplessness (see Maier & Jackson, 1979; Maier & Seligman, 1976; Maier, Seligman, & Solomon, 1969). Learned helplessness, to be specific, is the finding that learning is very poor following a period when the animal's behaviors and the reward presentations were unrelated. In fact, the inability to learn applies to many responses, not just the particular response that previously was uncorrelated with reward. Let us discuss a specific example of learned helplessness in order to illustrate this point.

In one early study by Seligman and Maier (1967), three groups of dogs were used as subjects. Animals in one group, called the Escape group, were restrained in a hammocklike device and given shock to their hind leg. These animals were allowed to terminate the shock by pressing a panel with their snout. In this regard, they were permitted to perform a normal escape response. Subjects in a second group, the Yoked animals, were also placed in the hammock and shocked whenever the first group received shock. They were not permitted to press the panel with their snout; they merely received shock whenever the first group was shocked. Finally, subjects in a third control group were placed in the hammock but were given no shock at any time in this part of the experiment.

The design of this experiment is especially important to the theory. Note that the shocks for the first two groups are identical in every respect. Each received the same intensity and duration of shock, and the pattern of shock was equivalent. In fact, the only difference between the first two groups was the fact that the first group of subjects, the Escape animals, could control the shock offset, whereas the second group of animals, the Yoked animals, could not. These Yoked animals depended upon the escape subjects for the termination of their shock. Whatever effects shock *per se* has, these two groups should be identical.

After this initial part of the experiment, all the subjects were given normal escape-avoidance training in a shuttlebox. Here, they were treated identically: they were placed in the box and, following a warning signal, were given shock. The median latency for the escape-avoidance response in the Escape subjects was 27.0 seconds; in the control group it was 25.9. In striking contrast, however, was the performance of the Yoked subjects. They showed virtually no improvement in their escape-avoidance behavior throughout training (their median latency was about 48.2 seconds).

The failure to perform the avoidance response by the Yoked subjects certainly could not have been due simply to the shock that they received in Phase 1 of the study. The reason is that the Escape group also received the same shock and yet they showed the normal learning pattern. The failure of the Yoked subjects to learn in Phase 2, therefore, must have been due to their inability to *control* shock in Phase 1. The important difference between those two groups was that in Phase 1 the Yoked subjects lacked control over the shock (they could not control its offset), whereas the Escape subjects had control. In other words, during Phase 1 the Yoked animals came to expect that their behavior was irrelevant to shock offset. No matter what they attempted, be it panel presses with their snout or struggling behaviors or freezing reactions, no behavior was consistently correlated with the offset of shock. This fact, in turn, led to the development of the general expectation that their behavior was irrelevant to reward. The notion that their behaviors are irrelevant to reward carried over to Phase 2 of the experiment and affected normal learning. The animals performed as if they were helpless; they, of course, could terminate shock during Phase 2, but they failed to do so. They failed to develop a new expectancy, namely, that responding does control reward, because they already had developed an expectancy that their responding was irrelevant to reward.

The concept of learned helplessness is enormously important. It has been demonstrated in many studies using a variety of species and various types of learning tasks. Perhaps most impressive is the fact that learned helplessness has been demonstrated in human subjects and, indeed, the theory serves as the basis for one explanation of human depression (see Seligman, 1975). The various studies using humans have all shown that lack of control over some outcome causes interference in future learning (Hiroto & Seligman, 1975; Roth & Bootzin, 1974; Thornton & Jacobs, 1971; Thornton & Powell, 1974). In the Thornton and Jacobs (1971) study, individuals in one group could avoid receiving a mild finger shock by pressing the correct button when a light was turned on. Subjects in a second group were yoked; that is, they received inescapable shock whenever the subjects in the first group were given shock. In the second phase of the study, all the subjects were required to learn another task, namely, to avoid shock by pressing the correct two buttons in an array of seven buttons. Failure to push the proper buttons in the correct order resulted in shock. The authors found that prior inescapable shock interfered with the acquisition of this simple task in the subjects who were yoked in Phase 1. The subjects who had been able to avoid shock during Phase 1, of course, learned this task easily. Thus, it appears that lack of control affects subsequent performance in humans just as it affects performance in lower species.

Principles of Learned Helplessness

The considerable research on learned helplessness has uncovered many important principles. One finding is that helplessness training seems to affect non-SSDRs. Although sufficient data have not been collected on this point, it has been shown that inescapable shock does not produce interference with learning when the learning involves very simple behaviors, such as SSDRs. If a rat is given inescapable shock and then required, in Phase 2, to make a simple escape response, like pressing a lever, no deficits are normally shown. However, if the animal is required to press the lever twice, or to shuttle in the shuttlebox twice, then the helplessness effect is demonstrated (see Maier, Albin, & Testa, 1973; Seligman & Beagley, 1975). Easy responses, of course, are SSDRs; they are essentially innate forms of behavior that are executed quite easily. These are not disrupted to any sufficient degree by helplessness training, but difficult responses are disrupted. Although no experiments have examined helplessness in terms of naturalistic versus less naturalistic behaviors, the point presumably still applies: the lower the response is on the SSDR hierarchy, the more susceptible it is to disruption by inescapable shock.

A second finding is that subjects may be immunized against the effects of inescapable shock by providing training with escapable shock (Seligman, Maier, & Greer, 1968; Seligman, Rosellini, & Kozak, 1975). In the Seligman et al. (1975) study, rats were first trained to jump onto a platform in order to escape shock. A second group was not given this experience. Both groups were then given the normal helplessness training, that is, inescapable shock. Finally, in the third phase of the study, both groups were given the escape training; that is, they were able to press a lever in order to terminate shock. A third control group was used but subjects in that group never received shock prior to the test; they were given neither the immunization training (escapable shock) nor the helplessness training (inescapable shock).

As shown in Figure 11–3, the latencies of the three groups varied considerably. Over the blocks of escape trials, the control animals showed a normal level of conditioning. In contrast, subjects in the inescapable group showed very poor learning; their mean latencies were very slow. This result illustrates the normal helplessness phenomenon. Most important, however, was the behavior of animals in the immunized group. They performed just like the control animals. Even though they received helplessness training, the fact that they had earlier received escapable shock immunized them against the effects of the helplessness training. Note that the immunized animals had more total shock than subjects in the helplessness group; however, being able to control shock during Phase 1 meant that the inescapable shock in Phase 2 did not create helplessness.

The immunization effect has been shown to operate even when the original training and the testing involved different responses. Williams and Maier (1977), for instance, allowed rats to escape shock by turning a wheel. Other animals did not receive this experience. Both groups were subsequently given inescapable shock (helplessness training), followed by a normal learning task during which the animals were placed in a shuttlebox and

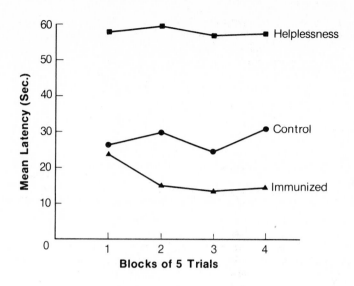

Figure 11-3. Mean response latency in seconds as a function of blocks of five escape trials for groups that received no shock (controls), helplessness training, or immunization training. (From Seligman, Rosellini, & Kozak, 1975)

trained to avoid shock by running from one side to the other and back again. The shuttle response is clearly different from the wheel turn behavior that the immunized subjects had previously experienced. The results showed that immunization occurred even when the two responses were vastly different. Having learned to turn the wheel to escape shock prevented helplessness from occurring. The major point, therefore, is that the escapable shock, regardless of what response is appropriate to terminate shock, leads to the expectation that responding does produce reward. It is this general cognition that immunizes the subjects and prevents helplessness from occurring later on. If the immunization procedure worked only because the original and subsequent responses were the same motor behaviors, then the argument that helplessness is a general cognitive state, would be considerably weakened. The fact that being able to control shock with one response prevents helplessness for a second response, (as shown in the Williams and Maier study) suggests that these changes are cognitive in nature (see discussion below).

A third important phenomenon concerning helplessness is the finding that the condition is reversible. Recall our previous discussion of flooding. We claimed that flooding permitted the animal to discover that the extinction conditions were in effect. There is a comparable therapy in the case of helplessness (see Seligman et al., 1968; Seligman et al., 1975). In the Seligman, Rosellini, and Kozak experiment, three groups of rats were used. All received inescapable shock in Phase 1 and all of them, when later tested on a bar press escape response, showed helplessness, that is, retardation of learning. Animals in one group were then forcibly dragged by the shock electrode wires attached to their tails to the area where the bar was located. By being placed on the bar, the animals were able to "discover" that shock could be terminated. This forced responding procedure had a therapeutic effect: those animals eventually learned the bar press escape response more quickly than animals that did not receive this therapy; the forced responding helped them to overcome the learning deficit caused by inescapable shock. In general, then, helplessness may be reversed by forcing the subject to respond appropriately. Presumably, such forced responding compels the animal to learn that its behavior is not irrelevant to reinforcement.

A fourth important discovery in the helplessness literature is the finding that response-reward independence can cause a learning deficit regardless of what behaviors are used later. More specifically, animals that learn that their behavior cannot control one kind of reward (helplessness training) show deficits in later learning even though a different type of response and reward are used (Altenor, Kay, & Richter, 1977; Braud, Wepman, & Russo, 1969; Goodkin, 1976; Rosellini & Seligman, 1975; Williams & Maier, 1977). For example, Altenor et al. (1977) gave rats inescapable shock in the first phase of the study. Such helplessness training, of course, later produces a deficit in shock escape learning. However, in this experiment, inescapable shock also produced interference in learning to escape from water. That is, when the testing situation involved an entirely different form of behavior and a different US, inescapable shock still caused retardation of learning.

Goodkin (1976) provided a more compelling demonstration of the notion that helplessness can transcend the particular situation in which it is tested. Three groups of rats were permitted to press a panel with their noses in order to obtain food, to pull a chain to get food, or to pull a chain to avoid shock. Those subjects, of course, could control their rewards. Other groups of subjects received either inescapable shock or free food (that is, food that was delivered independent of the animals' behavior). In Phase 2 of the study, all subjects were trained to press a small panel in order to escape or avoid shock. The groups that could control their rewards during Phase 1 learned quite readily, even those animals whose original training involved food reward. In contrast, the animals whose behavior was independent of reward in Phase 1, whether the reward was food or shock offset, failed to learn the simple escape-avoidance task in Phase 2 very well. The type of reward—food or shock offset—and the type of response—chain pulling, nose pressing, running, etc.—made no difference. Merely the lack of control was sufficient to produce the learning deficit. This is extremely powerful evidence for the theory that helplessness is a cognitive phenomenon, that the learning deficit is unrelated to the particulars of the behaviors and rewards.

Finally, a finding that is closely related to the one discussed above has been termed *learned laziness.* According to the helplessness notion, the independence of food and behaviors should produce a general cognition or expectancy that behaviors and reward are unrelated. This, as discussed above, interferes with future escape-avoidance learning. But it also interferes with future appetitive reward learning (for example, Tomie, Murphy, Fath, & Jackson, 1980; Wasserman & Molina, 1975). In these learned laziness studies, the CS and food are presented randomly to one group of pigeons; the control subjects do not receive such training. Later, in Phase 2, all the birds are trained to peck a lighted key via autoshaping. That is, they are given key light-food pairings. Normally, this procedure results in the acquisition of rapid key peck behavior in these animals (see Chapters 3 and 6). However, the animals that previously received the random CS-US pairings show retarded autoshaping. Not only do they take longer to begin pecking under these conditions, but they later show lower levels of key peck behavior once autoshaping has taken place (Wasserman & Molina, 1975). The learned laziness phenomenon, therefore, is a special demonstration of the helplessness principle: that the independence of behavior and rewards can lead in the future to a retardation of learning. Whereas helplessness normally focuses on aversive learning situations, the laziness paradigm has demonstrated similar effects using appetitive training procedures.

Theories of Helplessness

The first and most prominent theory of the interference effect is the helplessness theory, that is, the notion that subjects learn the expectancy that their behavior and reward are independent. This general cognition reduces their subsequent motivation for responding and, more importantly, their ability to learn a new expectation, namely, that their behavior now does indeed control reward. Quite obviously, the helplessness theory is a cognitive theory based upon the idea that the learning deficits reflect an inability to replace one cognition (the expectancy that behaviors and reward are independent) with a new cognition (the expectancy that behavior can control reward presentation).

Not all theorists, however, agree with the helplessness explanation (see Levis, 1976). Most of these investigators have suggested that the interference is due merely to response competition. That is to say, a motor response incompatibility develops in the helplessness subjects, an incompatibility between behaviors that occur during the inescapable shock phase and those that occur later on. Several forms of this competing response theory have been proposed. Bracewell and Black (1974) claimed that being active, moving around on the shock grids, involves more pain than being inactive. For this reason the subjects learn to be immobile during the inescapable shock phase. The helplessness animals later fail to respond because they are merely performing their previously learned response, namely, immobility. In other words, the theory claims that inescapable shock inadvertently causes the animal to freeze or be immobile. During Phase 2, when the animals are supposed to learn the active escape behavior, the freezing reaction competes with performance.

Other forms of the competing response hypothesis have been offered. Glazer and Weiss (1976a, 1976b) noted that shock naturally elicits locomotor activity for about 3 to 4 seconds in the rat followed by an immobility or freezing reaction. In many of the learned helplessness studies, each inescapable shock did indeed, last about 5 seconds. Therefore, Glazer and Weiss suggested that animals were simply becoming immobile toward the end of the 5-second period, and, furthermore, such immobility reactions were reinforced because they were consistently followed by shock offset.

Impressive evidence in favor of the competing response theory has been provided. In one study, Irwin, Suissa, and Anisman (1980) showed that inescapable shock caused a deficit in learning a water escape response. However, the deficit seemed to be caused primarily by the inability to perform the motor reaction, not by a cognition concerning a behavior-reward independence. Second, the work of Anisman and his colleagues also has provided convincing demonstrations of the competing response position (Anisman, deCatanzaro, & Remington, 1978; Anisman & Waller, 1971b, 1972, 1973). In one experiment, Anisman et al. (1978) showed that inescapable shock causes a reduction in activity produced by shock at some later time. Specifically, they placed mice in a shock box and delivered 60 inescapable shocks each 1 minute apart. The control subjects were placed in the box but were not given shock. The next day all animals were returned to the box and again given shock. This time, however, activity was measured. The authors found that inescapable shock (Phase 1) reduced activity that is normally elicited by shock (Phase 2). These results are illustrated in Figure 11–4. The mean activity score, computed in terms of the number of times the animals interrupted a photocell beam, was uniformly higher for the no-preshock subjects throughout the entire 6-second shock than for the shocked animals. If particular measures are taken to increase general activity, such as the administration of methamphetamine, then the interference effect is not observed (Anisman & Waller, 1971).

The notion that a change in activity is produced by inescapable shock seems to be well founded, but how this change is brought about is still unclear. It is possible that immobility is conditioned because it tends to occur just prior to shock offset in many experiments. Conversely, immobility reactions may result in a slightly less aversive shock and therefore be conditioned in that manner. Or it is possible that activity reductions occur

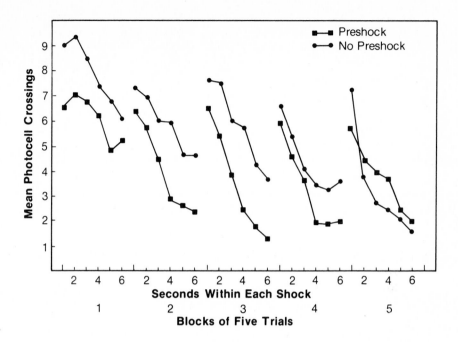

Figure 11-4. Mean activity scores as a function of blocks of 6-second shocks for groups that had (preshock) or had not (no preshock) received prior inescapable shock. (From Anisman, deCatanzaro, & Remington, 1978)

because of some neurochemical depletion which leaves the animal unable to respond actively. Regardless, it appears that the animal's behavior repertoire does change following inescapable shock and furthermore that such changes contribute to the learning deficit observed in the learned helplessness studies.

A good deal of work has been done to try to counteract the claims of the competing response theory. In one experiment, dogs were curarized during the inescapable shock phase. Curare is a drug that paralyzes the muscles but otherwise does not influence the animals. Since the muscles were paralyzed temporarily, the animals obviously were unable to perform any behaviors whatsoever. Even here helplessness was observed (Overmier & Seligman, 1967). The argument was that no systematic motor reaction could be learned during the inescapable shock phase since the animals were paralyzed, and therefore covert motor reactions could not have competed with the criterion response during the test.

Other studies have approached this problem somewhat differently. In one very important study, Maier (1970) deliberately trained the animals to remain passive in Phase 1 of the study. Yoked subjects were also used; they received the same intensity and pattern of shocks as the first group but remaining still did not control shock. If the competing response hypothesis were correct, then animals that were *deliberately* trained to be immobile should show the interference effect later on. On the other hand, if the learned helplessness theory were correct, then the animals that were deliberately trained to remain passive during Phase 1 should learn the active avoidance response in Phase 2 perfectly well. After all, they experienced control over the shock during Phase 1 and therefore should have no problem learning in Phase 2. Results favored the learned helplessness theory. Even deliberate training of an immobile reaction did not produce the interference effect (although under some conditions it may, in fact, cause a reduction in learning; see Anderson, Crowell, Cunningham, & Lupo, 1979).

Third, several experiments have shown that uncontrollable shock interferes with later learning even when the subsequent learning involves a passive reaction. Baker (1976) demonstrated that animals were deficient in learning to suppress their lever press behavior as a result of signaled punishment if they had previously received inescapable random shocks. The actual study involved presenting a noise CS and then a shock while the animals were bar pressing. Normally suppression of bar pressing was observed; this is the standard CER outcome (see Chapter 2). However, if the animals were given inescapable shock prior to the CER test, then they were quite poor at learning to suppress during the noise. The deficit could not have been due to the competition with immobility reactions. Such a theory would predict, if anything, that lever press suppression would be even *greater* following inescapable shock. That is, if interference with learning resulted merely from the presence of immobility or freezing reactions during the test (which were conditioned previously during inescapable shock), then one would predict that learning would be even more rapid when the test involved a passive reaction. Such was not the case, however.

Finally, the experiments on the immunization against helplessness and the experiments showing that the type of reward and response may differ between the inescapable shock phase and testing, all suggest that the competing response hypothesis is deficient. In the case of the immunization studies, immobility reactions cannot account for the phenomenon since the immunized subjects experience inescapable shock just like the helplessness animals; the competing response theory would predict that the immunized animals should also show interference effects. However they do not, suggesting that any freezing reactions which are present following the inescapable shock phase are not solely responsible for the interference affect.

Current theory and research (for example, Jackson, Alexander, & Maier, 1980; Jackson, Maier, & Rapaport, 1978; Maier & Jackson, 1979) have favored a compromise between the learned helplessness and competing response theories. The essence of the argument in these studies is that inescapable shock causes both effects to occur. Animals do become more immobile during inescapable shock, but they also develop helplessness. One of the studies that highlighted this was done by Jackson et al. (1980). Experimental animals were given inescapable shock followed by escape-avoidance training. The study differed from others in that Phase 2, the learning phase, was conducted in a Y maze. Animals not only had to leave the start box, but they also had to choose the correct alleyway in order to get to the goal box. Merely running away from the shock was not sufficient; the animals also had to make the correct choice. Inescapable shock, of course, produced the interference effect: the experimental animals ran more slowly and their choice performance was inferior to that of the control subjects. What was interesting was that by increasing the shock intensity, a procedure that eliminates immobility or freezing behavior, the speed of response was increased appropriately but the choice behavior was unaffected. In other words, the animals were demonstrating both helplessness and competing responses. The higher shock intensity eliminated the deficit caused by the competing immobility reaction, but it did not affect the deficit that was caused by the inappropriate cognition, the helplessness expectancy. In summary, the interference effect reflects two underlying processes. The first, and perhaps most interesting, is the fact that animals develop a cognition that their behavior is independent of reward. This cognition interferes with the learning of new expectancies later on. Second, inescapable shock also produces significant changes in the animal's motor behavior, notably in the immobility reactions the animal later demonstrates during shock.

The literature on learned helplessness is vast and complex and for this reason it is difficult to provide a simple conclusion to the research. It would appear, however, that various factors contribute to the occurrence of the phenomenon. The fact that responding does not control reward certainly leads to an expectancy that behavior is ineffective; this is the cognitive, or helplessness, component. In addition, it would seem reasonable to assume that animals are doing whatever they can to minimize shock during the helplessness training. Such efforts may include freezing, jumping, or any of a variety of reactions. These, in turn, may influence later learning. And so it seems that this very important phenomenon may have no single explanation; it may be produced under various conditions, all of which lead to the acquisition of helplessness and inappropriate motor behaviors.

On another level of analysis, it is highly appropriate to end with a discussion of helplessness because the phenomenon clearly illustrates several important themes that have been dominant throughout the text. First and most important, learned helplessness indicates how fundamental prediction is to animal learning and behavior. The ability of an animal to use stimuli or to use its own behavior to predict future events such as USs is critical to the animal's survival and central to our understanding of how learning enhances survival. When these predictive mechanisms fail, such as in the learned helplessness phenomenon, future learning (and thus future adaptive behavior) is jeopardized. Animals benefit from being able to predict threats or beneficial circumstances; they are debilitated when they come to expect that their behavior is irrelevant to the very commodities or threats that they must predict in order to survive. In short, learned helplessness research has highlighted the way in which expectancy learning, or cognition, is central to animal learning.

Also emphasized in the learned helplessness literature is a second theme that has been central in this text. This is the notion that species-typical behaviors are, ultimately, the ones that become expressed and modified by learning contingencies. More specifically, the finding that helplessness is, in part, explained by competing responses shows that the animals are indeed performing naturalistic reactions to the punishing stimuli; such reactions are the animals' only means of coping with the conditions. The fact that these reactions later influence avoidance learning (by competing with the proper avoidance response) is an interesting and important finding for the theory of helplessness. However, the more fundamental point is that the competing responses were executed in the first place. And they were the animals' attempt, if unsuccessful, at coping with great adversity. If one can imagine an animal trying, in vain, to cope with shock during the helplessness training by executing various species-typical defenses, while at the same time trying, also in vain, to detect a pattern to the events, a correlation between the external world (stimuli) or its own behavior (response) and the important US presentations, then one has grasped the essential focus of this text. Fortunately for animals that can learn, and that obviously includes human beings, the species-typical behavior potentials and the acquisition of expectancies do not usually occur in vain.

Summary

One problem encountered in the area of punishment and learning disorders occurs during avoidance extinction. If the animal responds quickly, as it did during acquisition, then it cannot discover that no aversive US is scheduled to be given. Accordingly, its fear will remain strong, leading to persistent responding.

One technique to counteract such an effect is to block the animal's response during a flooding session. This procedure decreases resistance to extinction. It appears that the more the animal experiences the CS without being able to respond, the quicker the extinction. One prominent theory suggests that flooding decreases fear which, in turn, leads to faster extinction. Other factors, however, have been shown to be important, including the development of freezing and other locomotor responses that compete with the avoidance CR.

A second technique that, in theory, should promote avoidance extinction involves punishment of the response. This procedure, however, tends to increase, rather than decrease, extinction responding. One reason appears to be that punishment maintains or enhances fear; this, in turn, leads to persistent responding during extinction. Conversely, animals may be "confused" about whether they are performing under the acquisition or extinction conditions; this confusion may cause them to persist longer.

Increased avoidance rates during extinction indicate maladaptive behavior in the sense that subjects are not behaving according to the contingencies in effect. Other disorders resulting from punishment include the development of ulcers and aggressive reactions.

Learned helplessness is an important phenomenon in modern learning research. If animals are given inescapable shocks, they later have great difficulty in learning simple escape-avoidance responses. Other subjects, that received the same amount of shock but could control its offset, did not show this learning disorder.

Responses most susceptible to helplessness treatment are non-SSDRs. In addition, subjects may be immunized against the effects of inescapable shock (by receiving prior escapable shock) or cured of the helplessness (through forced responding). The effects of helplessness are not limited to a single behavior or US; even noncontingent food presentations can later cause retarded avoidance learning.

The most important theory of helplessness claims that inescapable shock induces in the animal the expectancy that its behavior and rewards are independent. Other theories have stressed the idea that inescapable shock actually leads to the conditioning of covert responses that later interfere with the avoidance CR. Evidence seems to suggest that both factors underlie the helplessness phenomenon.

References

Adamec, R., & Melzack, R. The role of motivation and orientation in sensory preconditioning. *Canadian Journal of Psychology,* 1970, *24,* 230–39.

Adelman, H. M., & Maatsch, J. L. Resistance to extinction as a function of the type of response elicited by frustration. *Journal of Experimental Psychology,* 1955, *50,* 61–65.

Ahlers, R. H., & Best, P. J. Novelty vs. temporal contiguity in learned taste aversions. *Psychonomic Science,* 1971, *25,* 34–36.

Allen, J. D., Stein, G. W., & Long, C. J. The effect of food deprivation on responding for food odor in the rat. *Learning and Motivation,* 1972, *3,* 101–7.

Allison, J., Larson, D., & Jensen, D. D. Acquired fear, brightness preference, and one-way shuttle box performance. *Psychonomic Science,* 1967, *8,* 269–70.

Altenor, A., Kay, E., & Richter, M. The generality of learned helplessness in the rat. *Learning and Motivation,* 1977, *8,* 54–62.

Amsel, A. The role of frustrative nonreward in noncontinuous reward situations. *Psychological Bulletin,* 1958, *55,* 102–19.

Amsel, A. Frustrative nonreward in partial reinforcement and discrimination learning: Some recent history and a theoretical extension. *Psychological Review,* 1962, *69,* 306–28.

Amsel, A. Behavioral habituation, counter conditioning, and a general theory of persistence. In A. H. Black & W. F. Prokasy (Eds.), *Classical conditioning II: Current research and theory.* New York: Appleton-Century-Crofts, 1972.

Amsel, A., Hug, J. J., & Surridge, C. T. Number of food pellets, goal approaches, and the partial reinforcement effect after minimal acquisition. *Journal of Experimental Psychology,* 1968, *77,* 530–34.

Amsel, A., & Roussel, J. Motivational properties of frustration: I. Effect on a running response of the addition of frustration to the motivational complex. *Journal of Experimental Psychology,* 1952, *43,* 363–68.

Anderson, D. C., Crowell, C. R., Cunningham, C. L., & Lupo, J. V. Behavior during shock exposure as a determinant of subsequent interference with shuttle box escape-avoidance learning in the rat. *Journal of Experimental Psychology: Animal Behavior Processes,* 1979, *5,* 243–57.

Anderson, H. H. Comparison of different populations: Resistance to extinction and transfer. *Psychological Review,* 1963, *70,* 162–79.

Andrews, E. A., & Braveman, N. S. The combined effects of dosage level and interstimulus interval on the formation of one-trial poison-based aversions in rats. *Animal Learning and Behavior,* 1975, *3,* 287–89.

Anger, D. The role of temporal discrimination in the reinforcement of Sidman avoidance behavior. *Journal of the Experimental Analysis of Behavior,* 1963, *6,* 477–506.

Anisman, H. Time-dependent variations in aversively motivated behaviors: Nonassociative effects of cholinergic and catecholaminergic activity. *Psychological Review,* 1975, *82,* 359–85.

Anisman, H., deCatanzaro, D., & Remington, G. Escape performance following exposure to inescapable shock: Deficits in motor response maintenance. *Journal of Experimental Psychology: Animal Behavior Processes,* 1978, *4,* 197–218.

Anisman, H., & Waller, T. G. Effects of conflicting response requirements and shock-compartment confinement on the Kamin effect in rats. *Journal of Comparative and Physiological Psychology,* 1971, *77,* 240–44. (a)

Anisman, H., & Waller, T. G. Effects of methamphetamine and shock duration during inescapable shock exposure on subsequent active and passive avoidance. *Journal of Comparative and Physiological Psychology,* 1971, *77,* 143–51.(b)

Anisman, H., & Waller, T. G. Facilitative and disruptive effects of prior exposure to shock on subsequent avoidance performance. *Journal of Comparative and Physiological Psychology,* 1972, *78,* 113–22.

Anisman, H., & Waller T. G. Effects of inescapable shock on subsequent avoidance performance: Role of response repertoire changes. *Behavioral Biology,* 1973, *9,* 331–55.

Annau, Z., & Kamin, L. J. The conditioned emotional response as a function of intensity of the US. *Journal of Comparative and Physiological Psychology,* 1961, *54,* 428–32.

Ayres, C. E. Instinct and capacity: I. The instinct of belief-in-instincts. *Journal of Philosophy,* 1921, *18,* 561–66.

Ayres, J. J. B. Conditioned suppression and the information hypothesis. *Journal of Comparative and Physiological Psychology,* 1966, *62,* 21–25.

Ayres, J. J. B., & DeCosta, M. J. The truly random control as an extinction procedure. *Psychonomic Science,* 1971, *24,* 31–33.

Azrin, N. H., Hutchinson, R. R., & Hake, D. F. Pain-induced fighting in the squirrel monkey. *Journal of the Experimental Analysis of Behavior,* 1963, *6,* 620.

Azrin, N. H., Hutchinson, R. R., & Sallery, R. D. Pain-aggression toward inanimate objects. *Journal of the Experimental Analysis of Behavior,* 1964, *7,* 223–28.

Bacon, W. E. Partial-reinforcement extinction effect following different amounts of training. *Journal of Comparative and Physiological Psychology,* 1962, *55,* 998–1003.

Badia, P., & Defran, R. H. Orienting responses and GSR conditioning: A dilemma. *Psychological Review,* 1970, *77,* 171–81.

Badia, P., Harsh, J., & Abbott, B. Choosing between predictable and unpredictable shock conditions: Data and theory. *Psychological Bulletin,* 1979, *86,* 1107–31.

Baenninger, R., & Ulm, R. R. Overcoming the effects of prior punishment on interspecies aggression in the rat. *Journal of Comparative and Physiological Psychology,* 1969, *69,* 628–35.

Baker, A. G. Learned irrelevance and learned helplessness: Rats learn that stimuli, reinforcers, and responses are uncorrelated. *Journal of Experimental Psychology: Animal Behavior Processes,* 1976, *2,* 130–41.

Baker, A. G., & Mackintosh, N. J. Excitatory and inhibitory conditioning following uncorrelated presentations of CS and US. *Animal Learning and Behavior,* 1977, *5,* 315–19.

Baker, A. G., & Mackintosh, N. J. Preexposure to the CS alone, US alone, or CS and US uncorrelated: Latent inhibition, blocking by context or learned irrelevance? *Learning and Motivation,* 1979, *10,* 278–94.

Balagura, S. Conditioned glycemic responses in the control of food intake. *Journal of Comparative and Physiological Psychology,* 1968, *65,* 30–32.

Barker, L. M. CS duration, amount, and concentration effects in conditioning taste aversions. *Learning and Motivation,* 1976, *7,* 265–73.

Barker, L. M., Best, M. R., & Domjan, M. *Learning mechanisms in food selection.* Waco, Texas: Baylor University Press, 1977.

Barnes, G. W., & Kish, G. B. Reinforcing properties of the onset of auditory stimulation. *Journal of Experimental Psychology,* 1961, *62,* 164–70.

Baron, A. Delayed punishment of a runway response. *Journal of Comparative and Physiological Psychology,* 1965, *60,* 131–34.

Baron, A., Kaufman, A., & Fazzini, D. Density and delay of punishment of free-operant avoidance. *Journal of the Experimental Analysis of Behavior,* 1969, *12,* 1029–37.

Barrett, J. E. Schedules of electric shock presentation in the behavioral control of imprinted ducklings. *Journal of the Experimental Analysis of Behavior,* 1972, *18,* 305–21.

Barrett, R. J., Leith, N. J., & Ray, O. S. Kamin effect in rats: Index of memory or shock-induced inhibition? *Journal of Comparative and Physiological Psychology,* 1971, *77,* 234–39.

Barry, H. Effects of strength of drive on learning and on extinction. *Journal of Experimental Psychology,* 1958, *55,* 473–81.

Bateson, P. P. G. The characteristics and context of imprinting. *Biological Reviews,* 1966, *41,* 177–220.

Bateson, P. P. G., & Reese, E. P. The reinforcing properties of conspicuous stimuli in the imprinting situation. *Animal Behavior,* 1969, *17,* 692–99.

Batson, J. D., & Best, P. J. Drug-preexposure effects in flavor-aversion learning: Associative interference by conditioned environmental stimuli. *Journal of Experimental Psychology: Animal Behavior Processes,* 1979, *5,* 273–83.

Bauer, R. H. The effects of CS and US intensity on shuttlebox avoidance. *Psychonomic Science,* 1972, *27,* 266–68.

Baum, M. Extinction of an avoidance response following response prevention: Some parametric investigations. *Canadian Journal of Psychology,* 1969, *23,* 1–10. (a)

Baum, M. Extinction of an avoidance response motivated by intense fear: Social facilitation of the action of response prevention (flooding) in rats. *Behavior Research and Therapy,* 1969, *7,* 57–62. (b)

Baum, M. Extinction of avoidance responding through response prevention (flooding). *Psychological Bulletin,* 1970, *74,* 276–84.

Baum, M., & Myran, D. D. Response prevention (flooding) in rats: The effects of restricting exploration during flooding and of massed vs. distributed flooding. *Canadian Journal of Psychology,* 1971, *25,* 138–46.

Baum, W. M. Time allocation and negative reinforcement. *Journal of the Experimental Analysis of Behavior,* 1973, *20,* 313–22.

Baum, W. M. On two types of deviation from the matching law: Bias and undermatching. *Journal of the Experimental Analysis of Behavior,* 1974, *22,* 231–42.

Baum, W. M., & Rachlin, H. C. Choice as time allocation. *Journal of the Experimental Analysis of Behavior,* 1969, *12,* 861–74.

Baumeister, A., Hawkins, W., & Cromwell, R. Need states and activity level. *Psychological Bulletin,* 1964, *61,* 438–53.

Beck, S. B. Eyelid conditioning as a function of CS intensity, UCS intensity, and manifest anxiety scale score. *Journal of Experimental Psychology,* 1963, *66,* 429–38.

Beecroft, R. S. *Classical conditioning.* Goleta, Calif.: Psychonomic Press, 1966.

Beery, R. G. A negative contrast effect of reward delay in differential conditioning. *Journal of Experimental Psychology,* 1968, *77,* 429–34.

Behar, I., & LeBedda, J. M. Effects of differential pretraining on learning-set formation in rhesus monkeys. *Journal of Comparative and Physiological Psychology,* 1974, *87,* 277–83.

Bekhterev, V. M. *La psychologie objective.* Paris: Alcan, 1913.

Bender, L. Secondary punishment and self-punitive avoidance behavior in the rat. *Journal of Comparative and Physiological Psychology,* 1969, *69,* 261–66.

Benedict, J. O., & Ayres, J. J. B. Factors affecting conditioning in the truly random control procedure in the rat. *Journal of Comparative and Physiological Psychology,* 1972, *78,* 323–30.

Benson, H., Shapiro, D., Tursky, B., & Schwartz, G. F. Decreased systolic blood pressure through operant conditioning techniques in patients with essential hypertension. *Science,* 1971, *173,* 740–42.

Berlyne, D. E. The reward-value of indifferent stimulation. In J. T. Tapp (Ed.), *Reinforcement and behavior.* New York: Academic Press, 1969.

Bersh, P. J., & Keltz, J. R. Pavlovian reconditioning and the recovery of avoidance behavior in rats after extinction with response prevention. *Journal of Comparative and Physiological Psychology,* 1971, *76,* 262–66.

Bersh, P. J., & Miller, K. The influence of shock during response prevention upon resistance to extinction of an avoidance response. *Animal Learning and Behavior,* 1975, *3,* 140–42.

Bersh, P. J., & Paynter, W. E. Pavlovian extinction in rats during avoidance response prevention. *Journal of Comparative and Physiological Psychology,* 1972, *78,* 255–59.

Bertsch, G. J., & Leitenberg, H. A "Frustration Effect" following electric shock. *Learning and Motivation,* 1970, *1,* 150–56.

Best, M. R. Conditioned and latent inhibition in taste-aversion learning: Clarifying the role of learned safety. *Journal of Experimental Psychology: Animal Behavior Processes,* 1975, *1,* 97–113.

Best, M. R., Gemberling, G. A., & Johnson, P. E. Disrupting the conditioned stimulus preexposure effect in flavor-aversion learning: Effects of interoceptive distractor manipulations. *Journal of Experimental Psychology: Animal Behavior Processes,* 1979, *5,* 321–34.

Best, P. J., Best, M. R., & Lindsey, G. P. The role of cue additivity in salience in taste aversion conditioning. *Learning and Motivation,* 1976, *7,* 254–64.

Best, P. J., Best, M. R., & Mickley, G. A. Conditioned aversion to distinct environmental stimuli resulting from gastrointestinal distress. *Journal of Comparative and Physiological Psychology,* 1973, *85,* 250–57.

Bintz, J. Time-dependent memory deficits of aversively motivated behavior. *Learning and Motivation,* 1970, *1,* 382–90.

Bintz, J. Effects of detention in and removal from the training environment on retention of aversively motivated behavior. *Learning and Motivation,* 1972, *3,* 44–50.

Bintz, J., Braud, W. G., & Brown, J. S. An analysis of the role of fear in the Kamin effect. *Learning and Motivation,* 1970, *1,* 170–76.

Black, A. H., Carlson, N. J., & Solomon, R. L. Exploratory studies of the conditioning of autonomic responses in curarized dogs. *Psychological Monographs,* 1962, *76,* (Whole No. 548).

Black, A. H., & de Toledo, L. The relationship among classically conditioned responses: Heart rate and skeletal behavior. In A. H. Black & W. F. Prokasy (Eds.), *Classical conditioning II: Current research and theory.* New York: Appleton-Century-Crofts, 1972.

Blanchard, E. B., & Young, L. D. Self-control of cardiac functioning: A promise as yet unfulfilled. *Psychological Bulletin,* 1973, *79,* 145–63.

Blanchard, R., & Honig, W. K. Surprise value of food determines its effectiveness as a reinforcer. *Journal of Experimental Psychology: Animal Behavior Processes,* 1976, *2,* 67–74.

Blanchard, R. J., & Blanchard, D. C. Crouching as an index of fear. *Journal of Comparative and Physiological Psychology,* 1969, *67,* 370–75. (a)

Blanchard, R. J., & Blanchard, D. C. Passive and active reactions to fear-eliciting stimuli. *Journal of Comparative and Physiological Psychology,* 1969, *68,* 129–35. (b)

Blanchard, R. J., & Blanchard, D. C. Defensive reactions in the albino rat. *Learning and Motivation,* 1971, *2,* 351–62.

Blanchard, R. J., Kelley, M. J., & Blanchard, D. C. Defensive behaviors and exploratory behavior in rats. *Journal of Comparative and Physiological Psychology,* 1974, *87,* 1129–33.

Blough, D. S. Two-way generalization peak shift after two-key training in the pigeon. *Animal Learning and Behavior,* 1973, *1,* 171–74.

Bolles, R. C. Anticipatory general activity in thirsty rats. *Journal of Comparative and Physiological Psychology,* 1968, *65,* 511–13.

Bolles, R. C. Species-specific defense reactions and avoidance learning. *Psychological Review,* 1970, *77,* 32–48. (a)

Bolles, R. C. The cue value of illumination change in anticipatory general activity. *Learning and Motivation,* 1970, *1,* 177–85. (b)

Bolles, R. C. Reinforcement, expectancy, and learning. *Psychological Review,* 1972, *79,* 394–409.

Bolles, R. C. *Theory of motivation* (2nd ed.). New York: Harper & Row, 1975.

Bolles, R. C., & Collier, A. C. The effect of predictive cues on freezing in rats. *Animal Learning and Behavior,* 1976, *4,* 6–8.

Bolles, R. C., & Grossen, N. E. Effects of an informational stimulus on the acquisition of avoidance behavior in rats. *Journal of Comparative and Physiological Psychology,* 1969, *68,* 90–99.

Bolles, R. C., Grossen, N. E., Hargrave, G. E., & Duncan, P. M. Effects of conditioned appetitive stimuli on the acquisition and extinction of a runway response. *Journal of Experimental Psychology,* 1970, *85,* 138–40.

Bolles, R. C., Hayward, L., & Crandall, C. Conditioned taste preferences based on caloric density. *Journal of Experimental Psychology: Animal Behavior Processes,* 1981, *7,* 59–69.

Bolles, R. C., Holtz, R., Dunn, T., & Hill, W. Comparisons of stimulus learning and response learning in a punishment situation. *Learning and Motivation,* 1980, *11,* 78–96.

Bolles, R. C., & Moot, S. A. The rat's anticipation of two meals a day. *Journal of Comparative and Physiological Psychology,* 1973, *83,* 510–14.

Bolles, R. C., Moot, S. A., & Grossen, N. E. The extinction of shuttlebox avoidance. *Learning and Motivation,* 1971, *2,* 324–33.

Bolles, R. C., & Morlock, H. Some asymmetrical drive summation phenomena. *Psychological Report,* 1960, *6,* 373–78.

Bolles, R. C., & Riley, A. L. Freezing as an avoidance response: Another look at the operant-respondent distinction. *Learning and Motivation,* 1973, *4,* 268–75.

Bolles, R. C., Riley, A. L., Cantor, M. B., & Duncan, P. M. The rat's failure to anticipate regularly scheduled daily shock. *Behavioral Biology,* 1974, *11,* 365–72.

Bolles, R. C., & Seelbach, S. E. Punishing and reinforcing effects of noise onset and termination for different responses. *Journal of Comparative and Physiological Psychology,* 1964, *58,* 127–31.

Bolles, R. C., & Stokes, L. W. Rat's anticipation of diurnal and adiurnal feeding. *Journal of Comparative and Physiological Psychology,* 1965, *60,* 290–94.

Bolles, R. C., Stokes, L. W., & Younger, M. S. Does CS termination reinforce avoidance behavior? *Journal of Comparative and Physiological Psychology,* 1966, *62,* 201–7.

Bond, N., & Digiusto, E. Amount of solution drunk as a factor in the establishment of taste aversion. *Animal Learning and Behavior,* 1975, *3,* 81–84.

Bond, N., & Harland, W. Higher order conditioning of a taste aversion. *Animal Learning and Behavior,* 1975, *3,* 295–96.

Booth, D. A. Conditioned satiety in the rat. *Journal of Comparative and Physiological Psychology,* 1972, *81,* 457–71.

Boren, J. J., Sidman, M., & Herrnstein, R. J. Avoidance, escape, and extinction as functions of shock intensity. *Journal of Comparative and Physiological Psychology,* 1959, *52,* 420–25.

Bouton, M. E., & Bolles, R. C. Contextual control of the extinction of conditioned fear. *Learning and Motivation,* 1979, *10,* 445–66.

Bower, G. H. The influence of graded reductions in reward and prior frustrating events upon the magnitude of the frustration effect. *Journal of Comparative and Physiological Psychology,* 1962, *55,* 582–87.

Bower, G. H., Starr, R., & Lazarovitz, L. Amount of response-produced change in the CS and avoidance learning. *Journal of Comparative and Physiological Psychology,* 1965, *59,* 13–17.

Bowlby, J. *Attachment and loss* (Vol. 1: *Attachment*). New York: Basic Books, 1969.

Bracewell, R. J., & Black, A. H. The effects of restraint and noncontingent pre-shock on subsequent escape learning in the rat. *Learning and Motivation,* 1974, *5,* 53–69.

Braud, W., Wepman, B., & Russo, D. Task and species generality of the "helplessness" phenomenon. *Psychonomic Science,* 1969, *16,* 154–55.

Breland, K., & Breland, M. The misbehavior of organisms. *American Psychologist,* 1961, *16,* 681–84.

Brener, J., & Goesling, W. J. Avoidance conditioning of activity and immobility in rats. *Journal of Comparative and Physiological Psychology,* 1970, *70,* 276–80.

Brener, J., Phillips, K., & Connally, S. R. Oxygen consumption and ambulation during operant conditioning of heart rate increases and decreases in rats. *Psychophysiology,* 1977, *14,* 483–91.

Brodigan, D. L., & Peterson, G. B. Two-choice discrimination performance of pigeons as a function of reward expectancy, prechoice delay, and domesticity. *Animal Learning and Behavior,* 1976, *4,* 121–24.

Bronowski, M., & Bellugi, U. Language, name and concept. *Science,* 1970, *168,* 669–73.

Bronstein, P. M., Dworkin, T., Bilder, B. H., & Wolkoff, F. D. Repeated failures in reducing rat's spontaneous alternation through the intertrial disruption of spatial orientation. *Animal Learning and Behavior,* 1974, *2,* 207–9.

Brooks, C. I. Frustration to nonreward following limited reward experience. *Journal of Experimental Psychology,* 1969, *81,* 403–5.

Brooks, C. I. Primary frustration differences following brief partial-reinforcement acquisition under varying magnitudes of reward. *Animal Learning and Behavior,* 1975, *3,* 67–72.

Brower, L. P. Ecological chemistry. *Scientific American,* 1969, *220,* 22–29.

Brower, L. P., Ryerson, W. N., Coppinger, L. L., & Glazier, S. C. Ecological chemistry and the palatability spectrum. *Science,* 1968, *161,* 1349–51.

Brown, B. L. Stimulus generalization in salivary conditioning. *Journal of Comparative and Physiological Psychology,* 1970, *71,* 467–77.

Brown, J. S. Factors affecting self-punitive locomotor behavior. In B. A. Campbell & R. M. Church (Eds.), *Punishment and aversive behavior.* New York: Appleton-Century-Crofts, 1969.

Brown, J. S. Self-punitive behavior with a distinctively marked punishment zone. *Psychonomic Science,* 1970, *21,* 161–63.

Brown, J. S., Beier, E. M., & Lewis, R. W. Punishment-zone distinctiveness and self-punitive locomotor behavior in the rat. *Journal of Comparative and Physiological Psychology,* 1971, *77,* 513–20.

Brown, J. S., Martin, R. C., & Morrow, M. W. Self-punitive behavior in the rat: Facilitative effects of punishment on resistance to extinction. *Journal of Comparative and Physiological Psychology,* 1964, *57,* 127–33.

Brown, R. T., & Wagner, A. R. Resistance to punishment and extinction following training with shock or nonreinforcement. *Journal of Experimental Psychology,* 1964, *68,* 503–7.

Brownstein, A. J. Concurrent schedules of response-independent reinforcement: Duration of reinforcing stimulus. *Journal of the Experimental Analysis of Behavior,* 1971, *15,* 211–14.

Brush, F. R. Retention of aversively motivated behavior. In F. R. Brush (Ed.), *Aversive conditioning and learning.* New York: Academic Press, 1971.

Brush, F. R., & Levine, S. Adrenocortical activity and avoidance behavior as a function of time after fear conditioning. *Physiology and Behavior,* 1966, *1,* 309–11.

Bryan, R. G., & Spear, N. E. Forgetting of a discrimination after intervals of intermediate length: The Kamin effect with choice behavior. *Journal of Experimental Psychology: Animal Behavior Processes,* 1976, *2,* 221–34.

Bull, J. A. An interaction between appetitive Pavlovian CSs and instrumental avoidance responding. *Learning and Motivation,* 1970, *1,* 18–26.

Burdick, C. K., & James, J. P. Spontaneous recovery of conditioned suppression of licking by rats. *Journal of Comparative and Physiological Psychology,* 1970, *72,* 467–70.

Burkhardt, P. E., & Ayres, J. J. B. CS and US duration effects in one-trial simultaneous fear conditioning as assessed by conditioned suppression of licking in rats. *Animal Learning and Behavior,* 1978, *6,* 225–30.

Burr, D. E. S., & Thomas, D. R. Effect of proactive inhibition upon the post-discrimination generalization gradient. *Journal of Comparative and Physiological Psychology,* 1972, *81,* 441–48.

Burstein, K. R., & Moeser, S. The informational value of a distinctive stimulus associated with the imitation of acquisition trials. *Learning and Motivation,* 1971, *2,* 228–34.

Butler, R. A. Discrimination learning by rhesus monkeys to visual exploration motivation. *Journal of Comparative and Physiological Psychology,* 1953, *46,* 95–98.

Camp, D. S., Raymond, G. A., & Church, R. M. Temporal relationship between response and punishment. *Journal of Experimental Psychology,* 1967, *74,* 114–23.

Campbell, B. A. Effects of water deprivation on random activity. *Journal of Comparative and Physiological Psychology,* 1960, *53,* 240–41.

Campbell, B. A. Theory and research on the effects of water deprivation on random activity in the rat. In M. Wayner (Ed.), *Thirst—Proceedings of the 1st international symposium on thirst in the regulation of body water.* New York: Pergamon Press, 1964.

Campbell, B. A. Developmental studies of learning and motivation in infraprimate mammals. In H. W. Stevenson, E. H. Hess, & H. L. Rheingold (Eds.), *Early behavior: Comparative and developmental approaches.* New York: Wiley, 1967.

Campbell, B. A., & Campbell, E. H. Retention and extinction of learned fear in infant and adult rats. *Journal of Comparative and Physiological Psychology,* 1962, *55,* 1–8.

Campbell, B. A., & Cicala, G. A. Studies of water deprivation in rats as a function of age. *Journal of Comparative and Physiological Psychology,* 1962, *55,* 763–68.

Campbell, B. A., & Coulter, X. The ontogeny of learning and memory. In M. R. Rosenzweig & L. Bennett (Eds.), *Neural mechanisms of learning and memory.* Cambridge, Mass.: MIT Press, 1976.

Campbell, B. A., & Jaynes, J. Reinstatement. *Psychological Review,* 1966, *73,* 478–80.

Campbell, B. A., Jaynes, J., & Misanin, J. R. Retention of a light-dark discrimination in rats of different ages. *Journal of Comparative and Physiological Psychology,* 1968, *66,* 467–72.

Campbell, B. A., & Kraeling, D. Response strength as a function of drive level during training and extinction. *Journal of Comparative and Physiological Psychology,* 1954, *47,* 101–3.

Campbell, B. A., & Randall, P. The effect of reinstatement stimulus conditions on the maintenance of long-term memory. *Developmental Psychobiology,* 1976, *9,* 325–33.

Campbell, B. A., & Sheffield, F. D. Relation of random activity to food deprivation. *Journal of Comparative and Physiological Psychology,* 1953, *46,* 320–22.

Campbell, B. A., Smith, N. F., & Misanin, J. R. Effects of punishment on extinction of avoidance behavior: Avoidance-avoidance conflict or vicious circle behavior. *Journal of Comparative and Physiological Psychology,* 1966, *62,* 495–98.

Campbell, B. A., Smith, N. F., Misanin, J. R., & Jaynes, J. Species differences in activity during hunger and thirst. *Journal of Comparative and Physiological Psychology,* 1966, *61,* 123–27.

Campbell, B. A., & Spear, N. E. Ontogeny of memory. *Psychological Review,* 1972, *79,* 215–36.

Campbell, P. E. Partially delayed reward: Some effects of sequential reward magnitude in the rat. *Journal of Comparative and Physiological Psychology,* 1970, *71,* 152–59.

Campbell, P. E., Batsche, C. J., & Batsche, G. M. Spaced-trials reward magnitude effects in the rat: Single versus multiple food pellets. *Journal of Comparative and Physiological Psychology,* 1972, *81,* 360–64.

Cannon, D. S., Berman, R. F., Baker, T. B., & Atkinson, C. A. Effect of preconditioning unconditioned stimulus experience on learned taste aversions. *Journal of Experimental Psychology: Animal Behavior Processes,* 1975, *1,* 270–84.

Capaldi, E. D. Simultaneous shifts in reward magnitude and level of food deprivation. *Psychonomic Science,* 1971, *23,* 357–59.

Capaldi, E. D. Resistance to extinction in rats as a function of deprivation level and schedule of reward in acquisition. *Journal of Comparative and Physiological Psychology,* 1972, *79,* 90–98.

Capaldi, E. D. Effect of shifts in body weight on rats' straight alley performance as a function of reward magnitude. *Learning and Motivation,* 1973, *4,* 229–35.

Capaldi, E. D., & Hovancik, J. R. Effects of previous body weight level on rats' straight-alley performance. *Journal of Experimental Psychology,* 1973, *97,* 93–97.

Capaldi, E. D., Hovancik, J. R., & Davidson, T. L. Learning about water by hungry rats. *Learning and Motivation,* 1979, *10,* 58–72.

Capaldi, E. D., Hovancik, J. R., & Lamb, E. O. The effects of strong irrelevant thirst on food-rewarded instrumental performance. *Animal Learning and Behavior,* 1975, *3,* 172–78.

Capaldi, E. J. The effect of different amounts of training on the resistance to extinction of different patterns of partially reinforced responses. *Journal of Comparative and Physiological Psychology,* 1958, *51,* 367–71.

Capaldi, E. J. Effect of N-length, number of different N-lengths, and number of reinforcements on resistance to extinction. *Journal of Experimental Psychology,* 1964, *68,* 230–39.

Capaldi, E. J. Partial reinforcement: A hypothesis of sequential effects. *Psychological Review,* 1966, *73,* 459–77.

Capaldi, E. J. A sequential hypothesis of instrumental learning. In K. W. Spence & J. T. Spence (Eds.), *The psychology of learning and motivation* (Vol. 1). New York: Academic Press, 1967.

Capaldi, E. J., & Capaldi, E. D. Magnitude of partial reward, irregular reward schedules, and a 24-hour ITI: A test of several hypotheses. *Journal of Comparative and Physiological Psychology,* 1970, *72,* 203–9.

Capaldi, E. J., & Deutsch, E. A. Effects of severely limited acquisition training and pretraining on the partial reinforcement effect. *Psychonomic Science,* 1967, *9,* 171–72.

Capaldi, E. J., & Hart, D. Influence of a small number of partial reinforcement training trials on resistance to extinction. *Journal of Experimental Psychology,* 1962, *64,* 166–71.

Capaldi, E. J., Lanier, A. T., & Godbout, R. C. Reward schedule effects following severely limited acquisition training. *Journal of Experimental Psychology,* 1968, *78,* 521–24.

Capaldi, E. J., & Lynch, A. D. Magnitude of partial reward and resistance to extinction: Effect of N-R transitions. *Journal of Comparative and Physiological Psychology,* 1968, *65,* 179–81.

Capaldi, E. J., & Waters, R. W. Conditioning and nonconditioning interpretations of small-trial phenomena. *Journal of Experimental Psychology,* 1970, *84,* 518–22.

Caplan, H. J., Karpicke, J., & Rilling, M. Effects of extended fixed-interval training on reversed scallops. *Animal Learning and Behavior,* 1973, *1,* 293–96.

Carlson, J. G., & Wielkiewicz, R. M. Delay of reinforcement in instrumental discrimination learning in rats. *Journal of Comparative and Physiological Psychology,* 1972, *81,* 365–70.

Catania, A. C. Concurrent performances: A baseline for the study of reinforcement magnitudes. *Journal of the Experimental Analysis of Behavior,* 1963, *6,* 299–300.

Catania, A. C., & Reynolds, G. S. A quantitative analysis of the responding maintained by interval schedules of reinforcement. *Journal of the Experimental Analysis of Behavior,* 1968, *11,* 327–83.

Caul, W. F., Buchanan, D. C., & Hays, R. C. Effects of unpredictability of shock on incidence of gastric lesions and heart rate in immobilized rats. *Physiology and Behavior,* 1972, *8,* 669–72.

Chomsky, N. A. Review of *verbal behavior,* by B. F. Skinner. *Language,* 1959, *35,* 26–58.

Chung, S., & Hernstein, R. J. Choice and delay of reinforcement. *Journal of the Experimental Analysis of Behavior,* 1967, *10,* 67–74.

Church, R. M., Getty, D. J., & Lerner, N. D. Duration discrimination by rats. *Journal of Experimental Psychology: Animal Behavior Processes,* 1976, *2,* 303–12.

Church, R. M., LoLordo, V. M., Overmier, J. B., Solomon, R. L., & Turner, L. H. Cardiac responses to shock in curarized dogs. *Journal of Comparative and Physiological Psychology,* 1966, *62,* 1–7.

Church, R. M., Raymond, G. A., & Beauchamp, R. D. Response suppression as a function of intensity and duration of a punishment. *Journal of Comparative and Physiological Psychology,* 1967, *63,* 39–44.

Church, R. M., Wooten, C. L., & Matthews, T. J. Discriminative punishment and the conditioned emotional response. *Learning and Motivation,* 1970, *1,* 1–17.

Clarke, J. C., Westbrook, R. F. & Irwin, J. Potentiation instead of overshadowing in the pigeon. *Behavioral and Neural Biology,* 1979, *25,* 18–29.

Clements, M., & Lien, J. Effects of tactile stimulation on the initiation and maintenance of the following response in Japanese quail *(Coturnix coturnix japonica). Animal Learning and Behavior,* 1975, *3,* 301–4.

Coate, W. B. Effect of deprivation on post discrimination stimulus generalization in the rat. *Journal of Comparative and Physiological Psychology,* 1964, *57,* 134–38.

Colby, J. J., & Smith, N. F. The effect of three procedures for eliminating a conditioned taste aversion in the rat. *Learning and Motivation,* 1977, *8,* 404–13.

Coleman, S. R., & Gormezano, I. Classical conditioning of the rabbit's *(oryctolagus cuniculus)* nictitating membrane response under symmetrical CS-US interval shifts. *Journal of Comparative and Physiological Psychology,* 1971, *77,* 447–55.

Collerain, I., & Ludvigson, H. W. Hurdle-jump responding in the rat as a function of conspecific odor of reward and nonreward. *Animal Learning and Behavior,* 1977, *5,* 177–83.

Cook, R. G. Retroactive interference in pigeon short-term memory by a reduction in ambient illumination. *Journal of Experimental Psychology: Animal Behavior Processes,* 1980, *6,* 326–38.

Corson, J. A. Observational learning of a lever pressing response. *Psychonomic Science,* 1967, *7,* 197–98.

Coughlin, R. C. The frustration effect and resistance to extinction as a function of percentage of reinforcement. *Journal of Experimental Psychology,* 1970, *84,* 113–17.

Coulson, G., Coulson, V., & Gardner, L. The effect of two extinction procedures after acquisition on a Sidman avoidance contingency. *Psychonomic Science,* 1970, *18,* 309–10.

Coulter, X., Collier, A. C., & Campbell, B. A. Long-term retention of early Pavlovian fear conditioning in infant rats. *Journal of Experimental Psychology: Animal Behavior Processes,* 1976, *2,* 48–56.

Coulter, X., Riccio, D. C., & Page, H. A. Effects of blocking an instrumental avoidance response: Facilitated extinction but persistence of "fear." *Journal of Comparative and Physiological Psychology,* 1969, *68,* 377–81.

Cravens, R. W., & Renner, K. E. Conditioned hunger. *Journal of Experimental Psychology,* 1969, *81,* 312–16.

Cravens, R. W., & Renner, K. E. Conditioned appetitive drive states: Empirical evidence and theoretical status. *Psychological Bulletin,* 1970, *73,* 212–20.

Crawford, M. Brief "response prevention" in a novel place can facilitate avoidance extinction. *Learning and Motivation,* 1977, *8,* 39–53.

Crider, A., Schwartz, G. E., & Shnidman, S. On the criteria for instrumental autonomic conditioning: A reply to Katkin and Murray. *Psychological Bulletin,* 1969, *71,* 455–61.

Crowell, C. R., & Anderson, D. C. Variations in intensity, interstimulus interval, and interval between preconditioning CS exposures and conditioning with rats. *Journal of Comparative and Physiological Psychology,* 1972, *79,* 291–98.

Crowell, C. R., & Bernhardt, T. P. The feature-positive effect and sign-tracking behavior during the discrimination learning in the rat. *Animal Learning and Behavior,* 1979, *7,* 313–17.

Crum, J., Brown, W. L., & Bitterman, M. E. The effect of partial and delayed reinforcement on resistance to extinction. *American Journal of Psychology,* 1951, *64,* 228–37.

Daly, H. B. Excitatory and inhibitory effects of complete and incomplete reward reduction in the double runway. *Journal of Experimental Psychology,* 1968, *76,* 430–38.

Daly, H. B. Combined effects of fear and frustration on acquisition of a hurdle-jump response. *Journal of Experimental Psychology,* 1970, *83,* 89–93.

D'Amato, M. R. Delayed matching and short-term memory in monkeys. In G. H. Bower (Ed.), *The psychology of learning and motivation: Avoidances in theory and research* (Vol. 7). New York: Academic Press, 1973.

D'Amato, M. R. Derived motives. *Annual Review of Psychology,* 1974, *25,* 83–106.

D'Amato, M. R., & Cox, J. K. Delay of consequences and short-term memory in monkeys. In D. L. Medin, W. A. Roberts, & R. T. Davis (Eds.), *Processes of animal memory.* Hillsdale, N.J.: Lawrence Erlbaum, 1976.

D'Amato, M. R., & Fazzaro, J. Discriminated lever-press avoidance learning as a function of type and intensity of shock. *Journal of Comparative and Physiological Psychology,* 1966, *61,* 313–15.

D'Amato, M. R., Fazzaro, J., & Etkin, M. Anticipatory responding and avoidance discrimination as factors in avoidance conditioning. *Journal of Experimental Psychology,* 1968, *77,* 41–47.

D'Amato, M. R., & Meinrath, M. Bidirectional operant conditioning of heart rate in rats with sucrose reward. *Learning and Motivation,* 1979, *10,* 488–501.

D'Amato, M. R., & O'Neill, W. Effect of delay-interval illumination on matching behavior in the capuchin monkey. *Journal of the Experimental Analysis of Behavior,* 1971, *15,* 327–33.

D'Amato, M. R., & Schiff, D. Long-term discriminated avoidance performance in the rat. *Journal of Comparative and Physiological Psychology,* 1964, *57,* 123–26.

D'Amato, M. R., & Schiff, D. Overlearning and brightness-discrimination reversal. *Journal of Experimental Psychology,* 1965, *69,* 375–81.

D'Amato, M. R., & Worsham, R. W. Retrieval cues and short-term memory in capuchin monkeys. *Journal of Comparative and Physiological Psychology,* 1974, *86,* 274–82.

Davenport, J. W. Distribution of M and i parameters for rats trained under varying hunger drive levels. *Journal of Genetic Psychology,* 1965, *106,* 113–21.

Davis, H., & Kreuter, C. Conditioned suppression of an avoidance response by a stimulus paired with food. *Journal of the Experimental Analysis of Behavior,* 1972, *17,* 277–85.

Deane, G. E. Cardiac conditioning in the albino rabbit using three CS-UCS intervals. *Psychonomic Science,* 1965, *3,* 119–20.

Dearing, M. F., & Dickinson, A. Counterconditioning of shock by a water reinforcer in rabbits. *Animal Learning and Behavior,* 1979, *7,* 360–66.

Deets, A. C., Harlow, H. F., & Bloomquist, A. J. Effect of intertrial interval and trial 1 reward during acquisition of an object discrimination learning-set in monkeys. *Journal of Comparative and Physiological Psychology,* 1970, *73,* 501–5.

Delprato, D. J., & Carosio, L. A. Elimination of self-punitive behavior with a novel stimulus and safety signal. *Animal Learning and Behavior,* 1976, *4,* 210–12.

Delprato, D. J., & Holmes, P. A. Facilitation of discriminated lever-press avoidance by noncontingent shocks. *Learning and Motivation,* 1977, *8,* 238–46.

Del Russo, J. E. Observational learning of discriminative avoidance in hooded rats. *Animal Learning and Behavior,* 1975, *3,* 76–80.

Dember, W. N., & Fowler, H. Spontaneous alternation behavior. *Psychological Bulletin,* 1958, *55,* 412–28.

Denny, M. R. Elicitation theory applied to an analysis of the overlearning reversal effect. In J. H. Reynierse (Ed.), *Current issues in animal learning.* Lincoln: University of Nebraska Press, 1970.

DePaulo, P., Hoffman, H. S., Klein, S., & Gaioni, S. Effect of response-contingent vs. noncontingent shock on ducklings' preference for novel imprinting stimuli. *Animal Learning and Behavior,* 1978, *6,* 458–62.

Desiderato, O., Butler, B., & Meyer, C. Changes in fear generalization gradients as a function of delayed testing. *Journal of Experimental Psychology,* 1966, *72,* 678–82.

Desiderato, O., MacKinnon, J. R., & Hisson, H. Development of gastric ulcers in rats following stress termination. *Journal of Comparative and Physiological Psychology,* 1974, *87,* 208–14.

de Toledo, L. Changes in heart rate during conditioned suppression in rats as a function of US intensity and types of CS. *Journal of Comparative and Physiological Psychology,* 1971, *77,* 528–38.

Deutsch, R. Conditioned hypoglycemia: A mechanism for saccharin-induced sensitivity to insulin in the rat. *Journal of Comparative and Physiological Psychology,* 1974, *86,* 350–58.

Deutsch, R. Effects of CS amount on conditioned taste aversion at different CS-US intervals. *Animal Learning and Behavior,* 1978, *6,* 258–60.

de Villiers, P. A. The law of effect and avoidance: A quantitative relationship between response rate and shock-frequency reduction. *Journal of the Experimental Analysis of Behavior,* 1974, *21,* 223–35.

de Villiers, P. A. Choice in concurrent schedules and a quantitative formulation of the law of effect. In W. K. Honig & J. E. R. Staddon (Eds.), *Handbook of operant behavior.* Englewood Cliffs, N.J.: Prentice-Hall, 1977.

de Villiers, P. A., & Herrnstein, R. J. Toward a law of response strength. *Psychonomic Bulletin,* 1976, *83,* 1131–53.

Devine, J. V. Stimulus attributes and training procedures in learning-set formation of rhesus and cebus monkeys. *Journal of Comparative and Physiological Psychology,* 1970, *73,* 62–67.

Devine, J. V., Jones, L. C., Neville, J. W., & Sakai, D. J. Sample duration and type of stimuli in delayed matching-to-sample in rhesus monkeys. *Animal Learning and Behavior,* 1977, *5,* 57–62.

Deweer, B., Sara, S. J., & Hars, B. Contextual cues and memory retrieval in rats: Elimination of forgetting by a pretest exposure to background stimuli. *Animal Learning and Behavior,* 1980, *8,* 265–72.

Dews, P. B. Studies on responding under fixed-interval schedules of reinforcement: The effects on the pattern of responding of changes in requirement at reinforcement. *Journal of the Experimental Analysis of Behavior,* 1969, *12,* 191–99.

Dexter, W. R., & Merrill, H. K. Role of contextual discrimination in fear conditioning. *Journal of Comparative and Physiological Psychology,* 1969, *69,* 677–81.

Dickinson, A., Hall, G., & Mackintosh, N. J. Surprise and the attenuation of blocking. *Journal of Experimental Psychology: Animal Behavior Processes,* 1976, *2,* 313–22.

Dickinson, A., & Mackintosh, N. J. Reinforcer specificity in the enhancement of conditioning by posttrial surprise. *Journal of Experimental Psychology: Animal Behavior Processes,* 1979, *5,* 162–77.

Dickinson, A., & Pearce, J. M. Preference and response suppression under different correlations between shock and a positive reinforcer in rats. *Learning and Motivation,* 1976, *7,* 66–85.

Dickinson, A., & Pearce, J. M. Inhibitory interactions between appetitive and aversive stimuli. *Psychological Bulletin,* 1977, *84,* 690–711.

Domjan, M. Selective suppression of drinking during a limited period following aversive drug treatment in rats. *Journal of Experimental Psychology: Animal Behavior Processes,* 1977, *3,* 66–76.

Domjan, M., & Wilson, N. E. Contribution of ingestive behaviors to taste-aversion learning in the rat. *Journal of Comparative and Physiological Psychology,* 1972, *80,* 403–12. (a)

Domjan, M., & Wilson, N. E. Specificity of cue to consequence in aversion learning in the rat. *Psychonomic Science,* 1972, *26,* 143–45. (b)

Donahoe, J. W. Stimulus control within response sequences. In J. H. Reynierse (Ed.), *Current issues in animal learning: A colloquium.* Lincoln: University of Nebraska Press, 1970.

Douglas, R. J. Cues for spontaneous alternation. *Journal of Comparative and Physiological Psychology,* 1966, *62,* 171–83.

Douglas, R. J., Mitchell, D., & Del Valle, R. Angle between choice alleys as a critical factor in spontaneous alternation. *Animal Learning and Behavior,* 1974, *2,* 218–20.

Dragoin, W. B. Conditioning and extinction of taste aversions with variations in intensity of the CS and UCS in two strains of rats. *Psychonomic Science,* 1971, *22,* 303–4.

Dreyer, P., & Renner, K. W. Self-punitive behavior—Masochism or confusion? *Psychological Review,* 1971, *78,* 333–37.

Dunham, P. J. Contrasted conditions of reinforcement: A selective critique. *Psychological Bulletin,* 1968, *69,* 295–315.

Dyal, J. A., & Sytsma, D. Relative persistence as a function of order of reinforcement schedules. *Journal of Experimental Psychology: Animal Behavior Processes,* 1976, *2,* 370–75.

Egger, M. D., & Miller, N. E. Secondary reinforcement in rats as a function of information value and reliability of the stimulus. *Journal of Experimental Psychology,* 1962, *64,* 97–104.

Eibl-Eibesfeldt, I. *Ethology. The biology of behavior* (2nd ed.). New York: Holt, Rinehart, and Winston, 1975.

Eisenberger, R. Explanation of rewards that do not reduce tissue needs. *Psychological Bulletin,* 1972, *77,* 319–39.

Eisenberger, R., Frank, M., & Park, D. C. Incentive contrast of choice behavior. *Journal of Experimental Psychology: Animal Behavior Processes,* 1975, *1,* 346–54.

Eiserer, L. A., & Hoffman, H. S. Imprinting of ducklings to a second stimulus when a previously imprinted stimulus is occasionally presented. *Animal Learning and Behavior,* 1974, *2,* 123–25. (a)

Eiserer, L. A., & Hoffman, H. S. Acquisition of behavioral control by the auditory features of an imprinting object. *Animal Learning and Behavior,* 1974, *2,* 275–77. (b).

Ellison, G. D. Differential salivary conditioning to traces. *Journal of Comparative and Physiological Psychology,* 1964, *57,* 373–80.

Fallon, D. Increased resistance to extinction following punishment and reward: High frustration tolerance or low frustration magnitude? *Journal of Comparative and Physiological Psychology,* 1971, *77,* 245–55.

Feigley, D. A., & Spear, N. E. Effect of age and punishment condition on long-term retention by the rat of active- and passive-avoidance learning. *Journal of Comparative and Physiological Psychology,* 1970, *73,* 515–26.

Feldman, D. T., & Gordon, W. C. The alleviation of short-term retention decrements with reactivation. *Learning and Motivation,* 1979, *10,* 198–210.

Felton, M., & Lyon, D. O. The post-reinforcement pause. *Journal of the Experimental Analysis of Behavior,* 1966, *9,* 131–34.

Ferster, C. B., & Skinner, B. F. *Schedules of reinforcement.* New York: Appleton-Century-Crofts, 1957.

Filby, Y., & Appel, J. B. Variable-interval punishment during variable-interval reinforcement. *Journal of the Experimental Analysis of Behavior,* 1966, *9,* 521–27.

Fischer, G., Viney, W., Knight, J., & Johnson, N. Response decrement as a function of effort. *Quarterly Journal of Experimental Psychology,* 1968, *20,* 301–4.

Fiske, D. W., & Maddi, S. R. *Functions of varied experience.* Homewood, Ill.: Dorsey Press, 1961.

Fitzgerald, R. D. Effects of partial reinforcement with acid on the classically conditioned salivary response in dogs. *Journal of Comparative and Physiological Psychology,* 1963, *56,* 1056–60.

Fitzgerald, R. D., & Martin, G. K. Heart-rate conditioning in rats as a function of interstimulus interval. *Psychological Reports,* 1971, *29,* 1103–10.

Fitzgerald, R. D., & Teyler, T. J. Trace and delayed heart-rate conditioning in rats as a function of US intensity. *Journal of Comparative and Physiological Psychology,* 1970, *70,* 242–53.

Flaherty, C. F., & Largen, J. Within-subjects positive and negative contrast effects in rats. *Journal of Comparative and Physiological Psychology,* 1975, *88,* 653–64.

Fouts, R. S., & Rigby, R. L. Man-chimpanzee communication. In T. A. Sebeok (Ed.), *How animals communicate.* Bloomington: Indiana University Press, 1977.

Fowler, H. Satiation and curiosity. In K. W. Spence & J. T. Spence (Eds.), *The psychology of learning and motivation* (Vol. 1). New York: Academic Press, 1967.

Fowler, H., Fago, G. C., Domber, E. A., & Mochhauser, M. Signaling and affective functions in Pavlovian conditioning. *Animal Learning and Behavior,* 1973, *1,* 81–89.

Fowler, H., & Trapold, M. A. Escape performance as a function of delay of reinforcement. *Journal of Experimental Psychology,* 1962, *63,* 464–67.

Franchina, J. J. Escape behavior and shock intensity: Within-subject versus between-groups comparisons. *Journal of Comparative and Physiological Psychology,* 1969, *69,* 241–45.

Franchina, J. J., Agee, C. M., & Hauser, P. J. Response prevention and extinction of escape behavior: Duration, frequency, similarity, and retraining variables in rats. *Journal of Comparative and Physiological Psychology,* 1974, *87,* 354–63.

Freedman, P. E., Hennessy, J. W., & Groner, D. Effects of varying active/passive shock levels in shuttle box avoidance in rats. *Journal of Comparative and Physiological Psychology,* 1974, *86,* 79–84.

Frey, P. W., & Butler, C. S. Rabbit eyelid conditioning as a function of unconditioned stimulus duration. *Journal of Comparative and Physiological Psychology,* 1973, *85,* 289–94.

Friedman, H., & Guttman, N. Further analysis of the various effects of discrimination training upon stimulus generalization gradients. In D. I. Mostofsky (Ed.), *Stimulus generalization.* Stanford, Calif.: Stanford University Press, 1965.

Fuller, G. D. Current status of biofeedback in clinical practice. *American Psychologist,* 1978, *33,* 39–48.

Gage, F. H., Evans, S. H., & Olton, D. S. Multivariate analyses of performance in a DRL paradigm. *Animal Learning and Behavior,* 1979, *7,* 323–27.

Gaioni, S. J., Hoffman, H. S., DePaulo, P., & Stratton, V. N. Imprinting in older ducklings: Some tests of a reinforcement model. *Animal Learning and Behavior,* 1978, *6,* 19–26.

Galef, B. G. Aggression and timidity: Responses to novelty in Feral Norway rats. *Journal of Comparative and Physiological Psychology,* 1970, *70,* 370–81.

Galef, B. G., & Osborne, B. Novel taste facilitation of the association of visual cues with toxicosis in rats. *Journal of Comparative and Physiological Psychology,* 1978, *92,* 907–16.

Ganz, L., & Riesen, A. H. Stimulus generalization to hue in the dark-reared macaque. *Journal of Comparative and Physiological Psychology,* 1962, *55,* 92–99.

Garcia, J., Ervin, F. R., & Koelling, R. A. Learning with prolonged delay of reinforcement. *Psychonomic Science,* 1966, *5,* 121–22.

Garcia, J., Ervin, F. R., Yorke, C. H., & Koelling, R. A. Conditioning with delayed vitamin injections. *Science,* 1967, *155,* 716–18.

Garcia, J., Hawkins, W. G., & Rusiniak, K. W. Behavioral regulation of the milieu interne in man and rat. *Science,* 1974, *185,* 824–31.

Garcia, J., & Koelling, R. A. Relation of cue to consequence in avoidance learning. *Psychonomic Science,* 1966, *4,* 123–24.

Garcia, J., McGowan, B. K., Ervin, F. R., & Koelling, R. A. Cues: Their relative effectiveness as a function of the reinforcer. *Science,* 1968, *160,* 794–95.

Gardner, B. T., & Gardner, R. A. Two-way communication with an infant chimpanzee. In A. Schrier & F. Stollnitz (Eds.), *Behavior of nonhuman primates* (Vol. 4). New York: Academic Press, 1971.

Ghiselli, W. B., & Fowler, H. Signaling and affective functions of conditioned aversive stimuli in an appetitive choice discrimination: US intensity effects. *Learning and Motivation,* 1976, *7,* 1–16.

Gibbon, J., Baldock, M. D., Locurto, C., Gold, L., & Terrace, H. S. Trial and intertrial durations in autoshaping. *Journal of Experimental Psychology: Animal Behavior Processes.* 1977, *3,* 264–84.

Gilbert, R. M., & Sutherland, N. S. *Animal discrimination learning.* New York: Academic Press, 1969.

Gillan, D. J. Learned suppression of ingestion: Role of discriminative stimuli, ingestive responses, and aversive tastes. *Journal of Experimental Psychology: Animal Behavior Processes,* 1979, *5,* 258–72.

Gillan, D. J., & Domjan, M. Taste-aversion conditioning with expected versus unexpected drug treatment. *Journal of Experimental Psychology: Animal Behavior Processes,* 1977, *3,* 297–309.

Gillette, K., Martin, G. M., & Bellingham, W. P. Differential use of food and water cues in the formation of conditioned aversions by domestic chicks *(Gallus gallus). Journal of Experimental Psychology: Animal Behavior Processes,* 1980, *6,* 99–111.

Giulian, D., & Schmaltz, L. W. Enhanced discriminated bar-press avoidance in the rat through appetitive preconditioning. *Journal of Comparative and Physiological Psychology,* 1973, *83,* 106–12.

Glazer, H. I., & Weiss, J. M. Long-term and transitory interference effects. *Journal of Experimental Psychology: Animal Behavior Processes,* 1976, *2,* 191–201. (a)

Glazer, H. I., & Weiss, J. M. Long-term interference effect: An alternative to "learned helplessness." *Journal of Experimental Psychology: Animal Behavior Processes,* 1976, *2,* 202–13. (b)

Gliner, J. A. Predictable vs. unpredictable shock: Preference behavior and stomach ulceration. *Physiology and Behavior,* 1972, *9,* 693–98.

Goesling, W. J., & Brener, J. Effects of activity and immobility conditioning upon subsequent heart-rate conditioning in curarized rats. *Journal of Comparative and Physiological Psychology,* 1972, *81,* 311–17.

Goodkin, F. Rats learn the relationship between responding and environmental events: An expansion of the learned helplessness hypothesis. *Learning and Motivation,* 1976, *7,* 382–94.

Gordon, W. C., & Feldman, D. T. Reactivation-induced interference in a short-term retention paradigm. *Learning and Motivation,* 1978, *9,* 164–78.

Gordon, W. C., & Spear, N. E. The effect of reactivation of a previously acquired memory on the interaction between memories in the rat. *Journal of Experimental Psychology,* 1973, *99,* 349–55.

Gormezano, I. Investigations of defense and reward conditioning in the rabbit. In A. H. Black, & W. F. Prokasy (Eds.), *Classical conditioning II: Current theory and research.* New York: Appleton-Century-Crofts, 1972.

Gormezano, I., & Moore, J. W. Classical conditioning. In M. H. Marx (Ed.), *Learning: Processes.* New York: Macmillan Co., 1969.

Gottlieb, G. *Development of species identification in birds: An inquiry into the prenatal determinants of perception.* Chicago: University of Chicago Press, 1971.

Gottlieb, G. Neglected developmental variables in the study of species identification in birds. *Psychological Bulletin,* 1973, *79,* 362–72.

Graft, D. A., Lea, S. E. G., & Whitworth, T. L. The matching law in and within groups of rats. *Journal of the Experimental Analysis of Behavior,* 1977, *27,* 183–94.

Grant, D. S. Proactive interference in pigeon short-term memory. *Journal of Experimental Psychology: Animal Behavior Processes,* 1975, *1,* 207–20.

Grant, D. S., & Roberts, W. A. Sources of retroactive inhibition in pigeon short-term memory. *Journal of Experimental Psychology: Animal Behavior Processes,* 1976, *2,* 1–16.

Gray, J. A. Stimulus intensity dynamism. *Psychological Bulletin,* 1965, *63,* 180–96.

Gray, J. A. Sodium amobarbital and effects of frustrative nonreward. *Journal of Comparative and Physiological Psychology,* 1969, *69,* 55–64.

Green, K. F., Holmstrom, L. S., & Wollman, M. A. Relation of cue to consequence in rats: Effect of recuperation from illness. *Behavioral Biology,* 1974, *10,* 491–503.

Green, L., Bouzas, A., & Rachlin, H. Test of an electric-shock analog to illness-induced aversion. *Behavioral Biology,* 1972, *7,* 513–18.

Grice, G. R. The relation of secondary reinforcement to delayed reward in visual discrimination learning. *Journal of Experimental Psychology,* 1948, *38,* 1–16.

Grice, G. R. Stimulus intensity and response evocation. *Psychological Review,* 1968, *75,* 359–73.

Grossen, N. E., & Bolles, R. C. Effects of a classical conditioned "fear signal" and "safety signal" on nondiscriminated avoidance behavior. *Psychonomic Science,* 1968, *11,* 321–22.

Grossen, N. E., & Kelley, M. J. Species-specific behavior and acquisition of avoidance behavior in rats. *Journal of Comparative and Physiological Psychology,* 1972, *81,* 307–10.

Grossen, N. E., Kostansek, D. J., & Bolles, R. C. Effects of appetitive discriminative stimuli on avoidance behavior. *Journal of Experimental Psychology,* 1969, *81,* 340–43.

Guttman, N., & Kalish, H. I. Discriminability and stimulus generalization. *Journal of Experimental Psychology,* 1956, *51,* 79–88.

Gynther, M. D. Differential eyelid conditioning as a function of stimulus similarity and strength of response to the CS. *Journal of Experimental Psychology,* 1957, *53,* 408–16.

Habley, P., Gipson, M., & Hause, J. Acquisition and extinction in the runway as a joint function of constant reward magnitude and constant reward delay. *Psychonomic Science,* 1972, *29,* 133–36.

Halgren, C. R. Latent inhibition in rats: associative or nonassociative. *Journal of Comparative and Physiological Psychology,* 1974, *86,* 74–78.

Hall, G. Overtraining and reversal learning in the rat: Effects of stimulus salience and response strategies. *Journal of Comparative and Physiological Psychology,* 1973, *84,* 169–75.

Hall, G. Transfer effects produced by overtraining in the rat. *Journal of Comparative and Physiological Psychology,* 1974, *87,* 938–44.

Hammond, L. J. Increased responding to CS in differential CER. *Psychonomic Science,* 1966, *5,* 337–38.

Hammond, L. J. A traditional demonstration of the active properties of Pavlovian inhibition using differential CER. *Psychonomic Science,* 1967, *9,* 65–66.

Hanson, H. M. Effects of discrimination training on stimulus generalization. *Journal of Experimental Psychology,* 1959, *58,* 321–34.

Hanson, H. M., Stimulus generalization following three-stimulus discrimination training. *Journal of Comparative and Physiological Psychology,* 1961, *54,* 181–85.

Harlow, H. F. The formation of learning sets. *Psychological Review,* 1949, *56,* 51–65.

Harlow, H. F. Learning sets and error-factor theory. In S. Koch (Ed.), *Psychology: A study of a science* (Vol. 2). New York: McGraw-Hill, 1959.

Harris, A. H., & Brady, J. V. Animal learning—visceral and autonomic conditioning. *Annual Review of Psychology,* 1974, *25,* 107–33.

Hastings, S. E., & Obrist, P. A. Heart rate during conditioning in humans: Effect of varying interstimulus (CS-UCS) interval. *Journal of Experimental Psychology,* 1967, *74,* 431–42.

Hayes, C. *The ape in our house.* New York: Harper & Bros., 1951.

Hearst, E. Differential transfer of excitatory versus inhibitory pretraining to intradimensional discrimination learning in pigeons. *Journal of Comparative and Physiological Psychology,* 1971, *75,* 206–15.

Hearst, E. Stimulus relationships and feature selection in learning and behavior. In S. H. Hulse, H. Fowler, & W. K. Honig (Eds.), *Cognitive processes in animal behavior.* Hillsdale, N.J.: Lawrence Erlbaum, 1978.

Hearst, E., Besley, S., & Farthing, G. W. Inhibition and the stimulus control of operant behavior. *Journal of the Experimental Analysis of Behavior,* 1970, *14,* 373–409.

Hearst, E., & Jenkins, H. M. *Sign-tracking: The stimulus-reinforcer relation and directed action.* Austin, Texas: Psychonomic Society, 1974.

Hearst, E., & Koresko, M. B. Stimulus generalization and amount of prior training on variable-interval reinforcement. *Journal of Comparative and Physiological Psychology,* 1968, *66,* 133–38.

Hebb, D. O. On the nature of fear. *Psychological Review,* 1946, *53,* 259–76.

Hemmes, N. S., Eckerman, D. A., & Rubinsky, H. J. A functional analysis of collateral behavior under differential-reinforcement-of-low-rate schedules. *Animal Learning and Behavior,* 1979, *7,* 328–32.

Herrnstein, R. J. Relative and absolute strength of response as a function of frequency of reinforcement. *Journal of the Experimental Analysis of Behavior,* 1961, *4,* 267–74.

Herrnstein, R. J. Method and theory in the study of avoidance. *Psychological Review,* 1969, *76,* 49–69.

Herrnstein, R. J. On the law of effect. *Journal of the Experimental Analysis of Behavior,* 1970, *13,* 243–66.

Herrnstein, R. J. Formal properties of the matching law. *Journal of the Experimental Analysis of Behavior,* 1974, *21,* 159–64.

Herrnstein, R. J., & Hineline, P. N. Negative reinforcement as shock-frequency reduction. *Journal of the Experimental Analysis of Behavior,* 1966, *9,* 421–30.

Herrnstein, R. J., & Loveland, D. H. Maximizing and matching on concurrent ratio schedules. *Journal of the Experimental Analysis of Behavior,* 1975, *24,* 107–16.

Herzog, H. L., Grant, D. S., & Roberts, W. A. Effects of sample duration and spaced repetition upon delayed matching-to-sample in monkeys *(Macaca arctoides* and *Saimiri sciureus). Animal Learning and Behavior,* 1977, *5,* 347–54.

Hess, E. H. The relationship between imprinting and motivation. In M. R. Jones (Ed.), *Nebraska symposium on motivation* (Vol. 7). Lincoln: University of Nebraska Press, 1959. (a)

Hess, E. H. Imprinting. *Science,* 1959, *130,* 133–41. (b)

Heth, C. D., & Rescorla, R. A. Simultaneous and backward fear conditioning in the rat. *Journal of Comparative and Physiological Psychology,* 1973, *82,* 434–43.

Hickis, C. F., Robles, L., & Thomas, D. R. Contextual stimuli and memory retrieval in pigeons. *Animal Learning and Behavior,* 1977, *5,* 161–68.

Hill, W. F., & Spear, N. E. Extinction in a runway as a function of acquisition level and reinforcement percentage. *Journal of Experimental Psychology,* 1963, *65,* 495–500.

Hilton, A. Partial reinforcement of a conditioned emotional response in rats. *Journal of Comparative and Physiological Psychology,* 1969, *69,* 253–60.

Hinde, R. A., & Stevenson-Hinde, J. *Constraints on learning: Limitations and predispositions.* New York: Academic Press, 1973.

Hinson, R. E., & Siegel, S. Trace conditioning as an inhibitory procedure. *Animal Learning and Behavior,* 1980, *8,* 60–66.

Hiroto, D. S., & Seligman, M. E. P. Generality of learned helplessness in man. *Journal of Personality and Social Psychology*, 1975, *31*, 311–27.

Hoffman, H. S. Stimulus factors in conditioned suppression. In B. A. Campbell & R. M. Church (Eds.), *Punishment and aversive behavior*. New York: Appleton-Century-Crofts, 1969.

Hoffman, H. S., Eiserer, L. A., & Singer, D. Acquisition of behavioral control by a stationary imprinting stimulus. *Psychonomic Science*, 1972, *26*, 146–48.

Hoffman, H. S., & Ratner, A. M. A reinforcement model of imprinting: Implications for socialization in monkeys and men. *Psychological Review*, 1973, *80*, 527–44.

Hoffman, H. S., Ratner, A. M. & Eiserer, L. A. Role of visual imprinting in the emergence of specific filial attachments in ducklings. *Journal of Comparative and Physiological Psychology*, 1972, *81*, 399–409.

Hoffman, H. S., Searle, J. L., Toffey, S., & Kozma, F. Behavioral control by an imprinted stimulus. *Journal of the Experimental Analysis of Behavior*, 1966, *9*, 177–89.

Hoffman, H. S., & Solomon, R. L. An opponent-process theory of motivation: III. Some affective dynamics in imprinting. *Learning and Motivation*, 1974, *5*, 149–64.

Hoffman, H. S., Stratton, J. W., & Newby, V. Punishment by response-contingent withdrawal of an imprinting stimulus. *Science*, 1969, *163*, 702–4.

Hoffman, H. S., Stratton, J. W., Newby, V., & Barrett, J. E. Development of behavioral control by an imprinting stimulus. *Journal of Comparative and Physiological Psychology*, 1970, *71*, 229–36.

Holland, P. C. Conditioned stimulus as a determinant of the form of the Pavlovian conditioned response. *Journal of Experimental Psychology: Animal Behavior Processes*, 1977, *3*, 77–104.

Holland, P. C. Differential effects of omission contingencies on various components of Pavlovian appetitive conditioned responding in rats. *Journal of Experimental Psychology: Animal Behavior Processes*, 1979, *5*, 178–93. (a)

Holland, P. C. The effects of qualitative and quantitative variation in the US on individual components of Pavlovian appetitive conditioned behavior in rats. *Animal Learning and Behavior*, 1979, *7*, 424–32. (b)

Holland, P. C., & Rescorla, R. A. Second-order conditioning with food unconditioned stimulus. *Journal of Comparative and Physiological Psychology*, 1975, *88*, 459–67. (a)

Holland, P. C., & Rescorla, R. A. The effect of two ways of deviating the unconditioned stimulus after first- and second-order appetitive conditioning. *Journal of Experimental Psychology: Animal Behavior Processes*, 1975, *1*, 355–63. (b)

Holland, V., & Davison, M. C. Preference for qualitatively different reinforcers. *Journal of the Experimental Analysis of Behavior*, 1971, *16*, 375–80.

Holman, E. W. Temporal properties of gustatory spontaneous alternation in rats. *Journal of the Comparative and Physiological Psychology*, 1973, *85*, 536–39.

Hom, H. L., & Babb, H. Self-punitive responding in rats with goal shock and color change. *Animal Learning and Behavior*, 1975, *3*, 152–56.

Homzie, M. J. Nonreward anticipated: Effects on extinction runway performance in the rat. *Animal Learning and Behavior*, 1974, *2*, 77–79.

Honig, W. K., & James, P. H. R. *Animal memory*. New York: Academic Press, 1971.

Hooper, R. Variables controlling the overlearning reversal effect (ORE). *Journal of Experimental Psychology*, 1967, *73*, 612–19.

Hsiao, S. & Trankina, F. Thirst-hunger interaction: I. Effects of body-fluid restoration on food and water intake in water-deprived rats. *Journal of Comparative and Physiological Psychology*, 1969, *69*, 448–53.

Hull, C. L. *Principles of behavior*. New York: Appleton-Century-Crofts, 1943.

Hulse, S. H., & Campbell, C. E. "Thinking ahead" in rat discrimination learning. *Animal Learning and Behavior*, 1975, *3*, 305–11.

Hulse, S. H., & Dorsky, N. P. Structural complexity as a determinant of serial pattern learning. *Learning and Motivation,* 1977, *8,* 488–506.

Hutton, R. A., Woods, S. C., & Makous, W. L. Conditioned hypoglycemia: Pseudoconditioning controls. *Journal of Comparative and Physiological Psychology,* 1970, *71,* 198–201.

Hymowitz, N. Suppression of responding during signaled and unsignaled shock. *Psychological Bulletin,* 1979, *86,* 175–90.

Irwin, J., Suissa, A., & Anisman, H. Differential effects of inescapable shock on escape performance and discrimination learning in a water escape task. *Journal of Experimental Psychology: Animal Behavior Processes,* 1980, *6,* 21–40.

Ison, J. R. Experimental extinction as a function of number of reinforcements. *Journal of Experimental Psychology,* 1962, *64,* 314–17.

Ison, J. R., & Cook, P. E. Extinction performance as a function of incentive magnitude and number of acquisition trials. *Psychonomic Science,* 1964, *1,* 245–46.

Ison, J. R., & Pennes, E. S. Interaction of amobarbital sodium and reinforcement schedule in determining resistance to extinction of an instrumental running response. *Journal of Comparative and Physiological Psychology,* 1969, *68,* 215–19.

Jackson, R. L., Alexander, J. H., & Maier, S. F. Learned helplessness, inactivity, and associative deficits: Effects of inescapable shock on response choice escape learning. *Journal of Experimental Psychology: Animal Behavior Processes,* 1980, *6,* 1–20.

Jackson, R. L., Maier, S. F., & Rapaport, P. M. Exposure to inescapable shock produces both activity and associative deficits in the rat. *Learning and Motivation,* 1978, *9,* 69–98.

Jakubczak, L. F. Frequency, duration, and speed of wheel running of rats as a function of age and starvation. *Animal Learning and Behavior,* 1973, *1,* 13–16.

Jenkins, H. M., & Harrison, R. H. Effect of discrimination training on auditory generalization. *Journal of Experimental Psychology,* 1960, *59,* 246–53.

Jenkins, H. M., & Harrison, R. H. Generalization gradients of inhibition following auditory discrimination learning. *Journal of the Experimental Analysis of Behavior,* 1962, *5,* 435–41.

Jenkins, H. M., & Moore, B. A. The form of the auto-shaped response with food or water reinforcers. *Journal of the Experimental Analysis of Behavior,* 1973, *20,* 163–81.

Jenkins, H. M., & Sainsbury, R. S. Discrimination learning with the distinctive feature on positive or negative trials. In D. I. Mostofsky (Ed.), *Attention: Contemporary theory and analysis.* New York: Appleton-Century-Crofts, 1970.

John, E. R., Chesler, P., Bartlett, F., & Victor, I. Observation learning in cats. *Science,* 1968, *159,* 1489–91.

Johnson, N., & Viney, W. Resistance to extinction as a function of effort. *Journal of Comparative and Physiological Psychology,* 1970, *71,* 171–74.

Jones, R. B., & Nowell, N. W. A comparison of the aversive and female attractant properties of urine from dominant and subordinate male mice. *Animal Learning and Behavior,* 1974, *2,* 141–44.

Kalat, J. W. Taste salience depends on novelty, not concentration, in taste-aversion learning in the rat. *Journal of Comparative and Physiological Psychology,* 1974, *86,* 47–50.

Kalat, J. W. Should taste-aversion learning experiments control duration or volume of drinking on the training day? *Animal Learning and Behavior,* 1976, *4,* 96–98.

Kalat, J. W. Status of "learned safety" or "learned noncorrelation" as a mechanism in taste-aversion learning. In L. M. Barker, M. R. Best, & M. Domjan (Eds.), *Learning mechanisms in food selection.* Waco, Texas: Baylor University Press, 1977.

Kalat, J. W., & Rozin, P. "Salience": A factor which can override temporal contiguity in taste-aversion learning. *Journal of Comparative and Physiological Psychology,* 1970, *71,* 192–97.

Kalat, J. W., & Rozin, P. Role of interference in taste-aversion learning. *Journal of Comparative and Physiological Psychology*, 1971, *77*, 53–58.

Kalat, J. W., & Rozin, P. "Learned safety" as a mechanism in long-delay taste-aversion learning in rats. *Journal of Comparative and Physiological Psychology*, 1973, *83*, 198–207.

Kalish, H. I. Strength of fear as a function of the number of acquisition and extinction trials. *Journal of Experimental Psychology*, 1954, *55*, 637–44.

Kalish, H. I. Stimulus generalization. In M. H. Marx (Ed.), *Learning: Processes*. New York: Macmillan, 1969.

Kalish, H. I., & Haber, A. Generalization: I. Generalization gradients from single and multiple stimulus points. II. Generalization of inhibition. *Journal of Experimental Psychology*, 1963, *65*, 176–81.

Kalish, H. I., & Haber, A. Prediction of discrimination from generalization following variations in deprivation level. *Journal of Comparative and Physiological Psychology*, 1965, *60*, 125–28.

Kamil, A. C. The second-order conditioning of fear in rats. *Psychonomic Science*, 1968, *10*, 99–100.

Kamil, A. C. Some parameters of the second-order conditioning of fear in rats. *Journal of Comparative and Physiological Psychology*, 1969, *67*, 364–69.

Kamil, A. C., Lougee, M., & Shulman, R. I. Learning-set behavior in the learning-set experienced bluejay *(Cyanacitta cristata)*. *Journal of Comparative and Physiological Psychology*, 1973, *82*, 394–405.

Kamil, A. C., & Mauldin, J. E. Intraproblem retention during learning-set acquisition in bluejays *(Cyanacitta cristata)*. *Animal Learning and Behavior*, 1975, *3*, 125–30.

Kamin, L. J. Traumatic avoidance learning: The effects of CS-US interval with a trace conditioning procedure. *Journal of Comparative and Physiological Psychology*, 1954, *47*, 65–72.

Kamin, L. J. The effects of termination of the CS and avoidance of the US on avoidance learning. *Journal of Comparative and Physiological Psychology*, 1956, *49*, 420–24.

Kamin, L. J. The gradient of delay of secondary reward in avoidance learning. *Journal of Comparative and Physiological Psychology*, 1957, *50*, 445–49. (a)

Kamin, L. J. The gradient of delay of secondary reward in avoidance learning tested on avoidance trials only. *Journal of Comparative and Physiological Psychology*, 1957, *50*, 450–56. (b)

Kamin, L. J. The retention of an incompletely learned avoidance response. *Journal of Comparative and Physiological Psychology*, 1957, *50*, 457–60. (c)

Kamin, L. J. The delay-of-punishment gradient. *Journal of Comparative and Physiological Psychology*, 1959, *52*, 434–37.

Kamin, L. J. Temporal and intensity characteristics of the conditioned stimulus. In W. F. Prokasy (Ed.), *Classical conditioning: A symposium*. New York: Appleton-Century-Crofts, 1965.

Kamin, L. J. Predictability, surprise, attention, and conditioning. In B. A. Campbell & R. M. Church (Eds.), *Punishment and aversive behavior*. New York: Appleton-Century-Crofts, 1969.

Kamin, L. J., & Schaub, R. E. Effects of conditioned stimulus intensity on the conditioned emotional response. *Journal of Comparative and Physiological Psychology*, 1963, *56*, 502–7.

Kaplan, M., Jackson, B., & Sparer, R. Escape behavior under continuous reinforcement as a function of aversive light intensity. *Journal of the Experimental Analysis of Behavior*, 1965, *8*, 321–23.

Karpicke, J., Christoph, G., Peterson, G., & Hearst, E. Signal location and positive versus negative conditioned suppression in the rat. *Journal of Experimental Psychology: Animal Behavior Processes*, 1977, *3*, 105–18.

Katkin, E. S., & Murray, E. N. Instrumental conditioning of autonomically mediated behavior: Theoretical and methodological issues. *Psychological Bulletin,* 1968, *70,* 52–68.

Katzev, R. D., & Berman, J. S. Effect of exposure to conditioned stimulus and control of its termination in the extinction of avoidance behavior. *Journal of Comparative and Physiological Psychology,* 1974, *87,* 347–53.

Keehn, J. D. On the nonclassical nature of avoidance behavior. *American Journal of Psychology,* 1959, *72,* 243–47.

Keith-Lucas, T., & Guttman, N. Robust single-trial delayed backward conditioning. *Journal of Comparative and Physiological Psychology,* 1975, *88,* 468–76.

Keller, J. V., & Gollub, L. R. Duration and rate of reinforcement as determinants of concurrent responding. *Journal of the Experimental Analysis of Behavior,* 1977, *28,* 145–53.

Keller, R. J., Ayres, J. J. B., & Mahoney, W. J. Brief versus extended exposure to truly random control procedures. *Journal of Experimental Psychology: Animal Behavior Processes,* 1977, *3,* 53–65.

Kello, J. E. The reinforcement-omission effect on fixed-interval schedules: Frustration or inhibition? *Learning and Motivation,* 1972, *3,* 138–47.

Kendall, S. B. Some effects of response-dependent clock stimuli in a fixed-interval schedule. *Journal of the Experimental Analysis of Behavior,* 1972, *17,* 161–68.

Kerr, L. M., Ostapoff, E. M., & Rubel, E. W. Influence of acoustic experience on the ontogeny of frequency generalization gradients in the chicken. *Journal of Experimental Psychology: Animal Behavior Processes,* 1979, *5,* 97–115.

Killeen, P. Reinforcement frequency and contingency as factors in fixed-ration behavior. *Journal of the Experimental Analysis of Behavior,* 1969, *12,* 391–95.

Kimble, G. A. Behavior strength as a function of the intensity of the hunger drive. *Journal of Experimental Psychology,* 1951, *41,* 341–48.

Kimmel, H. D. Instrumental inhibitory factors in classical conditioning. In W. F. Prokasy (Ed.), *Classical conditioning.* New York: Appleton-Century-Crofts, 1965.

Kimmel, H. D., Brennan, A. F., McLeod, D. C., Raich, M. S., & Schonfeld, L. I. Instrumental electrodermal conditioning in the monkey *(Cebus albifrons):* Acquisition and long-term retention. *Animal Learning and Behavior,* 1979, *7,* 447–51.

Kintsch, W. Runway performance as a function of drive strength and magnitude of reinforcement. *Journal of Comparative and Physiological Psychology,* 1962, *55,* 882–87.

Kirby, R. H. Acquisition, extinction, and retention of an avoidance response in rats as a function of age. *Journal of Comparative and Physiological Psychology,* 1963, *56,* 158–62.

Kish, G. B. Learning when the onset of illumination is used as reinforcing stimulus. *Journal of Comparative and Physiological Psychology,* 1955, *48,* 261–64.

Kish, G. B. Studies of sensory reinforcement. In W. K. Honig (Ed.), *Operant behavior: Areas of research and application.* New York: Appleton-Century-Crofts, 1966.

Klare, W. F. Conditioned fear and post shock emotionality in vicious circle behavior of rats. *Journal of Comparative and Physiological Psychology,* 1974, *87,* 364–72.

Klein, S. B. Adrenal-pituitary influence in reactivation of avoidance-learning memory in the rat after intermediate intervals. *Journal of Comparative and Physiological Psychology,* 1972, *79,* 341–59.

Klein, S. B., & Spear, N. E. Forgetting by the rat after intermediate intervals ("Kamin effect") as retrieval failure. *Journal of Comparative and Physiological Psychology,* 1970, *71,* 165–70. (a)

Klein, S. B., & Spear, N. E. Reactivation of avoidance-learning memory in the rat after intermediate retention intervals. *Journal of Comparative and Physiological Psychology,* 1970, *72,* 498–504. (b)

Knouse, S. B., & Campbell, P. E. Partially delayed reward in the rat: A parametric study of delay duration. *Journal of Comparative and Physiological Psychology,* 1971, *75,* 116–19.

Kohler, E. A., & Ayres, J. J. B. The Kamin blocking effect with variable-duration CSs. *Animal Learning and Behavior,* 1979, *7,* 347–50.

Kohn, B. Observation and discrimination learning in the rat: Effects of stimulus substitution. *Learning and Motivation,* 1976, *7,* 303–12.

Kohn, B., & Dennis, M. Observation and discrimination learning in the rat: Specific and nonspecific effects. *Journal of Comparative and Physiological Psychology,* 1972, *78,* 292–96.

Koppenaal, R. J., Jagoda, E., & Cruce, J. A. F. Recovery from ECS-produced amnesia following a reminder. *Psychonomic Science,* 1967, *9,* 293–94.

Kraeling, D. Analysis of amount of reward as a variable in learning. *Journal of Comparative and Physiological Psychology,* 1961, *54,* 560–65.

Kramer, T. J., & Rilling, M. Differential reinforcement of low rates: A selective critique. *Psychological Review,* 1970, *74,* 225–54.

Krebs, J. R. Optimal foraging: Decision rules for predators. In J. R. Krebs & N. B. Davies (Eds.), *Behavioural ecology, an evolutionary approach.* Oxford: Blackwell Scientific Pub., 1978.

Krebs, J. R., Erichsen, J. T., Webber, M. I., & Charnov, E. L. Optimal prey selection in the great tit *(Parus Major). Animal Behavior,* 1977, *25,* 30–38.

Krebs, J. R., Kacelnik, A., & Taylor, P. Test of optimal sampling by foraging great tits. *Nature,* 1978, *275,* 27–31.

Kremer, E. F. Truly random and traditional control procedures in CER conditioning in the rat. *Journal of Comparative and Physiological Psychology,* 1971, *76,* 441–48.

Kremer, E. F. The truly random control procedure: Conditioning to static cues. *Journal of Comparative and Physiological Psychology,* 1974, *86,* 700–707.

Kremer, E. F. The Rescorla-Wagner model: Losses in associative strength in compound conditioned stimuli. *Journal of Experimental Psychology: Animal Behavior Processes,* 1978, *4,* 22–36.

Kremer, E. F. Effect of post-trial episodes on conditioning in compound conditioned stimuli. *Journal of Experimental Psychology: Animal Behavior Processes,* 1979, *5,* 130–41.

Kremer, E. F., & Kamin, L. J. The truly random procedure: Associative or nonassociative effects in rats. *Journal of Comparative and Physiological Psychology,* 1971, *74,* 203–10.

Kulkarni, A. S., & Job, W. M. Instrumental response pretraining and avoidance acquisition in rats. *Journal of Comparative and Physiological Psychology,* 1970, *70,* 254–57.

LaBarbera, J. D., & Caul, W. F. An opponent-process interpretation of post-shock bursts in appetitive responding. *Animal Learning and Behavior,* 1976, *4,* 386–90. (a)

LaBarbera, J. D., & Caul, W. F. Decrement in distress to an aversive event during a conditioned positive opponent-process. *Animal Learning and Behavior,* 1976, *4,* 485–89. (b)

Larkin, S., & McFarland, D. J. The cost of changing from one activity to another. *Animal Behavior,* 1978, *26,* 1237–46.

Larsen, J. D., & Hyde, T. S. A comparison of learned aversions to gustatory and exteroceptive cues in rats. *Animal Learning and Behavior,* 1977, *5,* 17–20.

Lashley, K. S., & Wade, M. The Pavlovian theory of generalization. *Psychological Review,* 1946, *53,* 72–87.

Laties, V. C., Weiss, B., Clark, R. L., & Reynolds, M. D. Overt "mediating" behavior during temporally spaced responding. *Journal of the Experimental Analysis of Behavior,* 1965, *8,* 107–16.

Laties, V. C., Weiss, B., & Weiss, A. B. Further observations on overt "mediating" behavior and the discrimination of time. *Journal of the Experimental Analysis of Behavior,* 1969, *12,* 43–57.

Lavin, M. J. The establishment of flavor-flavor associations using a sensory preconditioning training procedure. *Learning and Motivation,* 1976, *7,* 173–83.

Lawrence, D. H. The transfer of a discrimination along a continuum. *Journal of Comparative and Physiological Psychology,* 1952, *45,* 511–16.

Lea, S. E. G. Foraging and reinforcement schedules in the pigeon: Optimal and non-optimal aspects of choice. *Animal Behavior,* 1979, *27,* 875–86.

Leach, D. A. Rats' extinction performance as a function of deprivation level during training and partial reinforcement. *Journal of Comparative and Physiological Psychology,* 1971, *75,* 317–23.

Leitenberg, H., Rawson, R. A., & Bath, K. Reinforcement of competing behavior during extinction. *Science,* 1970, *169,* 301–3.

Leitenberg, H., Rawson, R. A., & Mulick, J. A. Extinction and reinforcement of alternative behavior. *Journal of Comparative and Physiological Psychology,* 1975, *88,* 640–52.

Lenderhendler, I., & Baum, M. Mechanical facilitation of the action of response prevention (flooding) in rats. *Behavior Research and Therapy,* 1970, *8,* 43–48.

Lenneberg, E. H. *Biological foundations of language.* New York: Wiley, 1967.

Leonard, D. W. Amount and sequence of reward in partial and continuous reinforcement. *Journal of Comparative and Physiological Psychology,* 1969, *67,* 204–11.

Leonard, D. W. Partial reinforcement effects in classical aversive conditioning in rabbits and human beings. *Journal of Comparative and Physiological Psychology,* 1975, *88,* 596–608.

Lett, B. T. Taste potentiates color-sickness associations in pigeons and quail. *Animal Learning and Behavior.* 1980, *8,* 193–98.

Levine, M. A model of hypothesis behavior in discrimination learning set. *Psychological Review,* 1959, *66,* 353–66.

Levine, M. Hypothesis behavior. In A. M. Schrier, H. F. Harlow, & F. Stollnitz (Eds.), *Behavior of nonhuman primates: Modern research trends.* New York: Academic Press, 1965.

Levine, S. UCS intensity and avoidance learning. *Journal of Experimental Psychology,* 1966, *71,* 163–64.

Levinthal, C. F. The CS-US interval function in rabbit nictitating membrane response conditioning: Single vs. multiple trials per conditioning session. *Learning and Motivation,* 1973, *4,* 259–67.

Levis, D. J. Learned helplessness: A reply and an alternative S-R interpretation. *Journal of Experimental Psychology: General,* 1976, *105,* 47–65.

Lewis, D. J. Partial reinforcement: A selective review of the literature since 1950. *Psychological Bulletin,* 1960, *57,* 1–28.

Lewis, D. J. Sources of experimental amnesia. *Psychological Review,* 1969, *76,* 461–72.

Lewis, D. J., Misanin, J. R., & Miller, R. R. Recovery of memory following amnesia. *Nature,* 1968, *220,* 704–5.

Lewis, M. Psychological effect of effort. *Psychological Bulletin,* 1965, *64,* 183–90.

Libby, M. E., & Church, R. M. Fear gradients as a function of the temporal interval between signal and aversive event in the rat. *Journal of Comparative and Physiological Psychology,* 1975, *88,* 911–16.

Lieberman, P. H., Klatt, D. H., & Wilson, W. H. Vocal tract limitations on the vowel repertoires of rhesus monkeys and other nonhuman primates. *Science,* 1969, *164,* 1185–87.

Linden, D. R., & Hallgren, S. O. Transfer of approach responding between punishment and frustrative nonreward sustained through continuous reinforcement. *Learning and Motivation,* 1973, *4,* 207–17.

Linton, J., Riccio, D. C., Rohrbaugh, M., & Page, H. A. The effects of blocking an instrumental avoidance response: Fear reduction or enhancement? *Behavior Research and Therapy,* 1970, *8,* 267–72.

Logan, F. A. A micromolar approach to behavior. *Psychological Review,* 1956, *63,* 63–73.

Logan, F. A. Transfer of discrimination. *Journal of Experimental Psychology,* 1966, *71,* 616–18.

Logue, A. W. Taste aversion and the generality of the laws of learning. *Psychological Bulletin,* 1979, *86,* 276–96.

LoLordo, V. M., McMillan, J. C. & Riley, A. L. The effects upon food-reinforced pecking and treadle-pressing of auditory and visual signals for response-independent food. *Learning and Motivation,* 1974, *5,* 24–41.

Long, C. J., & Tapp, J. T. Reinforcing properties of odors for the albino rat. *Psychonomic Science,* 1967, *7,* 17–18.

Looney, T. A., & Griffin, R. W. A sequential feature-positive effect using tone as the distinguishing feature in an auto-shaping procedure. *Animal Learning and Behavior,* 1978, *6,* 401–5.

Lorenz, K. Z. The companion in the bird's world. *Auk,* 1937, *54,* 245–73.

Lovejoy, E. Analysis of the overlearning reversal effect. *Psychological Review,* 1966, *73,* 87–103.

Lubow, R. E. Latent inhibition: Effects of frequency of nonreinforced preexposure of the CS. *Journal of Comparative and Physiological Psychology,* 1965, *60,* 454–57.

Lubow, R. E. Latent inhibition. *Psychological Bulletin,* 1973, *79,* 398–407.

Lubow, R. E., Rifkin, B., & Alek, M. The context effect: The relationship between stimulus preexposure and environmental preexposure determines subsequent learning. *Journal of Experimental Psychology: Animal Behavior Processes,* 1976, *2,* 38–47.

Lubow, R. E., Schnur, P., & Rifkin, B. Latent inhibition and conditioned attention theory. *Journal of Experimental Psychology: Animal Behavior Processes,* 1976, *2,* 163–74.

Lynch, J. J. Pavlovian inhibition of delay in cardiac and somatic responses in dogs: Schizokinesis. *Psychological Reports,* 1973, *32,* 1339–46.

Lyons, J., Klipec, W. D., & Steinsultz, G. The effect of chlorpromazine on discrimination performance and the peak shift. *Physiological Psychology,* 1973, *1,* 121–24.

Mackintosh, N. J. Extinction of a discrimination habit as a function of overtraining. *Journal of Comparative and Physiological Psychology,* 1963, *56,* 842–47. (a)

Mackintosh, N. J. The effect of irrelevant cues on reversal learning in the rat. *British Journal of Psychology,* 1963, *54,* 127–34. (b)

Mackintosh, N. J. Overtraining and transfer within and between dimensions in the rat. *Quarterly Journal of Experimental Psychology,* 1964, *16,* 250–56.

Mackintosh, N. J. Overtraining, reversal, and extinction in rats and chicks. *Journal of Comparative and Physiological Psychology,* 1965, *59,* 31–36.

Mackintosh, N. J. Further analysis of the overtraining reversal effect. *Journal of Comparative and Physiological Psychology, Monograph Supplement,* 1969, *67,* part 2, 1–18.

Mackintosh, N. J. Blocking of conditioned suppression: Role of the first compound trial. *Journal of Experimental Psychology: Animal Behavior Processes,* 1975, *1,* 335–45. (a)

Mackintosh, N. J. A theory of attention: Variation of the associability of stimuli with reinforcement. *Psychological Review,* 1975, *82,* 276–98. (b)

Mackintosh, N. J. Overshadowing and stimulus intensity. *Animal Learning and Behavior,* 1976, *4,* 186–92.

Mackintosh, N. J., & Honig, W. K. Blocking and attentional enhancement in pigeons. *Journal of Comparative and Physiological Psychology,* 1970, *73,* 78–85.

Mackintosh, N. J., & Little, L. Intradimensional and extradimensional shift learning by pigeons. *Psychonomic Science,* 1969, *14,* 5–6.

Mackintosh, N. J., & Little, L. An analysis of transfer along a continuum. *Canadian Journal of Psychology,* 1970, *24,* 362–69.

Mackintosh, N. J., Little, L., & Lord, J. Some determinants of behavioral contrast in pigeons and rats. *Learning and Motivation,* 1972, *3,* 148–61.

Mackintosh, N. J., & Lord, J. Simultaneous and successive contrast with delay of reward. *Animal Learning and Behavior,* 1973, *1,* 283–86.

Mackintosh, N. J., & Turner, C. Blocking as a function of novelty of CS and predictability of UCS. *Quarterly Journal of Experimental Psychology,* 1971, *23,* 359–66.

MacPhail, E. M. Avoidance responding in pigeons. *Journal of the Experimental Analysis of Behavior,* 1968, *11,* 625–32.

Mahoney, W. J., & Ayres, J. J. B. One-trial simultaneous and backward fear conditioning as reflected in conditioned suppression of licking in rats. *Animal Learning and Behavior,* 1976, *4,* 357–62.

Maier, S. F. Failure to escape traumatic shock: Incompatible skeletal motor response or learned helplessness? *Learning and Motivation,* 1970, *1,* 157–70.

Maier, S. F., Albin, R. W., & Testa, T. J. Failure to learn to escape in rats previously exposed to inescapable shock depends on nature of escape response. *Journal of Comparative and Physiological Psychology,* 1973, *85,* 581–92.

Maier, S. F., & Jackson, R. L. Learned helplessness: All of us were right (and wrong): Inescapable shock has multiple effects. In G. H. Bower (Ed.), *The psychology of learning and motivation* (Vol. 13). New York: Academic Press, 1979.

Maier, S. F., & Seligman, M. E. P. Learned helplessness: Theory and evidence. *Journal of Experimental Psychology: General,* 1976, *105,* 3–46.

Maier, S. F., Seligman, M. E. P., & Solomon, R. L. Pavlovian fear conditioning and learned helplessness: Effects on escape and avoidance behavior of (a) the CS-US contingency and (b) the independence of the US and voluntary responding. In B. A. Campbell & R. M. Church (Eds.), *Punishment and aversive behavior.* New York: Appleton-Century-Crofts, 1969.

Maisiak, R., & Frey, P. W. Second-order conditioning: The importance of stimulus overlap on second-order trials. *Animal Learning and Behavior,* 1977, *5,* 309–14.

Maki, W. S. Pigeon's short-term memories for surprising vs. expected reinforcement and nonreinforcement. *Animal Learning and Behavior,* 1979, *7,* 31–37.

Maki, W. S., & Leith, C. R. Shared attention in pigeons. *Journal of the Experimental Analysis of Behavior,* 1973, *19,* 345–49.

Maki, W. S., Moe, J. C., & Bierley, C. M. Short-term memory for stimuli, responses, and reinforcers. *Journal of Experimental Psychology: Animal Behavior Processes,* 1977, *3,* 156–77.

Maki, W. S., Riley, D. A. & Leith, C. R. The role of test stimuli in matching to compound samples by pigeons. *Animal Learning and Behavior,* 1976, *4,* 13–21.

Maleske, R. T., & Frey, P. W. Blocking in eyelid conditioning: Effect of changing the CS-US interval and introducing an intertrial stimulus. *Animal Learning and Behavior,* 1979, *7,* 452–56.

Manning, A. A., Schneiderman, N., & Lordahl, D. S. Delay vs. trace heart rate classical discrimination conditioning in rabbits as a function of ISI. *Journal of Experimental Psychology,* 1969, *80,* 225–30.

Marchant, H. G., & Moore, J. W. Blocking of the rabbit's conditioned nictitating membrane response in Kamin's two-stage paradigm. *Journal of Experimental Psychology,* 1973, *101,* 155–58.

Marchant, H. G., & Moore, J. W. Below-zero conditioned inhibition of the rabbit's nictitating membrane response. *Journal of Experimental Psychology,* 1974, *102,* 350–52.

Marler, P. A comparative approach to vocal learning: Song development in white-crowned sparrows. *Journal of Comparative and Physiological Psychology Monograph,* 1970, *71* (2, Pt. 2), 1–25.

Marrazo, M. J., Riccio, D. C., & Riley, J. Effects of Pavlovian conditioned stimulus—unconditioned stimulus pairings during avoidance response-prevention trials in rats. *Journal of Comparative and Physiological Psychology,* 1974, *86,* 96–100.

Marsh, G. An evaluation of three explanations for the transfer of discrimination effect. *Journal of Comparative and Physiological Psychology,* 1969, *68,* 268–75.

Marsh, G. Prediction of the peak shift in pigeons from gradients of excitation and inhibition. *Journal of Comparative and Physiological Psychology,* 1972, *81,* 262–66.

Martin, L. K., & Riess, D. Effects of US intensity during previous discrete delay conditioning on conditioned acceleration during avoidance extinction. *Journal of Comparative and Physiological Psychology,* 1969, *69,* 196–200.

Marwine, A. G., & Collier, G. Instrumental and consummatory behavior as a function of rate of weight loss and weight maintenance schedule. *Journal of Comparative and Physiological Psychology,* 1971, *74,* 441–47.

Marx, M. H. Positive contrast in instrumental learning from qualitative shift in incentive. *Psychonomic Science,* 1969, *16,* 254–55.

Marx, M. H., & Murphy, W. W. Resistance to extinction as a function of the presentation of a motivating cue in the start box. *Journal of Comparative and Physiological Psychology,* 1961, *54,* 207–10.

Mason, W. A. Motivation factors in psychosocial development. In W. J. Arnold & M. M. Page (Eds.), *Nebraska symposium on motivation: 1970.* Lincoln: University of Nebraska Press, 1970.

Masterson, F. A. Escape from noise. *Psychological Reports,* 1969, *24,* 484–86.

Masterson, F. A., Crawford, M., & Bartter, W. D. Brief escape from a dangerous place: The role of reinforcement in the rat's one-way avoidance acquisition. *Learning and Motivation,* 1978, *9,* 141–63.

McAllister, D. E., & McAllister, W. R. Forgetting of acquired fear. *Journal of Comparative and Physiological Psychology,* 1968, *65,* 352–55.

McAllister, D. E., McAllister, W. R., Brooks, C. I., & Goldman, J. A. Magnitude and shift of reward in instrumental aversive learning in rats. *Journal of Comparative and Physiological Psychology,* 1972, *80,* 490–501.

McAllister, D. E., McAllister, W. R., & Dieter, S. E. Reward magnitude and shock variables (continuity and intensity) in shuttlebox-avoidance learning. *Animal Learning and Behavior,* 1976, *4,* 204–9.

McAllister, W. R., & McAllister, D. E. Increase over time in the stimulus generalization of acquired fear. *Journal of Experimental Psychology,* 1963, *65,* 576–82.

McAllister, W. R., McAllister, D. E., & Douglass, W. K. The inverse relationship between shock intensity and shuttle-box avoidance learning in rats: A reinforcement explanation. *Journal of Comparative and Physiological Psychology,* 1971, *74,* 426–33.

McCain, G. Partial reinforcement effects following a small number of acquisition trials. *Psychonomic Monograph Supplement,* 1966, *1,* 251–70.

McFarland, D. J. Interaction of hunger and thirst in the Barbary dove. *Journal of Comparative and Physiological Psychology,* 1964, *58,* 174–79.

McFarland, D. J., & Lloyd, I. H. Time-shared feeding and drinking. *Quarterly Journal of Experimental Psychology,* 1973, *25,* 48–61.

McFarland, D. J., & Sibly, R. M. The behavioural final common path. *Philosophical Translations of the Royal Society* (B), 1975, *207,* 265–93.

McHose, J. H., & Peters, D. P. Partial reward, the negative contrast effect, and incentive averaging. *Animal Learning and Behavior,* 1975, *3,* 239–44.

McHose, J. H., & Tauber, L. Changes in delay of reinforcement in simple instrumental conditioning. *Psychonomic Science,* 1972, *27,* 291–92.

McKearney, J. W. Fixed-interval schedules of electric shock presentation: Extinction and recovery of performance under different shock intensities and fixed-interval durations. *Journal of the Experimental Analysis of Behavior,* 1969, *12,* 301–13.

McNeill, D. *The acquisition of language.* New York: Harper & Row, 1970.

Medin, D. L. Role of reinforcement in discrimination learning set in monkeys. *Psychological Bulletin,* 1972, *77,* 305–18.

Medin, D. L., Roberts, W. A., & Davis, R. T. *Processes of animal memory.* Hillsdale, N.J.: Laurence Erlbaum, 1976.

Mednick, S. A., & Freedman, J. L. Stimulus generalization. *Psychological Bulletin,* 1960, *57,* 169–200.

Mellgren, R. L. Positive and negative contrast effects using delayed reinforcement. *Learning and Motivation,* 1972, *3,* 185–93.

Mellgren, R. L., Seybert, J. A., & Dyck, D. G. The order of continuous, partial and nonreward trials and resistance to extinction. *Learning and Motivation,* 1978, *9,* 359–71.

Mellgren, R. L., Wrather, D. M., & Dyck, D. G. Differential conditioning and contrast effects in rats. *Journal of Comparative and Physiological Psychology,* 1972, *80,* 478–83.

Melvin, K. B., & Martin, R. C. Facilitative effects of two modes of punishment on resistance to extinction. *Journal of Comparative and Physiological Psychology,* 1966, *62,* 491–94.

Mikhail, A. A. Stress and ulceration in the glandular and nonglandular portions of the rat's stomach. *Journal of Comparative and Physiological Psychology,* 1973, *85,* 636–42.

Mikulka, P. J., Leard, B., & Klein, S. B. Illness-alone exposure as a source of interference with the acquisition and retention of a taste aversion. *Journal of Experimental Psychology: Animal Behavior Processes,* 1977, *3,* 189–201.

Miles, R. C. Discrimination-learning sets. In A. M. Schrier, H. F. Harlow, & F. Stollnitz (Eds.), *Behavior of nonhuman primates: Modern research trends.* New York: Academic Press, 1965.

Millenson, J. R., & de Villiers, P. A. Motivational properties of conditioned suppression. *Learning and Motivation,* 1972, *3,* 125–37.

Miller, H. L. Matching-based hedonic scaling in the pigeon. *Journal of the Experimental Analysis of Behavior,* 1976, *26,* 335–47.

Miller, H. L., & Loveland, D. H. Matching when the number of response alternatives is large. *Animal Learning and Behavior,* 1974, *2,* 106–10.

Miller, N. E. Studies of fear as an acquirable drive: I. Fear as motivation and fear reduction as reinforcement in the learning of new responses. *Journal of Experimental Psychology,* 1948, *38,* 89–101.

Miller, N. E. Learnable drives and rewards. In *Handbook of experimental psychology.* S. S. Stevens (Ed.), New York: Wiley, 1951.

Miller, N. E. Learning of visceral and glandular responses. *Science,* 1969, *163,* 434–45.

Miller, N. E. Biofeedback and visceral learning. *Annual Review of Psychology,* 1978, *29,* 373–404.

Miller, N. E., & Dworkin, B. R. Visceral learning: Recent difficulties with curarized rats and significant problems for human research. In P. A. Obrist, A. H. Black, J. Brener, & L. V. DiCara (Eds.), *Cardiovascular psychophysiology: Current issues in response mechanisms, biofeedback, and methodology.* Chicago: Aldine, 1974.

Miller, R. R., Ott, C. A., Berk, A. M., & Springer, A. D. Appetitive memory restoration after electroconvulsive shock in the rat. *Journal of Comparative and Physiological Psychology,* 1974, *87,* 717–23.

Miller, R. R., & Springer, A. D. Induced recovery of memory in rats following electroconvulsive shock. *Physiology and Behavior,* 1972, *8,* 645–51.

Miller, R. R., & Springer, A. D. Amnesia, consolidation, and retrieval. *Psychological Review,* 1973, *80,* 69–79.

Mineka, S. Some new perspectives on conditioned hunger. *Journal of Experimental Psychology: Animal Behavior Processes,* 1975, *1,* 134–48.

Mineka, S. Effects of flooding an irrelevant response on the extinction of avoidance responses. *Journal of Experimental Psychology: Animal Behavior Processes,* 1976, *2,* 142–53.

Mineka, S. The role of fear in theories of avoidance learning, flooding, and extinction. *Psychological Bulletin,* 1979, *86,* 985–1010.

Mineka, S., & Gino, A. Some further tests of the brief confinement effect and the SSDR account of flooding. *Learning and Motivation,* 1979, *10,* 98–115. (a)

Mineka, S., & Gino, A. Dissociative effects of different types and amounts of nonreinforced CS exposure on avoidance extinction and the CER. *Learning and Motivation,* 1979, *10,* 141–60. (b)

Mineka, S., Seligman, M. E. P., Hetrick, M., & Zuelzer, K. Poisoning and conditioned drinking. *Journal of Comparative and Physiological Psychology,* 1972, *79,* 377–84.

Mis, F. W., & Moore, J. W. Effect of preacquisition UCS exposure on classical conditioning of the rabbit's nictitating membrane response. *Learning and Motivation,* 1973, *4,* 108–14.

Misanin, J. R., Campbell, B. A., & Smith, N. F. Duration of punishment and the delay of punishment gradient. *Canadian Journal of Psychology,* 1966, *20,* 407–12.

Misanin, J. R., Nagy, Z. M., Keiser, E. F., & Bowen, W. Emergence of long-term memory in the neonatal rat. *Journal of Comparative and Physiological Psychology,* 1971, *77,* 188–99.

Mistler-Lachman, J. L., & Lachman, R. Language in man, monkeys, and machines. *Science,* 1974, *185,* 871–72.

Mitchell, D., Kirschbaum, E. H., & Perry, R. L. Effects of neophobia and habituation on the poison-induced avoidance of exteroceptive stimuli in the rat. *Journal of Experimental Psychology: Animal Behavior Processes,* 1975, *1,* 47–55.

Modaresi, H. A. Facilitating effects of a safe platform on two-way avoidance learning. *Journal of Experimental Psychology: Animal Behavior Processes,* 1978, *4,* 83–94.

Mollenauer, S. O. Shifts in deprivation level: Different effects depending on amount of preshift training. *Learning and Motivation,* 1971, *2,* 58–66.

Moltz, H., & Stettner, L. J. The influence of patterned-light deprivation on the critical period for imprinting. *Journal of Comparative and Physiological Psychology,* 1961, *54,* 279–83.

Montgomery, K. C. The role of exploratory drive in learning. *Journal of Comparative and Physiological Psychology,* 1954, *47,* 60–64.

Moore, J. W. Differential eyelid conditioning as a function of the frequency and intensity of auditory CSs. *Journal of Experimental Psychology,* 1964, *68,* 250–59.

Moot, S. A., Cebulla, R. P., & Crabtree, J. M. Instrumental control and ulceration in rats. *Journal of Comparative and Physiological Psychology,* 1970, *71,* 405–10.

Morrison, G. R., & Collyer, R. Taste-mediated conditioned aversion to an exteroceptive stimulus following LiCl poisoning. *Journal of Comparative and Physiological Psychology,* 1974, *86,* 51–55.

Mostofsky, D. *Stimulus generalization.* Stanford, Calif.: Stanford University Press, 1965.

Mountjoy, P. P., & Malott, M. K. Wave-length generalization curves for chickens reared in restricted portions of the spectrum. *Psychological Record,* 1968, *18,* 575–83.

Mowrer, O. H. On the dual nature of learning: A reinterpretation of "conditioning" and "problem-solving." *Harvard Educational Review,* 1947, *17,* 102–48.

Mowrer, O. H. *Learning theory and behavior.* New York: Wiley, 1960.

Moyer, K. E., & Korn, J. H. Effect of UCS intensity on the acquisition and extinction of an avoidance response. *Journal of Experimental Psychology,* 1964, *67,* 352–59.

Myers, A. K. Shock intensity and warning signal effects on several measures of operant avoidance acquisition. *Animal Learning and Behavior,* 1977, *5,* 51–56.

Nachman, M., & Ashe, J. H. Learned taste aversions in rats as a function of dosage, concentration, and route of administration of LiCl. *Physiology and Behavior,* 1973, *10,* 73–78.

Nachman, M., & Jones, D. R. Learned taste aversions over long delays in rats: The role of learned safety. *Journal of Comparative and Physiological Psychology,* 1974, *86,* 949–56.

Nageishi, Y., & Imada, H. Suppression of licking behavior in rats as a function of predictability of shock and probability of conditioned-stimulus-shock pairings. *Journal of Comparative and Physiological Psychology,* 1974, *87,* 1165–73.

Nagy, Z. M., Misanin, J. R., Newman, J. A., Olsen, P. L., & Hinderliter, C. F. Ontogeny of memory in the neonatal mouse. *Journal of Comparative and Physiological Psychology*, 1972, *81*, 380–93.

Nagy, Z. M., & Mueller, P. W. Effect of amount of original training upon onset of a 24-hour memory capacity in neonatal mice. *Journal of Comparative and Physiological Psychology*, 1973, *85*, 151–59.

Nation, J. R., Wrather, D. M., & Mellgren, R. L. Contrast effects in escape conditioning of rats. *Journal of Comparative and Physiological Psychology*, 1974, *86*, 69–73.

Neuringer, A. J., & Schneider, B. A. Separating the effects of interreinforcement time and number of interreinforcement responses. *Journal of the Experimental Analysis of Behavior*, 1968, *11*, 661–67.

Nevin, J. A. Interval reinforcement of choice behavior in discrete trials. *Journal of the Experimental Analysis of Behavior*, 1969, *12*, 875–85.

Nevin, J. A. Overall matching versus momentary maximizing: Nevin (1969) revisited. *Journal of Experimental Psychology: Animal Behavior Processes*, 1979, *5*, 300–306.

Newman, J. R., & Grice, G. R. Stimulus generalization as a function of drive level, and the relation between two measures of response strength. *Journal of Experimental Psychology*, 1965, *69*, 357–62.

Nottebohm, F. The origins of vocal learning. *American Naturalist*, 1972, *106*, 116–40.

Obrist, P. A. The cardiovascular-behavioral interaction—as it appears today. *Psychophysiology*, 1976, *13*, 95–107.

Olton, D. S. Discrimination behavior in the rat: Differential effects of reinforcement and nonreinforcement. *Journal of Comparative and Physiological Psychology*, 1972, *79*, 284–90.

Olton, D. S., & Samuelson, R. J. Remembrance of places passed: Spatial memory in rats. *Journal of Experimental Psychology: Animal Behavior Processes*. 1976, *2*, 97–116.

O'Neil, H. F., Skeen, L. C., & Ryan, F. J. Prevention of vicious circle behavior. *Journal of Comparative and Physiological Psychology*, 1970, *70*, 281–85.

Osborne, S. R. A quantitative analysis of the effects of amount of reinforcement on two response classes. *Journal of Experimental Psychology: Animal Behavior Processes*, 1978, *4*, 297–317.

Ost, J. W. P., & Lauer, D. W. Some investigations of classical salivary conditioning in the dog. In W. F. Prokasy (Ed.), *Classical conditioning: A symposium*. New York: Appleton-Century-Crofts, 1965.

Overmier, J. B., & Bull, J. A. Influences of appetitive Pavlovian conditioning upon avoidance behavior. In J. H. Reynierse (Ed.), *Current issues in animal learning: A colloquium*. Lincoln: University of Nebraska Press, 1970.

Overmier, J. B., Bull, J. A., & Pack, K. On instrumental response interaction as explaining the influences of Pavlovian CS+ upon avoidance behavior. *Learning and Motivation*, 1971, *2*, 103–12.

Overmier, J. B., & Lawry, J. A. Pavlovian conditioning and the mediation of behavior. In G. H. Bower (Ed.), *The psychology of learning and motivation* (Vol. 13). New York: Academic Press, 1979.

Overmier, J. B., & Schwarzkopf, K. H. Summation of food and shock based responding. *Learning and Motivation*, 1974, *5*, 42–52.

Overmier, J. B., & Seligman, M. E. P. Effects of inescapable shock upon subsequent escape and avoidance responding. *Journal of Comparative and Physiological Psychology*, 1967, *63*, 28–33.

Owen, J. W., Cicala, G. A., & Herdegen, R. T. Fear inhibition and species specific defense reaction termination may contribute independently to avoidance learning. *Learning and Motivation*, 1978, *9*, 297–313.

Pacitti, W. A., & Smith, N. F. A direct comparison of four methods for eliminating a response. *Learning and Motivation,* 1977, *8,* 229–37.

Parsons, P. J., & Spear, N. E. Long-term retention of avoidance learning by immature and adult rats as a function of environmental enrichment. *Journal of Comparative and Physiological Psychology,* 1972, *80,* 297–303.

Pavlik, W. B., & Reynolds, W. F. Effects of deprivation schedule and reward magnitude on acquisition and extinction performance. *Journal of Comparative and Physiological Psychology,* 1963, *56,* 452–55.

Pavlov, I. P. [*Conditioned reflexes*] (G. V. Anrep, Trans.). London: Oxford University Press, 1927.

Pearce, J. M. The relationship between shock magnitude and passive avoidance learning. *Animal Learning and Behavior,* 1978, *6,* 341–45.

Pearce, J. M., & Dickinson, A. Pavlovian counterconditioning: Changing the suppressive properties of shock by association with food. *Journal of Experimental Psychology: Animal Behavior Processes,* 1975, *1,* 170–77.

Pearce, J. M., & Hall, G. Overshadowing the instrumental conditioning of a lever-press response by a more valid predictor of the reinforcer. *Journal of Experimental Psychology: Animal Behavior Processes,* 1978, *4,* 356–67.

Peckham, R. H., & Amsel, A. The within-S demonstration of a relationship between frustration and magnitude of reward in a differential magnitude of reward discrimination. *Journal of Experimental Psychology,* 1967, *73,* 187–95.

Perin, C. T. A quantitative investigation of the delay-of-reinforcement gradient. *Journal of Experimental Psychology,* 1943, *32,* 37–51.

Perkins, C. C., & Weyant, R. G. The interval between training and test trials as a determiner of the slope of generalization gradients. *Journal of Comparative and Physiological Psychology,* 1958, *51,* 596–600.

Peterson, G. B., Wheeler, R. L., & Armstrong, G. D. Expectancies as mediators in the differential-reward conditioned discrimination performance of pigeons. *Animal Learning and Behavior,* 1978, *6,* 279–85.

Peterson, G. B., Wheeler, R. L., & Trapold, M. A. Enhancement of pigeons' conditional discrimination performance by expectancies of reinforcement and nonreinforcement. *Animal Learning and Behavior,* 1980, *8,* 22–30.

Peterson, N. Effect of monochromatic rearing on the control of responding by wavelength. *Science,* 1962, *136,* 774–75.

Pinel, J. P. J., Corcoran, M. E., & Malsbury, C. W. Incubation effect in rats: Decline of footshock-produced activation. *Journal of Comparative and Physiological Psychology,* 1971, *77,* 271–76.

Pinel, J. P. J., & Mucha, R. F. Incubation and Kamin effects in the rat: Changes in activity and reactivity after foot shock. *Journal of Comparative and Physiological Psychology,* 1973, *84,* 661–68.

Plotkin, H. C., & Oakley, D. A. Backward conditioning in the rabbit *(oryctologus cuniculus). Journal of Comparative and Physiological Psychology,* 1975, *88,* 586–90.

Powell, R. W. The effect of punishment shock intensity upon responding under multiple schedules. *Journal of the Experimental Analysis of Behavior,* 1970, *14,* 201–11.

Powell, R. W. Analysis of warm-up effects during avoidance in wild and domesticated rodents. *Journal of Comparative and Physiological Psychology,* 1972, *78,* 311–16. (a)

Powell, R. W. Some effects of response-independent shocks after unsignaled avoidance conditioning in rats. *Learning and Motivation,* 1972, *3,* 420–41. (b)

Premack, D. *Intelligence in ape and man.* Hillsdale, N.J.: Lawrence Erlbaum Publishers, 1976.

Prewitt, E. P. Number of preconditioning trials in sensory preconditioning using CER training. *Journal of Comparative and Physiological Psychology,* 1967, *64,* 360–62.

Prokasy, W. F., & Chambliss, D. J. Temporal conditioning: Negative results. *Psychological Reports,* 1960, *7,* 539–42.

Prokasy, W. F., & Hall, J. F. Primary stimulus generalization. *Psychological Review,* 1963, *70,* 310–22.

Prokasy, W. F., Hall, J. F., & Fawcett, J. T. Adaptation sensitization, forward and backward conditioning and pseudoconditioning of the GSR. *Psychological Reports,* 1962, *10,* 103–6.

Prokasy, W. F., & Whaley, F. L. The intertrial interval in classical conditioning. *Journal of Experimental Psychology,* 1961, *62,* 560–64.

Prokasy, W. F., & Whaley, F. L. Intertrial interval range shift in classical eyelid conditioning. *Psychological Reports,* 1963, *12,* 55–58.

Pubols, B. H. The facilitation of visual and spatial discrimination reversal by overlearning. *Journal of Comparative and Physiological Psychology,* 1956, *49,* 243–48.

Purdy, J. E., & Cross, H. A. The role of R-S* expectancy in discrimination and discrimination reversal learning. *Learning and Motivation,* 1979, *10,* 211–27.

Purtle, R. B. Peak shift: A review. *Psychological Bulletin,* 1973, *80,* 408–21.

Pyke, G. H., Pulliam, H. R., & Charnov, E. L. Optimal foraging: A selective review of theory and tests. *Quarterly Review of Biology,* 1977, *52,* 137–54.

Rabin, J. S. Effects of varying sucrose reinforcers and anobarbital sodium on positive contrast in rats. *Animal Learning and Behavior,* 1975, *3,* 290–94.

Rachlin, H., & Baum, W. M. Effects of alternative reinforcement: Does the source matter? *Journal of the Experimental Analysis of Behavior,* 1972, *18,* 231–41.

Rajecki, D. W. Imprinting in precocial birds: Interpretation, evidence, and evaluation. *Psychological Bulletin,* 1973, *79,* 48–58.

Randall, P. K., & Riccio, D. C. Fear and punishment as determinants of passive-avoidance responding. *Journal of Comparative and Physiological Psychology,* 1969, *69,* 550–53.

Randich, A., & LoLordo, V. M. Associative and nonassociative theories of the UCS preexposure phenomenon: Implications for Pavlovian conditioning. *Psychological Bulletin,* 1979, *86,* 523–48. (a)

Randich, A., & LoLordo, V. M. Preconditioning exposure to the unconditioned stimulus affects the acquisition of a conditioned emotional response. *Learning and Motivation,* 1979, *10,* 245–77. (b)

Ratliff, R. G., & Ratliff, A. R. Runway acquisition and extinction as a joint function of magnitude of reward and percentage of rewarded acquisition trials. *Learning and Motivation,* 1971, *2,* 289–95.

Ratner, A. M. Modification of ducklings' filial behavior by aversive stimulation. *Journal of Experimental Psychology: Animal Behavior Processes,* 1976, *2,* 266–84.

Ratner, A. M., & Hoffman, H. S. Evidence for a critical period for imprinting in Khaki Campbell ducklings *(Anas platyrhynchos domesticus). Animal Behavior,* 1974, *22,* 249–55.

Rawson, R. A., & Leitenberg, H. Reinforced alternative behavior during punishment and extinction with rats. *Journal of Comparative and Physiological Psychology,* 1973, *85,* 593–600.

Rawson, R. A., Leitenberg, H., Mulick, J. A., & Lefebvre, M. F. Recovery of extinction responding in rats following discontinuation of reinforcement of alternative behavior: A test of two explanations. *Animal Learning and Behavior,* 1977, *5,* 415–20.

Reese, H. W. Discrimination learning set in rhesus monkeys. *Psychological Bulletin,* 1964, *61,* 321–40.

Reid, L. S. The development of noncontinuity behavior through continuity learning. *Journal of Experimental Psychology,* 1953, *46,* 107–12.

Reiss, S., & Wagner, A. R. CS habituation produces a "latent inhibition effect" but no active "conditioned inhibition". *Learning and Motivation,* 1972, *3,* 237–45.

Renner, K. E. Delay of reinforcement: An historical review. *Psychological Bulletin*, 1964, *61*, 341–61.

Renner, K. E. Delay of reinforcement and resistance to extinction: A supplementary report. *Psychological Reports*, 1965, *16*, 197–98.

Rescorla, R. A. Inhibition of delay in Pavlovian fear conditioning. *Journal of Comparative and Physiological Psychology*, 1967, *64*, 114–20. (a)

Rescorla, R. A. Pavlovian conditioning and its proper control procedures. *Psychological Review*, 1967, *74*, 71–80. (b)

Rescorla, R. A. Probability of shock in the presence and absence of CS in fear conditioning. *Journal of Comparative and Physiological Psychology*, 1968, *66*, 1–5.

Rescorla, R. A. Pavlovian conditioned inhibition. *Psychological Bulletin*, 1969, *72*, 77–94.

Rescorla, R. A. Summation and retardation tests of latent inhibition. *Journal of Comparative and Physiological Psychology*, 1971, *75*, 77–81. (a)

Rescorla, R. A. Variation in the effectiveness of reinforcement and nonreinforcement following prior inhibitory conditioning. *Learning and Motivation*, 1971, *2*, 113–23. (b)

Rescorla, R. A. Informational variables in Pavlovian conditioning. In G. H. Bower (Ed.), *The psychology of learning and motivation* (Vol. 6). London: Academic Press, 1972.

Rescorla, R. A. Effect of US habituation following conditioning. *Journal of Comparative and Physiological Psychology*, 1973, *82*, 137–43. (a)

Rescorla, R. A. Second-order conditioning: Implications for theories of learning. In F. J. McGuigan & D. B. Lumsden (Eds.), *Contemporary approaches to conditioning and learning.* Washington, D.C.: V. H. Winston & Sons, 1973. (b)

Rescorla, R. A. Effect of inflation of the unconditioned stimulus value following conditioning. *Journal of Comparative and Physiological Psychology*, 1974, *86*, 101–6.

Rescorla, R. A. Second-order conditioning of Pavlovian conditioned inhibition. *Learning and Motivation*, 1976, *7*, 161–72.

Rescorla, R. A., & Furrow, D. R. Stimulus similarity as a determinant of Pavlovian conditioning. *Journal of Experimental Psychology: Animal Behavior Processes*, 1977, *3*, 203–15.

Rescorla, R. A., & LoLordo, V. M. Inhibition of avoidance behavior. *Journal of Comparative and Physiological Psychology*, 1965, *59*, 406–12.

Rescorla, R. A., & Skucy, J. C. Effect of response-independent reinforcers during extinction. *Journal of Comparative and Physiological Psychology*, 1969, *67*, 381–89.

Rescorla, R. A., & Solomon, R. L. Two-process learning theory: Relationships between Pavlovian conditioning and instrumental learning. *Psychological Review*, 1967, *74*, 151–82.

Rescorla, R. A., & Wagner, A. R. A theory of Pavlovian conditioning: Variations in the effectiveness of reinforcement and nonreinforcement. In A. H. Black & W. F. Prokasy (Eds.), *Classical conditioning II: Current research and theory.* New York: Appleton-Century-Crofts, 1972.

Revusky, S. H., & Bedarf, E. W. Association of illness with prior ingestion of novel foods. *Science*, 1967, *155*, 219–20.

Revusky, S. H., & Parker, L. A. Aversions to unflavored water and cup drinking produced by delayed sickness. *Journal of Experimental Psychology: Animal Behavior Processes*, 1976, *2*, 342–53.

Reynierse, J. H., & Rizley, R. C. Stimulus and response contingencies in extinction of avoidance by rats. *Journal of Comparative and Physiological Psychology*, 1970, *73*, 86–92.

Reynolds, T. J., & Medin, D. L. Strength vs. temporal-order information in delayed-matching-to-sample performance by monkeys. *Animal Learning and Behavior*, 1979, *7*, 294–300.

Richter, C. P. A behavioristic study of the activity of the rat. *Comparative Psychological Monograph*, 1922, *1*, 2.

Riess, D. Sidman avoidance in rats as a function of shock intensity and duration. *Journal of Comparative and Physiological Psychology*, 1970, *73*, 481–85.

Riess, D., & Farrar, C. H. Shock intensity, shock duration, Sidman avoidance acquisition, and the "all or nothing" principle in rats. *Journal of Comparative and Physiological Psychology, 1972, 81,* 347–55.

Riess, D., & Farrar, C. H. UCS duration and conditioned suppression: Acquisition and extinction between-groups and terminal performance within-subjects. *Learning and Motivation,* 1973, *4,* 366–73.

Riley, A. L., & Clarke, C. M. Conditioned taste aversions: A bibliography. In L. M. Barker, M. R. Best, & M. Domjan (Eds.), *Learning mechanisms in food selection.* Waco, Texas: Baylor University Press, 1977.

Riley, D. A., & Leuin, T. C. Stimulus-generalization gradients in chickens reared in monochromatic light and tested with a single wavelength value. *Journal of Comparative and Physiological Psychology,* 1971, *75,* 399–402.

Rizley, R. C., & Rescorla, R. A. Associations in second-order conditioning and sensory preconditioning. *Journal of Comparative and Physiological Psychology, 1972, 81,* 1–11.

Robbins, D. Partial reinforcement: A selective review of the alleyway literature since 1960. *Psychological Bulletin, 1971, 76,* 415–31.

Roberts, W. A. Resistance to extinction following partial and consistent reinforcement with varying magnitudes of reward. *Journal of Comparative and Physiological Psychology,* 1969, *67,* 395–400.

Roberts, W. A., & Grant, D. S. Short-term memory in the pigeon with presentation time precisely controlled. *Learning and Motivation, 1974, 5,* 393–408.

Roberts, W. A., & Grant, D. S. Studies of short-term memory in the pigeon using the delayed matching to sample procedure. In D. L. Medin, W. A. Roberts, & R. T. Davis (Eds.), *Processes of animal memory.* Hillsdale, N.J.: Lawrence Erlbaum, 1976.

Roberts, W. A., & Grant, D. S. An analysis of light-induced retroactive inhibition in pigeon short-term memory. *Journal of Experimental Psychology: Animal Behavior Processes,* 1978, *4,* 219–36. (a)

Roberts, W. A., & Grant, D. S. Interaction of sample and comparison stimuli in delayed matching to sample with the pigeon. *Journal of Experimental Psychology: Animal Behavior Processes, 1978, 4,* 68–82. (b)

Rodgers, W. L. Specificity of specific hungers. *Journal of Comparative and Physiological Psychology,* 1967, *64,* 49–58.

Rodgers, W., & Rozin, P. Novel food preferences in Thiamine-deficient rats. *Journal of Comparative and Physiological Psychology,* 1966, *61,* 1–4.

Rosellini, R. A., & Seligman, M. E. P. Frustration and learned helplessness. *Journal of Experimental Psychology: Animal Behavior Processes,* 1975, *1,* 149–57.

Ross, L. E., & Hartman, T. F. Human-eyelid conditioning: The recent experimental literature. *Genetic Psychology Monographs, 1965, 71,* 177–220.

Ross, S. M., & Ross, L. E. Comparison of trace and delay classical eyelid conditioning as a function of interstimulus interval. *Journal of Experimental Psychology, 1971, 91,* 165–67.

Roth, S., & Bootzin, R. R. Effects of experimentally induced expectancies of external control: An investigation of learned helplessness. *Journal of Personality and Social Psychology,* 1974, *29,* 253–64.

Rozin, P. Specific aversions as a component of specific hungers. *Journal of Comparative and Physiological Psychology,* 1967, *64,* 237–42.

Rozin, P. Specific aversions and neophobia resulting from vitamin deficiency or poisoning in half-wild and domestic rats. *Journal of Comparative and Physiological Psychology,* 1968, *66,* 82–88.

Rozin, P. Adaptive food sampling patterns in vitamin deficient rats. *Journal of Comparative and Physiological Psychology,* 1969, *69,* 126–32.

Rozin, P., & Kalat, J. W. Specific hungers and poison avoidance as adaptive specializations of learning. *Psychological Review, 1971, 78,* 459–86.

Rozin, P., Wells, C., & Mayer, J. Thiamine specific hunger: Vitamin in water versus vitamins in food. *Journal of Comparative and Physiological Psychology,* 1964, *57,* 78–84.

Rudolph, R. I., Honig, W. K., & Gerry, J. E. Effects of monochromatic rearing on the acquisition of stimulus control. *Journal of Comparative and Physiological Psychology,* 1969, *67,* 50–57.

Rudy, J. W. Sequential variables as determiners of the rat's discrimination of reinforcement events: Effects on extinction performance. *Journal of Comparative and Physiological Psychology,* 1971, *77,* 476–81.

Rudy, J. W., Rosenberg, L., & Sandell, J. H. Disruption of a taste familiarity effect by novel exteroceptive stimulation. *Journal of Experimental Psychology: Animal Behavior Processes,* 1977, *3,* 26–36.

Rumbaugh, D., Gill, T. V., & von Glasersfeld, E. Reading and sentence completion by a chimpanzee *(Pan). Science,* 1973, *182,* 731–33.

Russell, A., & Glow, P. H. Some effects of short-term immediate prior exposure to light change on responding for light change. *Animal Learning and Behavior,* 1974, *2,* 262–66.

Sadler, E. W. A within- and between-subjects comparison of partial reinforcement in classical salivary conditioning. *Journal of Comparative and Physiological Psychology,* 1968, *66,* 695–98.

Sahakian, B. J., Robbins, T. W., & Iverson, S. D. The effects of isolation rearing on exploration in the rat. *Animal Learning and Behavior,* 1977, *5,* 193–98.

Sainsbury, R. S. Effect of proximity of elements on the feature-positive effect. *Journal of the Experimental Analysis of Behavior,* 1971, *16,* 315–25.

Salafia, W. R., Lambert, R. W., Host, K. C., Chiaia, N. L., & Ramirez, J. J. Rabbit nictitating membrane conditioning: Lower limit of the effective interstimulus interval. *Animal Learning and Behavior,* 1980, *8,* 85–91.

Salafia, W. R., Mis, F. W., Terry, W. S., Bartosiak, R. S., & Daston, A. P. Conditioning of the nictitating membrane response of the rabbit *(Oryctolagus cuniculus)* as a function of the length and degree of variation of the intertrial interval. *Animal Learning and Behavior,* 1973, *1,* 109–15.

Scavio, M. J. Classical-classical transfer: Effects of prior aversive conditioning upon appetitive conditioning in rabbits. *Journal of Comparative and Physiological Psychology,* 1974, *86,* 107–15.

Scheuer, C., & Sutton, C. O. Discriminated vs. motivational interpretations of avoidance extinction: Extensions to learned helplessness. *Animal Learning and Behavior,* 1973, *1,* 193–97.

Schiff, R., Smith, N., & Prochaska, J. Extinction of avoidance in rats as a function of duration and number of blocked trials. *Journal of Comparative and Physiological Psychology,* 1972, *81,* 356–59.

Schneider, B. A. A two-state analysis of fixed-interval responding in the pigeon. *Journal of the Experimental Analysis of Behavior,* 1969, *12,* 677–87.

Schneiderman, N. Interstimulus interval function of the nictitating membrane response of the rabbit under delay versus trace conditioning. *Journal of Comparative and Physiological Psychology,* 1966, *62,* 397–402.

Schnur, P., & Ksir, C. J. Latent inhibition in human eyelid conditioning. *Journal of Experimental Psychology,* 1969, *80,* 388–89.

Schoenfeld, W. N. An experimental approach to anxiety, escape, and avoidance behavior. In P. H. Hock & J. Zubin (Eds.), *Anxiety.* New York: Grune & Stratton, 1950.

Schoenfeld, W. N. *Theory of reinforcement schedules.* New York: Appleton-Century-Crofts, 1970.

Schrier, A. M. Transfer between the repeated reversal and learning set tasks: A reexamination. *Journal of Comparative and Physiological Psychology,* 1974, *87,* 1004–10.

Schulenburg, C. J., Riccio, D. C., & Stikes, E. R. Acquisition and retention of a passive-avoidance response as a function of age in rats. *Journal of Comparative and Physiological Psychology*, 1971, *74*, 75–83.

Schusterman, R. J. Successive discrimination-reversal training and multiple discrimination training in one-trial learning by chimpanzees. *Journal of Comparative and Physiological Psychology*, 1964, *58*, 153–56.

Schwartz, B. Maintenance of key pecking in pigeons by a food avoidance but not by a shock avoidance contingency. *Animal Learning and Behavior*, 1973, *1*, 164–66.

Schwartz, B., & Williams, D. R. Discrete-trials spaced responding in the pigeon: The dependence of efficient performance on the availability of a stimulus for collateral pecking. *Journal of the Experimental Analysis of Behavior*, 1971, *16*, 155–60.

Schwartz, G. E. Voluntary control of human cardiovascular integration and differentiation through feedback and reward. *Science*, 1972, *175*, 90–93.

Schwartz, G. E. Biofeedback as therapy: Some theoretical and practical issues. *American Psychologist*, 1973, *28*, 666–73.

Schwartz, G. E. Biofeedback, self-regulation, and the patterning of physiological processes. *American Scientist*, 1975, *63*, 314–24.

Schwartz, G. E., & Beatty, J. *Biofeedback: Theory and research*. New York: Academic Press, 1977.

Scobie, S. R. Interaction of an aversive Pavlovian conditioned stimulus with aversively and appetitively motivated operants in rats. *Journal of Comparative and Physiological Psychology*, 1972, *79*, 171–88.

Sebeok, T. A. *How animals communicate*. Bloomington: Indiana University Press, 1977.

Seger, K. A., & Scheuer, C. The informational properties of S1, S2, and the S1-S2 sequence on conditioned suppression. *Animal Learning and Behavior*, 1977, *5*, 39–41.

Seidel, R. J. A review of sensory preconditioning. *Psychological Bulletin*, 1959, *56*, 58–73.

Seligman, M. E. P. CS redundancy and secondary punishment. *Journal of Experimental Psychology*, 1966, *72*, 546–50.

Seligman, M. E. P. On the generality of the laws of learning. *Psychological Review*, 1970, *77*, 406–18.

Seligman, M. E. P. *Helplessness: On depression, development, and death*. San Francisco: W. H. Freeman Company, 1975.

Seligman, M. E. P., & Beagley, G. Learned helplessness in the rat. *Journal of Comparative and Physiological Psychology*, 1975, *88*, 534–41.

Seligman, M. E. P., & Hager, J. *Biological boundaries of learning*. New York: Appleton-Century-Crofts, 1972.

Seligman, M. E. P., Ives, C. E., Ames, H., & Mineka, S. Conditioned drinking and its failure to extinguish: Avoidance, preprocedures, or functional autonomy. *Journal of Comparative and Physiological Psychology*, 1970, *71*, 411–19.

Seligman, M. E. P., & Maier, S. F. Failure to escape traumatic shock. *Journal of Experimental Psychology*, 1967, *74*, 1–9.

Seligman, M. E. P., Maier, S. F., & Greer, J. The alleviation of learned helplessness in the dog. *Journal of Abnormal and Social Psychology*, 1968, *73*, 256–62.

Seligman, M. E. P., Mineka, S., & Fillit, H. Conditioned drinking produced by procaine, NaCl, and angiotensin. *Journal of Comparative and Physiological Psychology*, 1971, *77*, 110–21.

Seligman, M. E. P., Rosellini, R. A., & Kozak, M. J. Learned helplessness in the rat: Time course, immunization, and reversibility. *Journal of Comparative and Physiological Psychology*, 1975, *88*, 542–47.

Senkowski, P. C., & Vogel, V. A. Effects of number of CS-US pairings on the strength of conditioned frustration in rats. *Animal Learning and Behavior*, 1976, *4*, 421–26.

Seraganian, P. Extra dimensional transfer in the easy-to-hard effect. *Learning and Motivation*, 1979, *10*, 39–57.

Sgro, J. A. Complete removal and delay of sucrose reward in the double alleyway. *Journal of Comparative and Physiological Psychology,* 1969, *69,* 442–47.

Sgro, J. A., Dyal, J. A., & Anastasio, E. J. Effects of constant delay of reinforcement on acquisition asymptote and resistance to extinction. *Journal of Experimental Psychology,* 1967, *73,* 634–36.

Sgro, J. A., Glotfelty, R. A., & Moore, B. D. Delay of reward in the double alleyway: A within-subjects versus between-groups comparison. *Journal of Experimental Psychology,* 1970, *84,* 82–87.

Shanab, M. E., & Biller, J. D. Positive contrast in the runway obtained following a shift in both delay and magnitude of reward. *Learning and Motivation,* 1972, *3,* 179–84.

Shanab, M. E., & Birnbaum, D. W. Durability of the partial reinforcement and partial delay of reinforcement extinction effects after minimal acquisition training. *Animal Learning and Behavior,* 1974, *2,* 81–85.

Shanab, M. E., Sanders, R., & Premack, D. Positive contrast in the runway obtained with delay of reward. *Science,* 1969, *164,* 724–25.

Sheafor, P. J. "Pseudoconditioned" jaw movements of the rabbit reflect associations conditioned to contextual background cues. *Journal of Experimental Psychology: Animal Behavior Processes,* 1975, *1,* 245–60.

Sheafor, P. J., & Gormezano, I. Conditioning the rabbit's *(Oryctolagus cuniculus)* jaw-movement response: US magnitude effects on URs, CRs, and pseudo-CRs. *Journal of Comparative and Physiological Psychology,* 1972, *81,* 449–56.

Sheffield, F. D., & Campbell, B. A. The role of experience in the "spontaneous" activity of hungry rats. *Journal of Comparative and Physiological Psychology,* 1954, *47,* 97–100.

Shepp, B. E., & Eimas, P. D. Intradimensional and extradimensional shifts in the rat. *Journal of Comparative and Physiological Psychology,* 1964, *57,* 357–61.

Shepp, B. E., & Schrier, A. M. Consecutive intradimensional and extradimensional shifts in monkeys. *Journal of Comparative and Physiological Psychology,* 1969, *67,* 199–203.

Sherman, J. E. US inflation with trace and simultaneous fear conditioning. *Animal Learning and Behavior,* 1978, *6,* 463–68.

Sherman, J. E., & Maier, S. F. The decrement in conditioned fear with increased trials of simultaneous conditioning is not specific to the simultaneous procedure. *Learning and Motivation,* 1978, *9,* 31–53.

Shettleworth, S. J. Constraints on learning. In D. S. Lehrman, R. A. Hinde, & E. Shaw (Eds.), *Advances in the study of behavior* (Vol. 4). New York: Academic Press, 1972.

Shettleworth, S. J. Reinforcement and the organization of behavior in golden hamsters: Hunger, environment, and food reinforcement. *Journal of Experimental Psychology: Animal Behavior Processes,* 1975, *1,* 56–87.

Shettleworth, S. J. Reinforcement and the organization of behavior in golden hamsters: Sunflower seed and nest paper reinforcers. *Animal Learning and Behavior,* 1978, *6,* 352–62. (a)

Shettleworth, S. J. Reinforcement and the organization of behavior in golden hamsters: Punishment of three action patterns. *Learning and Motivation,* 1978, *9,* 99–123. (b)

Shettleworth, S. J. Reinforcement and the organization of behavior in golden hamsters: Pavlovian conditioning with food and shock unconditioned stimuli. *Journal of Experimental Psychology: Animal Behavior Processes,* 1978, *4,* 152–69. (c)

Shimp, C. P. Probabilistically reinforced choice behavior in pigeons. *Journal of the Experimental Analysis of Behavior,* 1966, *9,* 443–55.

Shimp, C. P. The reinforcement of short interresponse times. *Journal of the Experimental Analysis of Behavior,* 1967, *10,* 425–34.

Shipley, R. H. Extinction of conditioned fear in rats as a function of several parameters of CS exposure. *Journal of Comparative and Physiological Psychology,* 1974, *87,* 699–707.

Shipley, R. H., Mock, L. A., & Levis, D. J. Effects of several response prevention procedures on activity, avoidance responding, and conditioned fear in rats. *Journal of Comparative and Physiological Psychology*, 1971, *77*, 256–70.

Sibly, R. M. How incentive and deficit determine feeding tendency. *Animal Behavior*, 1975, *23*, 437–46.

Sibly, R. M., & McCleery, R. H. The dominance boundary method of determining motivational state. *Animal Behavior*, 1976, *24*, 108–24.

Sidman, M. Avoidance conditioning with brief shock and no exteroceptive warning signal. *Science*, 1953, *118*, 157–58. (a)

Sidman, M. Two temporal parameters of the maintenance of avoidance behavior by the white rat. *Journal of Comparative and Physiological Psychology*, 1953, *46*, 253–61. (b).

Sidman, M. Some properties of the warning stimulus in avoidance behavior. *Journal of Comparative and Physiological Psychology*, 1955, *48*, 444–50.

Sidman, M. Stimulus generalization in an avoidance situation. *Journal of the Experimental Analysis of Behavior*, 1961, *4*, 157–69.

Siegel, S. Conditioning of insulin-induced glycemia. *Journal of Comparative and Physiological Psychology*, 1972, *78*, 233–41. (a)

Siegel, S. Latent inhibition and eyelid conditioning. In A. H. Black & W. F. Prokasy (Eds.), *Classical conditioning II: Current theory and research*. New York: Appleton-Century-Crofts, 1972. (b)

Siegel, S. Flavor preexposure and "learned safety." *Journal of Comparative and Physiological Psychology*, 1974, *87*, 1073–82.

Siegel, S. & Nettleton, N. Conditioning of insulin-induced hyperphagia. *Journal of Comparative and Physiological Psychology*, 1970, *72*, 390–93.

Silberberg, A., Hamilton, B., Ziriax, J. M., & Casey, J. The structure of choice. *Journal of Experimental Psychology: Animal Behavior Processes*, 1978, *4*, 368–98.

Silvestri, R., Rohrbaugh, M. J., & Riccio, D. C. Conditions influencing the retention of learned fear in young rats. *Developmental Psychology*, 1970, *2*, 389–95.

Singh, D., & Wickens, D. D. Disinhibition in instrumental conditioning. *Journal of Comparative and Physiological Psychology*, 1968, *66*, 557–59.

Singh, P. J., Sakellaris, P. C., & Brush, F. R. Retention of active and passive avoidance responses tested in extinction. *Learning and Motivation*, 1971, *2*, 305–23.

Skinner, B. F. Are theories of learning necessary? *Psychological Review*, 1950, *57*, 193–216.

Skinner, B. F. Behaviorism at fifty. *Science*, 1963, *140*, 951–58.

Slotnick, B. M., & Katz, H. M. Olfactory learning-set formation in rats. *Science*, 1974, *185*, 796–98.

Sluckin, W. *Imprinting and early Learning*. London: Methuen, 1964.

Smith, M. C. CS-US interval and US intensity in classical conditioning of the rabbit's nictitating membrane responses. *Journal of Comparative and Physiological Psychology*, 1968, *66*, 679–87.

Smith, M. C., Coleman, S. R., & Gormezano, I. Classical conditioning of the rabbit's nictitating membrane response at backward, simultaneous, and forward CS-US intervals. *Journal of Comparative and Physiological Psychology*, 1969, *69*, 226–31.

Smith, N. F. Effects of interpolated learning on the retention of an escape response in rats as a function of age. *Journal of Comparative and Physiological Psychology*, 1968, *65*, 422–26.

Smith, N. F., Misanin, J. R., & Campbell, B. A. Effect of punishment on extinction of an avoidance response: Facilitation or inhibition? *Psychonomic Science*, 1966, *4*, 271–72.

Smith, R. D., Dickson, A. L., & Sheppard, L. Review of flooding procedures (implosion) in animals and man. *Perceptual and Motor Skills*, 1973, *37*, 351–74.

Solomon, R. L. An opponent-process theory of acquired motivation: IV. The affective dynamics of addiction. In J. D. Maser & M. E. P. Seligman (Eds.), *Psychopathology: Experimental models*. San Francisco: W. H. Freeman, 1977.

Solomon, R. L., & Corbit, J. D. An opponent-process theory of motivation: I. Temporal dynamics of affect. *Psychological Review, 1974, 81,* 119–45.

Solomon, R. L., Kamin, L. J., & Wynne, L. C. Traumatic avoidance learning: The outcomes of several extinction procedures with dogs. *Journal of Abnormal and Social Psychology,* 1953, *48,* 291–302.

Solomon, R. L., & Wynne, L. C. Traumatic avoidance learning: The principles of anxiety conservation and partial irreversibility. *Psychological Review, 1954, 61,* 353–85.

Spear, N. E. Forgetting as retrieval failure. In W. K. Honig & P. H. R. James (Eds.), *Animal memory.* New York: Academic Press, 1971.

Spear, N. E. Retrieval of memory in animals. *Psychological Review, 1973, 80,* 163–94.

Spear, N. E. *The processing of memories: Forgetting and retention.* Hillsdale, N.J.: Lawrence Erlbaum, 1978.

Spear, N. E., Gordon, W. C., & Martin, P. A. Warm-up decrement as failure in memory retrieval in the rat. *Journal of Comparative and Physiological Psychology, 1973, 85,* 601–14.

Spear, N. E., Klein, S. B., & Riley, E. P. The Kamin effect as "state-dependent learning." *Journal of Comparative and Physiological Psychology, 1971, 74,* 416–25.

Spear, N. E., & Parsons, P. J. Analysis of a reactivation treatment: Ontogenetic determinants of alleviated forgetting. In D. L. Medin, W. A. Roberts, & R. T. Davis (Eds.), *Processes of animal memory.* Hillsdale, N.J.: Lawrence Erlbaum, 1976.

Spence, K. W. The nature of discrimination learning in animals. *Psychological Review, 1936, 43,* 427–49.

Spence, K. W. The role of secondary reinforcement in delayed-reward learning. *Psychological Review, 1947, 54,* 1–8.

Spence, K. W., & Norris, E. B. Eyelid conditioning as a function of the intertrial interval. *Journal of Experimental Psychology, 1950, 40,* 716–20.

Spence, K. W., & Platt, J. R. UCS intensity and performance in eyelid conditioning. *Psychological Bulletin, 1966, 65,* 1–10.

Sperling, S. E. Reversal learning and resistance to extinction: A review of the rat literature. *Psychological Bulletin, 1965, 63,* 281–97. (a)

Sperling, S. E. Reversal learning and resistance to extinction: A supplementary report. *Psychological Bulletin, 1965, 64,* 310–12. (b)

Spevack, A. A., & Suboski, M. D. Retrograde effects of electroconvulsive shock on learned responses. *Psychological Bulletin, 1969, 72,* 66–76.

Spiker, C. C. An extension of Hull-Spence discrimination learning theory. *Psychological Review, 1970, 77,* 496–515.

Springer, A. D., & Miller, R. R. Retrieval failure induced by electroconvulsive shock: Reversal with dissimilar training and recovery agents. *Science, 1972, 177,* 628–30.

Starr, M. D. An opponent-process theory of motivation: VI. Time and intensity variables in the development of separation-induced distress calling in ducklings. *Journal of Experimental Psychology: Animal Behavior Processes, 1978, 4,* 338–55.

Staveley, H. E. Effect of escape duration and shock intensity on the acquisition and extinction of an escape response. *Journal of Experimental Psychology, 1966, 72,* 698–703.

St. Claire-Smith, R. The overshadowing of instrumental conditioning by a stimulus that predicts reinforcement better than the response. *Animal Learning and Behavior, 1979, 7,* 224–28.

Steranka, L. R., & Barrett, R. J. Kamin effect in rats: Differential retention or differential acquisition of an active-avoidance response? *Journal of Comparative and Physiological Psychology, 1973, 85,* 324–30.

Still, A. W. Memory and spontaneous alternation in the rat. *Nature, 1966, 210,* 401–2.

Stimmel, D. T., & Adams, P. C. The magnitude of the frustration effect as a function of the number of previously reinforced trials. *Psychonomic Science, 1969, 16,* 31–32.

Stoffer, G. R., & Zimmerman, R. R. Airblast avoidance learning sets in rhesus monkeys. *Animal Learning and Behavior,* 1973, *1,* 211–14.

Suiter, R. D., & LoLordo, V. M. Blocking of inhibitory Pavlovian conditioning in the conditioned emotional response procedure. *Journal of Comparative and Physiological Psychology,* 1971, *76,* 137–44.

Sutherland, N. S., & Mackintosh, N. J. *Mechanisms of animal discrimination learning.* New York: Academic Press, 1971.

Sytsma, D., & Dyal, J. A. Effects of varied reward schedules on generalized persistence of rats. *Journal of Comparative and Physiological Psychology,* 1973, *85,* 179–85.

Szakmary, G. A. Second-order conditioning of the conditioned emotional response: Some methodological considerations. *Animal Learning and Behavior,* 1979, *7,* 181–84.

Tapp, J. T., Zimmerman, R. S., & D'Encarnacao, P. S. Intercorrelational analysis of some common measures of rat activity. *Psychological Reports,* 1968, *23,* 1047–50.

Tarpy, R. M. Reinforcement difference limen (RDL) for delay in shock escape. *Journal of Experimental Psychology,* 1969, *79,* 116–21.

Tarpy, R. M., & Koster, E. D. Stimulus facilitation of delayed-reward learning in the rat. *Journal of Comparative and Physiological Psychology,* 1970, *71,* 147–51.

Tarpy, R. M., & Sawabini, F. L. Reinforcement delay: A selective review of the last decade. *Psychological Bulletin,* 1974, *81,* 984–97.

Taylor, G. T. Stimulus change and complexity in exploratory behavior. *Animal Learning and Behavior,* 1974, *2,* 115–18.

Teghtsoonian, R., & Campbell, B. A. Random activity of the rat during food deprivation as a function of environmental conditions. *Journal of Comparative and Physiological Psychology,* 1960, *53,* 242–44.

Terlecki, L. J., Pinel, J. P. J., & Treit, D. Conditioned and unconditioned defensive burying in the rat. *Learning and Motivation,* 1979, *10,* 337–50.

Terrace, H. S. Discrimination learning with and without "errors." *Journal of the Experimental Analysis of Behavior,* 1963, *6,* 1–27.

Terrace, H. S., Gibbon, J., Farrell, L., & Baldock, M. D. Temporal factors influencing the acquisition and maintenance of an autoshaped key peck. *Animal Learning and Behavior.* 1975, *3,* 53–62.

Terry, W. S. & Wagner, A. R. Short-term memory for "surprising" versus "expected" unconditioned stimuli in Pavlovian conditioning. *Journal of Experimental Psychology: Animal Behavior Processes,* 1975, *1,* 122–33.

Testa, T. J. Causal relationships and the acquisition of avoidance responses. *Psychological Review,* 1974, *81,* 491–505.

Testa, T. J. Effects of similarity of location and temporal intensity pattern of conditioned and unconditioned stimuli on the acquisition of conditioned suppression in rats. *Journal of Experimental Psychology: Animal Behavior Processes,* 1975, *1,* 114–21.

Theios, J. The partial reinforcement sustained through blocks of continuous reinforcement. *Journal of Experimental Psychology,* 1962, *64,* 1–6.

Theios, J. Drive stimulus generalization increments. *Journal of Comparative and Physiological Psychology,* 1963, *56,* 691–95.

Theios, J., & Blosser, D. Overlearning reversal effect and magnitude of reward. *Journal of Comparative and Physiological Psychology,* 1965, *59,* 252–57.

Theios, J., Lynch, A. D., & Lowe, W. F. Differential effects of shock intensity on one-way and shuttle avoidance conditioning. *Journal of Experimental Psychology,* 1966, *72,* 294–99.

Theios, J., & McGinnis, R. W. Partial reinforcement before and after continuous reinforcement. *Journal of Experimental Psychology,* 1967, *73,* 479–81.

Thomas, D. R. The effects of drive and discrimination training on stimulus generalization. *Journal of Experimental Psychology,* 1962, *64,* 24–28.

Thomas, D. R., & Lopez, L. J. The effects of delayed testing on generalization slope. *Journal of Comparative and Physiological Psychology,* 1962, *55,* 541–44.

Thomas, E., & Wagner, A. R. Partial reinforcement of the classically conditioned eyelid response in the rabbit. *Journal of Comparative and Physiological Psychology,* 1964, *58,* 157–58.

Thompson, R., & Dean, W. A further study on the retroactive effect of electroconvulsive shock. *Journal of Comparative and Physiological Psychology,* 1955, *48,* 488–91.

Thompson, R. F. Sensory preconditioning. In R. F. Thompson & J. S. Voss (Eds.), *Topics in Learning and Performance.* New York: Academic Press, 1972.

Thorndike, E. L. Animal intelligence. An experimental study of the associative process in animals. *Psychological Monographs,* 1898, *2,* No. 8.

Thornton, J. W., & Jacobs, P. D. Learned helplessness in human subjects. *Journal of Experimental Psychology,* 1971, *87,* 367–72.

Thornton, J. W., & Powell, G. D. Immunization to and alleviation of learned helplessness in man. *American Journal of Psychology,* 1974, *87,* 351–67.

Tolman, E. C., & Honzik, C. H. Introduction and removal of reward and maze performance in rats. *University of California Publications in Psychology,* 1930, *4,* 257–75.

Tombaugh, J. W., & Tombaugh, T. N. Effects of delay of reinforcement and cues upon acquisition and extinction performance. *Psychological Reports,* 1969, *25,* 931–34.

Tombaugh, T. N. Resistance to extinction as a function of the interaction between training and extinction delays. *Psychological Reports,* 1966, *19,* 791–98.

Tombaugh, T. N. A comparison of the effects of immediate reinforcement, constant delay of reinforcement and partial delay of reinforcement on performance. *Canadian Journal of Psychology,* 1970, *24,* 276–88.

Tombaugh, T. N., & Tombaugh, J. W. Effects on performance of placing a visual cue at different temporal locations within a constant delay interval. *Journal of Experimental Psychology,* 1971, *87,* 220–24.

Tomie, A., Murphy, A. L., Fath, S., & Jackson, R. L. Retardation of autoshaping following pretraining with unpredictable food: Effects of changing the context between pretraining and testing. *Learning and Motivation,* 1980, *11,* 117–34.

Tortora, D. F., & Denny, M. R. Flooding as a function of shock level and length of confinement. *Learning and Motivation,* 1973, *4,* 276–83.

Tracy, W. K. Wavelength generalization and preference in monochromatically reared ducklings. *Journal of the Experimental Analysis of Behavior,* 1970, *13,* 163–78.

Trapold, M. A. Are expectancies based upon different positive reinforcing events discriminably different? *Learning and Motivation,* 1970, *1,* 129–40.

Trapold, M. A., & Fowler, H. Instrumental escape performance as a function of the intensity of noxious stimulation. *Journal of Experimental Psychology,* 1960, *60,* 323–26.

Traupmann, K. L. Drive, reward, and training parameters, and the overlearning-extinction effect (OEE). *Learning and Motivation,* 1972, *3,* 359–68.

Treichler, F. R., & Hall, J. F. The relationship between deprivation weight loss and several measures of activity. *Journal of Comparative and Physiological Psychology,* 1962, *55,* 346–49.

Turney, T. H. The easy-to-hard effect: Transfer along the dimension of orientation in the rat. *Animal Learning and Behavior,* 1976, *4,* 363–66.

Uhl, C. N. Eliminating behavior with omission and extinction after varying amounts of training. *Animal Learning and Behavior,* 1973, *1,* 237–40.

Uhl, C. N., & Garcia, E. E. Comparison of omission with extinction in response elimination in rats. *Journal of Comparative and Physiological Psychology,* 1969, *69,* 554–62.

Ulrich, R. E., & Azrin, N. H. Reflexive fighting in response to aversive stimulation. *Journal of the Experimental Analysis of Behavior,* 1962, *5,* 511–20.

Vardaris, R. M. Partial reinforcement and extinction of heart rate deceleration in rats with the US interpolated on nonreinforced trials. *Learning and Motivation,* 1971, *2,* 280–88.

Vardaris, R. M., & Fitzgerald, R. R. Effects of partial reinforcement on a classically conditioned eyeblink response in dogs. *Journal of Comparative and Physiological Psychology,* 1969, *67,* 531–34.

Verplanck, W. S., & Hayes, J. R. Eating and drinking as a function of maintenance schedule. *Journal of Comparative and Physiological Psychology,* 1953, *46,* 327–33.

Wagner, A. R. Effects of amount and percentage of reinforcement and number of acquisition trials on conditioning and extinction. *Journal of Experimental Psychology,* 1961, *62,* 234–42.

Wagner, A. R. Frustrative nonreward: A variety of punishment? In B. A. Campbell & R. A. Church (Eds.), *Punishment and aversive behavior.* New York: Appleton-Century-Crofts, 1969. (a)

Wagner, A. R. Stimulus selection and a "modified continuity theory." In G. H. Bower & J. T. Spence (Eds.), *The psychology of learning and motivation* (Vol. 3). New York: Academic Press, 1969. (b)

Wagner, A. R., Rudy, J. W., & Whitlow, J. W. Rehearsal in animal conditioning. *Journal of Experimental Psychology Monograph,* 1973, *97,* No. 3, 407–26.

Wagner, A. R., Siegel, S., Thomas, E., & Ellison, G. D. Reinforcement history and the extinction of a conditioned salivary response. *Journal of Comparative and Physiological Psychology,* 1964, *58,* 354–58.

Waller, T. G. Effects of irrelevant cues on discrimination acquisition and transfer in rats. *Journal of Comparative and Physiological Psychology,* 1971, *73,* 477–80.

Waller, T. G. Effect of consistency of reward during runway training on subsequent discrimination performance in rats. *Journal of Comparative and Physiological Psychology,* 1973, *83,* 120–23. (a)

Waller, T. G. The effect of overtraining on a visual discrimination on transfer to a spatial discrimination. *Animal Learning and Behavior,* 1973, *1,* 65–67. (b)

Wang, G. H. Relation between "spontaneous" activity and oestrus cycle in the white rat. *Comparative Psychological Monograph,* 1923, *2,* No. 6.

Warden, C. J. *Animal motivation: Experimental studies on the albino rat.* New York: Columbia University Press, 1931.

Warren, J. M. Primate learning in comparative perspective. In A. M. Schrier, H. F. Harlow, & F. Stollnitz (Eds.), *Behavior of nonhuman primates: Modern research trends.* New York: Academic Press, 1965.

Warren, J. M. Irrelevant cues and shape discrimination learning by cats. *Animal Learning and Behavior,* 1976, *4,* 22–24.

Wasserman, E. A., & Molina, E. J. Explicitly unpaired key light and food presentations: Interference with subsequent autoshaped key pecking in pigeons. *Journal of Experimental Psychology: Animal Behavior Processes,* 1975, *1,* 30–38.

Weisinger, R. S., & Woods, S. C. Formalin-like sodium appetite and thirst elicited by a conditioned stimulus in rats. *Journal of Comparative and Physiological Psychology,* 1972, *80,* 413–21.

Weisman, R. G., & Litner, J. S. Positive conditioned reinforcement of Sidman avoidance behavior in rats. *Journal of Comparative and Physiological Psychology,* 1969, *68,* 597–603. (a)

Weisman, R. G., & Litner, J. S. The course of Pavlovian excitation and inhibition of fear in rats. *Journal of Comparative and Physiological Psychology,* 1969, *69,* 667–72. (b)

Weisman, R. G., & Palmer, J. A. Factors influencing inhibitory stimulus control: Discrimination training and prior nondifferential reinforcement. *Journal of the Experimental Analysis of Behavior,* 1969, *12,* 229–37.

Weiss, K. M., & Strongman, K. T. Shock-induced response bursts and suppression. *Psychonomic Science*, 1969, *15*, 238–40.

Welker, R. L., & McAuley, K. Reductions in resistance to extinction and spontaneous recovery as a function of changes in transportational and contextual stimuli. *Animal Learning and Behavior*, 1978, *6*, 451–57.

Welker, R. L., & Wheatley, K. L. Differential acquisition of conditioned suppression in rats with increased and decreased luminance levels as CS+s. *Learning and Motivation*, 1977, *8*, 247–62.

White, C. T., & Schlosberg, H. Degree of conditioning of the GSR as a function of the period of delay. *Journal of Experimental Psychology*, 1952, *43*, 357–62.

Wielkiewicz, R. M. Effects of CSs for food and water upon rats bar pressing for different magnitudes of food reinforcement. *Animal Learning and Behavior*, 1979, *7*, 246–50.

Wike, E. L., & Atwood, M. E. The effects of sequences of reward magnitude, delay, and delay-box confinement upon runway performance. *Psychological Record*, 1970, *20*, 51–56.

Wike, E. L., Cour, C., & Mellgren, R. L. Establishment of a learned drive with hunger. *Psychological Reports*, 1967, *20*, 143–45.

Wike, E. L., & King, D. D. Sequences of reward magnitude and runway performance. *Animal Learning and Behavior*, 1973, *1*, 175–78.

Wike, E. L., Mellgren, R. L., & Wike, S. S. Runway performance as a function of delayed reinforcement and delay-box confinement. *Psychological Record*, 1968, *18*, 9–18.

Wilcoxon, H. C., Dragoin, W. B., & Kral, P. A. Illness-induced aversions in rat and quail: Relative salience of visual and gustatory cues. *Science*, 1971, *171*, 826–28.

Wilkie, D. M. Behavioral contrast produced by a signaled decrease in local rate of reinforcement. *Learning and Motivation*, 1977, *8*, 182–93.

Williams, B. A. Information effects on the response-reinforcer association. *Animal Learning and Behavior*, 1978, *6*, 371–79.

Williams, B. A. Contrast, component duration, and the following schedule of reinforcement. *Journal of Experimental Psychology: Animal Behavior Processes*, 1979, *5*, 379–96.

Williams, J. L., & Maier, S. F. Transituational immunization and therapy of learned helplessness in the rat. *Journal of Experimental Psychology: Animal Behavior Processes*, 1977, *3*, 240–53.

Williams, S. B. Resistance to extinction as a function of the number of reinforcements. *Journal of Experimental Psychology*, 1938, *23*, 506–22.

Willis, R. D. The partial reinforcement of conditioned suppression. *Journal of Comparative and Physiological Psychology*, 1969, *68*, 289–95.

Willner, J. A. Blocking of a taste aversion by prior pairings of exteroceptive stimuli with illness. *Learning and Motivation*, 1978, *9*, 125–40.

Wilson, G. T. Counterconditioning versus forced exposure in extinction of avoidance responding and conditioned fear in rats. *Journal of Comparative and Physiological Psychology*, 1973, *82*, 105–14.

Wilson, J. J. Level of training and goal-box movements as parameters of the partial reinforcement effect. *Journal of Comparative and Physiological Psychology*, 1964, *57*, 211–13.

Wilson, R. S. Cardiac response: Determinants of conditioning. *Journal of Comparative and Physiological Psychology Monograph*, 1969, *68*, (Pt. 2).

Wolfe, J. B. The effect of delayed reward upon learning in the white rat. *Journal of Comparative Psychology*, 1934, *17*, 1–21.

Wong, P. T. P. A behavioral field approach to instrumental learning in the rat: I. Partial reinforcement and sex difference. *Animal Learning and Behavior*, 1977, *5*, 5–13.

Wong, P. T. P. A behavioral field approach to general activity: Sex differences and food deprivation in the rat. *Animal Learning and Behavior*, 1979, *7*, 111–18.

Woods, P. J., & Bolles, R. C. Effects of current hunger and prior eating habits on exploratory behavior. *Journal of Comparative and Physiological Psychology*, 1965, *59*, 141–43.

Woods, P. J., Davidson, E. H., & Peters, R. J. Instrumental escape conditioning in a water tank: Effects of variations in drive stimulus intensity and reinforcement magnitude. *Journal of Comparative and Physiological Psychology*, 1964, *57*, 466–70.

Woods, S. C. Conditioned hypoglycemia. *Journal of Comparative and Physiological Psychology*, 1976, *90*, 1164–68.

Woods, S. C., Makous, W., & Hutton, R. A. Temporal parameters of conditioned hypoglycemia. *Journal of Comparative and Physiological Psychology*, 1969, *69*, 301–7.

Worsham, R. W. Temporal discrimination factors in the delayed matching-to-sample task in monkeys. *Animal Learning and Behavior*, 1975, *3*, 93–97.

Worsham, R. W., & D'Amato, M. R. Ambient light, white noise, and monkey vocalization as sources of interference in visual short-term memory of monkeys. *Journal of Experimental Psychology*, 1973, *99*, 99–105.

Wright, J. H. Test for a learned drive based on the hunger drive. *Journal of Experimental Psychology*, 1965, *70*, 580–84.

Wynne, J. D., & Brogden, W. J. Supplementary report: Effect upon sensory preconditioning of backward, forward, and trace preconditioning training. *Journal of Experimental Psychology*, 1962, *64*, 422–23.

Yelen, D. Magnitude of the frustration effect and number of training trials. *Psychonomic Science*, 1969, *15*, 137–38.

Young, G. A., & Black, A. H. A comparison of operant licking and lever pressing in the rat. *Learning and Motivation*, 1977, *8*, 387–403.

Zahorik, D. M. Conditioned physiological changes associated with learned aversions to tastes paired with Thiamine deficiency in the rat. *Journal of Comparative and Physiological Psychology*, 1972, *79*, 189–200.

Zahorik, D. M., & Bean, C. A. Resistance of "recovery" flavors to later association with illness. *Bulletin of Psychonomic Society*, 1975, *6*, 309–12.

Zahorik, D. M., & Maier, S. F. Appetitive conditioning with recovery from Thiamine deficiency as the unconditioned stimulus. *Psychonomic Science*, 1969, *17*, 309–10.

Zahorik, D. M., Maier, S. F., & Pies, R. W. Preferences for tastes paired with recovery from Thiamine deficiency in rats: Appetitive conditioning or learned safety? *Journal of Comparative and Physiological Psychology*, 1974, *87*, 1083–91.

Zamble, E. Conditioned motivation patterns in instrumental responding of rats. *Journal of Comparative and Physiological Psychology*, 1969, *69*, 536–43.

Zamble, E. Pavlovian appetitive conditioning under curare in rats. *Animal Learning and Behavior*, 1974, *2*, 101–5.

Zaretsky, H. H. Runway performance during extinction as a function of drive and incentive. *Journal of Comparative and Physiological Psychology*, 1965, *60*, 463–64.

Zentall, T. R., & Hogan, D. E. Short-term proactive inhibition in the pigeon. *Learning and Motivation*, 1977, *8*, 367–86.

Ziff, D. R., & Capaldi, E. J. Amytal and the small trial partial reinforcement effect: Stimulus properties of early trial nonrewards. *Journal of Experimental Psychology*, 1971, *87*, 263–69.

Zimmer-Hart, C. L., & Rescorla, R. A. Extinction of Pavlovian conditioned inhibition. *Journal of Comparative and Physiological Psychology*, 1974, *86*, 837–45.

Author Index

Abbott, B., 336
Adamec, R., 75
Adams, P. C., 215
Adelman, H. M., 216
Agee, C. M., 328
Ahlers, R. H., 174
Albin, R. W., 340
Alek, M., 81, 82
Alexander, J. H., 346
Allen, J. D., 106, 300
Allison, J., 61
Altenor, A., 342
Ames, H., 308
Amsel, A., 214, 215, 226, 227
Anastasio, E. J., 207
Anderson, D. C., 78, 346
Anderson, H. H., 219
Andrews, E. A., 58, 59, 64, 172
Anger, D., 138–39
Anisman, H., 245, 246, 344, 345
Annau, Z., 63, 148
Appel, J. B., 98
Armstrong, G. D., 274
Ashe, J. H., 64, 65
Atkinson, C. A., 89
Atwood, M. E., 205
Ayres, C. E., 295
Ayres, J. J. B., 54, 55, 56, 87, 196, 202
Azrin, N. H., 336

Babb, H., 335
Bacon, W. E., 221
Badia, P., 57, 336
Baenninger, R., 336
Baker, A. G., 88, 175, 346
Baker, T. B., 89
Balagura, S., 306, 307
Baldock, M. D., 69
Barker, L. M., 60, 166
Barnes, G. W., 105
Baron, A., 104, 105
Barrett, J. E., 183, 184
Barrett, R. J., 246
Barry, H., 300
Bartlett, F., 277
Bartosiak, R. S., 69
Bartter, W. D., 102
Bateson, P. P. G., 183, 184, 185
Bath, K., 202
Batsche, C. J., 204
Batsche, G. M., 204
Batson, J. D., 89, 90
Bauer, R. H., 100
Baum, M., 326, 328, 329, 331
Baum, W. M., 125, 126, 127
Baumeister, A., 299
Beagley, G., 340

Bean, C. A., 165
Beatty, J., 123
Beauchamp, R. D., 98
Beck, S. B., 59
Bedarf, E. W., 174
Beecroft, R. S., 53
Beery, R. G., 120
Behar, I., 289
Beier, E. M., 335
Bekhterev, V. M., 26, 35–36, 46
Bellingham, W. P., 169
Bellugi, U., 188–89
Bender, L., 334
Benedict, J. O., 55
Benson, H., 124
Berk, A. M., 236
Berlyne, D. E., 105
Berman, J. S., 327, 328
Berman, R. F., 89
Bernhardt, T. P., 274
Bersh, P. J., 327, 330, 331
Bertsch, G. J., 215
Besley, S., 258
Best, M. R., 61, 79, 80, 166, 168, 175
Best, P. J., 61, 89, 90, 168, 174
Bierley, C. M., 242
Bilder, B. H., 319
Biller, J. D., 120
Bintz, J., 245, 247
Birnbaum, D. W., 207
Bitterman, M. E., 207
Black, A. H., 52, 57, 123, 302, 343
Blanchard, D. C., 102, 178, 180, 318
Blanchard, E. B., 124
Blanchard, R., 87
Blanchard, R. J., 178, 180, 318
Bloomquist, A. J., 289
Blosser, D., 282
Blough, D. S., 269
Bolles, R. C., 21, 42, 43, 98, 102, 138, 140,
    142, 147, 150, 159, 175, 177, 178, 202,
    203, 212, 295, 298, 299, 300, 304
Bond, N., 60, 70
Booth, D. A., 307
Bootzin, R. R., 339
Boren, J. J., 100
Bouton, M. E., 212
Bouzas, A., 166
Bower, G. H., 140, 141, 215
Bowlby, J., 186
Bracewell, R. J., 343
Brady, J. V., 123
Braud, W., 342
Braud, W. G., 245
Braveman, N. S., 58, 59, 64, 172
Breland, K., 159, 160
Breland, M., 159, 160

Brener, J., 123, 178, 179
Brennan, A. F., 123, 124
Brodigan, D. L., 274
Brogden, W. J., 74
Bronowski, M., 188
Bronstein, P. M., 319
Brooks, C. I., 120, 215, 227
Brower, L. P., 176
Brown, B. L., 261
Brown, J. S., 245, 332, 333, 334, 335
Brown, R. T., 215
Brown, W. L., 207
Brownstein, A. J., 126
Brush, F. R., 245, 248
Bryan, R. G., 246
Buchanan, D. C., 336
Bull, J. A., 144, 150, 151, 152
Burdick, C. K., 194, 195
Burkhardt, P. E., 54
Burr, D. E. S., 262
Burstein, K. R., 56
Butler, B., 262
Butler, C. S., 66
Butler, R. A., 317

Camp, D. S., 98, 104
Campbell, B. A., 98, 105, 249–50, 251, 253, 298, 299, 334, 359
Campbell, C. E., 107
Campbell, E. H., 249–50
Campbell, P. E., 204, 207
Cannon, D. S., 89
Cantor, M. B., 159
Capaldi, E. D., 225, 301, 302
Capaldi, E. J., 219, 220, 221, 225, 226, 227
Caplan, H. J., 109, 110
Carlson, J. G., 275, 276
Carlson, N. J., 52
Carosio, L. A., 334
Casey, J., 127
Catania, A. C., 113, 126
Caul, W. F., 316, 336
Cebullà, R. P., 336
Chambliss, D. J., 69
Charnov, E. L., 97
Chesler, P., 277
Chiaia, N. L., 57
Chomsky, N. A., 187
Christoph, G., 199
Chung, S., 126
Church, R. M., 43, 56, 98, 109, 312, 313, 314
Cicala, G. A., 299
Clark, R. L., 115
Clarke, C. M., 58, 166
Clarke, J. C., 172
Clements, M., 183
Coate, W. B., 261
Colby, J. J., 197
Coleman, S. R., 57
Collier, A. C., 102, 178, 250, 251
Collier, G., 300
Collyer, R., 172
Connally, S. R., 123

Cook, P. E., 206
Cook, R. G., 242
Coppinger, L. L., 176
Corbit, J. D., 312, 315
Corcoran, M. E., 245
Corson, J. A., 277
Coughlin, R. C., 221
Coulson, G., 203
Coulson, V., 203
Coulter, X., 249, 250, 251, 330
Cour, C., 306
Cox, J. K., 239, 243
Crabtree, J. M., 336
Crandall, C., 175
Cravens, R. W., 306
Crawford, M., 102, 331
Crider, A., 123
Cromwell, R., 299
Cross, H. A., 284
Crowell, C. R., 78, 274, 346
Cruce, J. A. F., 236
Crum, J., 277
Cunningham, C. L., 346

Daly, H. B., 215, 216
D'Amato, M. R., 123, 140, 177, 239, 240, 241, 243, 244, 282, 305, 306
Darwin, Charles, 3–4
Daston, A. P., 69
Davenport, J. W., 296
Davidson, E. H., 98
Davidson, T. L., 302
Davis, H., 151
Davis, R. T., 238
Davison, M. C., 127
Dean, W., 234, 235
Deane, G. E., 57
Dearing, M. F., 66
deCatanzaro, D., 344, 345
DeCosta, M. J., 196, 202
Deets, A. C., 289
Defran, R. H., 57
Delprato, D. J., 177, 334
Del Russo, J. E., 277
Del Valle, R., 319
Dember, W. N., 319
D'Encarnacao, P. S., 299
Dennis, M., 277
Denny, M. R., 282, 329
DePaulo, P., 183, 184
Desiderato, O., 262, 336
de Toledo, L., 63, 123
Deutsch, E. A., 220
Deutsch, R., 60, 311
de Villiers, P. A., 125, 126, 127, 302
Devine, J. V., 240, 288
Deweer, B., 238
Dews, P. B., 114
Dexter, W. R., 81, 82
Dickinson, A., 66, 87, 88, 152
Dickson, A. L., 326
Digiusto, E., 60
Domber, A., 144, 153
Domjam, M., 52–53, 84, 166, 167
Donahoe, J. W., 109

Dorsky, N. P., 107
Douglas, R. J., 319
Douglass, W. K., 102
Dragoin, W. B., 60, 169, 170
Dreyer, P., 334, 335
Duncan, P. M., 150, 159
Dunham, P. J., 118
Dunn, T., 42, 43
Dworkin, B. R., 123
Dworkin, T., 319
Dyal, J. A., 204, 207, 221, 222, 223
Dyck, D. G., 118, 119, 221

Eckerman, D. A., 115
Egger, M. D., 55–56
Eibl-Eibesfeldt, I., 160
Eimas, P. D., 285
Eisenberger, R., 105, 121, 317
Eiserer, L. A., 183
Ellison, G. D., 63
Erichsen, J. T., 97
Ervin, F. R., 58, 165, 166, 172
Etkin, M., 140
Evans, S. H., 115

Fago, G. C., 144, 153
Fallon, D., 215
Farrar, C. H., 66, 100
Farrell, L., 69
Farthing, G. W., 258
Fath, S., 343
Fawcett, J. T., 57
Fazzaro, J., 140, 177
Fazzini, D., 105
Feigley, D. A., 250
Feldman, D. T., 238
Felton, M., 111, 112
Ferster, C. B., 113
Filby, Y., 98
Fillit, H., 308
Fischer, G., 209
Fiske, D. W., 318
Fitzgerald, R. D., 57, 63
Fitzgerald, R. R., 67, 218
Flaherty, C. F., 120
Fouts, R. S., 187
Fowler, H., 98, 103, 104, 105, 144, 153,
  318, 319
Franchina, J. J., 98–99, 328
Frank, M., 121
Freedman, J. L., 258
Freedman, P. E., 101
Frey, P. W., 66, 72, 87
Friedman, H., 261, 265
Fuller, G. D., 124
Furrow, D. R., 72

Gaioni, S., 183, 184, 185
Galef, B. G., 165, 172
Ganz, L., 263
Garcia, J., 26, 58, 165, 166, 167, 168, 172,
  199, 200
Gardner, B. T., 187
Gardner, L., 203
Gardner, R. A., 187

Gemberling, G. A., 79
Gerry, J. E., 263
Ghiselli, W. B., 144, 153
Gibbon, J., 69
Gilbert, R. M., 257
Gill, T. V., 188
Gillan, D. J., 84, 176
Gillette, K., 169, 170, 171, 172
Gino, A., 331
Gipson, M., 207
Giulian, D., 177
Glazer, H. I., 344
Glazier, S. C., 176
Gliner, J. A., 336
Glotfelty, R. A., 215
Godbout, R. C., 219
Goesling, W. J., 123, 178, 179
Gold, L., 69
Goldman, J. A., 120
Gollub, L. R., 126
Goodkin, F., 342
Gordon, W. C., 238
Gormezano, I., 54, 57, 63
Gottlieb, G., 183, 185
Graft, D. A., 127
Grant, D. S., 239, 240, 241, 242, 243
Gray, J. A., 59, 227
Green, K. F., 166
Green, L., 166
Greer, J., 340
Grice, G. R., 59, 103, 261
Griffin, R. W., 274
Groner, D., 101
Grossen, N. E., 140, 147, 150, 151, 178,
  179, 202
Guttman, N., 63, 258, 259, 261, 265
Gynther, M. D., 278

Haber, A., 261
Habley, P., 207
Hager, J., 96, 159
Hake, D. F., 336
Halgren, C. R., 80
Hall, G., 44, 87, 196, 284
Hall, J. F., 57, 258, 299
Hallgren, S. O., 215
Hamilton, B., 127
Hammond, L. T., 29, 30, 150
Hanson, H. M., 265, 266, 267, 268
Hargrave, G. E., 150
Harland, W., 70
Harlow, H. F., 286, 287, 288, 289
Harris, A. H., 123
Harrison, R. H., 258, 265
Hars, B., 238
Harsh, J., 336
Hart, D., 221
Hartman, T. F., 67
Hastings, S. E., 57
Hause, J., 207
Hauser, P. J., 328
Hawkins, W., 299
Hawkins, W. G., 26, 58
Hayes, C., 187
Hayes, J. R., 303

Hayes, R. C., 336
Hayward, L., 175
Hearst, E., 199, 258, 261, 262, 272, 273, 274
Hebb, D. O., 306
Hemmes, N. S., 115
Hennessy, J. W., 101
Herrnstein, R. J., 100, 125, 126, 127, 138, 142
Herzog, H. L., 240
Hess, E. H., 181, 182, 183
Heth, C. D., 54
Hetrick, M., 308
Hickis, C. F., 238
Hill, W., 42, 43
Hill, W. F., 220
Hilton, A., 218
Hinde, R. A., 159
Hinderliter, C. F., 250
Hineline, P. N., 142
Hinson, R. E., 52, 75
Hiroto, D. S., 339
Hisson, H., 336
Hoffman, H. S., 183, 184, 185, 269, 316
Hogan, D. E., 243
Holland, P. C., 61, 67, 72, 73, 207
Holland, V., 127
Holman, E. W., 318
Holmes, P. A., 177
Holmstrom, L. S., 166
Holtz, R., 42, 43
Hom, H. L., 335
Homzie, M. J., 212
Honig, W. K., 87, 238, 263, 265
Honzik, C. H., 14, 15
Hooper, R., 282
Host, K. C., 57
Hovancik, J. R., 302
Hsiao, S., 303
Hug, J. J., 227
Hull, C. L., 278, 295, 296, 297
Hulse, S. H., 107
Hutchinson, R. R., 336
Hutton, R. A., 309, 310
Hyde, T. S., 166
Hymowitz, N., 336

Imada, H., 56
Irwin, J., 172, 344
Ison, J. R., 206, 227
Iverson, S. D., 317
Ives, C. E., 308

Jackson, B., 98
Jackson, R. L., 338, 343, 346, 347
Jacobs, P. D., 339
Jagoda, E., 236
Jakubczak, L. F., 298
James, J. J. B., 194, 195
James, P. H. R., 238
Jaynes, J., 249, 253
Jenkins, H. M., 68, 258, 265, 273, 274
Jensen, D. D., 61
Job, W. M., 177
John, E. R., 277

Johnson, N., 209
Johnson, P. E., 79
Jones, D. R., 174, 175
Jones, L. C., 240
Jones, R. B., 105

Kacelnik, A., 97
Kalat, J. W., 58, 60, 61, 164, 172, 173, 174
Kalish, H. I., 194, 257, 258, 259, 261
Kamil, A. C., 70, 289
Kamin, L. J., 55, 59, 63, 83, 87, 105, 138, 139, 140, 148, 244, 245, 325, 326
Kaplan, M., 98
Karpicke, J., 109, 110, 199
Katkin, E. S., 123
Katz, H. M., 288
Katzev, R. D., 327, 328
Kaufman, A., 105
Kay, E., 342
Keehn, J. D., 139
Keiser, E. F., 251
Keith-Lucas, T., 54, 63
Keller, J. V., 126
Keller, R. J., 55
Kelley, M. J., 178, 179, 318
Kello, J. E., 110
Keltz, J. R., 327
Kendall, S. B., 109
Kerr, L. M., 263, 264
Killeen, P., 111, 114
Kimble, G. A., 300
Kimmel, H. D., 76, 123, 124
King, D. D., 205, 206
Kintsch, W., 96
Kirby, R. H., 250
Kirschbaum, E. H., 168
Kish, G. B., 105
Klare, W. F., 334
Klatt, D. H., 187
Klein, S. B., 89, 247, 248, 249
Klein, S., 184
Klipec, W. D., 269
Knight, J., 209
Knouse, S. B., 207, 208
Koelling, R. A., 58, 165, 166, 167, 168, 172
Kohler, E. A., 87
Kohn, B., 277
Koppenaal, R. J., 236
Koresko, M. B., 261, 262
Korn, J. H., 100
Kostansek, D. J., 150, 151
Koster, E. D., 103
Kozak, M. J., 340, 341, 342
Kozman, F., 185
Kraeling, D., 96, 300
Kral, P. A., 169, 170
Kramer, T. J., 115
Krebs, J. R., 97, 98
Kremer, E. F., 55, 88
Kreuter, C., 151
Ksir, C. J., 80
Kulkarni, A. S., 177

La Barbera, J. D., 316
Lachman, R., 189
Lamb, E. O., 303
Lambert, R. W., 57
Lanier, A. T., 219
Largen, J., 120
Larkin, S., 305
Larsen, J. D., 166
Larson, D., 61
Lashley, K. S., 265
Laties, V. C., 115
Lauer, D. W., 63
Lavin, M. J., 74
Lawrence, D. H., 270
Lawry, J. A., 152
Lazarovitz, L., 140, 141
Lea, S. E. G., 97, 127
Leach, D. A., 301
Leard, B., 89
LeBedda, J. M., 289
Lefebvre, M. F., 200
Leitenberg, H., 200, 201, 202, 215
Leith, C. R., 241
Leith, N. J., 246
Lenderhendler, I., 329
Lenneberg, E. H., 189
Leonard, D. W., 218, 219
Lett, B. T., 172
Leuin, T. C., 263
Levine, M., 289
Levine, S., 100, 248
Levinthal, C. F., 57
Levis, D. J., 330, 343
Lewis, D. J., 95, 221, 234,
   236
Lewis, R. W., 335
Libby, M. E., 56
Lieberman, P. H., 187
Lien, J., 183
Linden, D. R., 215
Lindsey, G. P., 61
Linton, J., 330
Litner, J. S., 147, 148, 149
Little, L., 120, 270, 285
Lloyd, I. H., 304
Locurto, C., 69
Logan, F. A., 271, 301
Logue, A. W., 166
LoLordo, V. M., 84, 89, 147, 150, 312,
   313
Long, C. J., 106, 300
Looney, T. A., 274
Lopez, L. J., 262
Lord, J., 120
Lordahl, D. S., 52
Lorenz, K. Z., 181
Lougee, M., 289
Lovejoy, E., 282
Loveland, D. H., 126
Lowe, W. F., 100
Lubow, R. E., 78, 79, 80, 81, 82
Lupo, J. V., 346
Lynch, A. D., 100, 225
Lynch, J. J., 76, 77

Lyon, D. O., 111, 112
Lyons, J., 269

Maatsch, J. L., 216
McAllister, D. E., 102, 120, 252, 262
McAllister, W. R., 102, 120, 252, 262
McAuley, K., 212, 213
McCain, G., 220
McCleery, R. H., 304
McFarland, D. J., 303, 304, 305
McGinnis, R. W., 222
McGowan, B. K., 166
McHose, J. H., 120, 122
McKearney, J. W., 203
Mackinnon, J. R., 336
Mackintosh, N. J., 81, 87, 88–89, 120,
   175, 265, 279, 282, 283, 284, 285, 337
McLeod, D. C., 123, 124
McMillan, J. C., 150
McNeill, D., 189
MacPhail, E. M., 177
Maddi, S. R., 318
Mahoney, W. J., 54, 55
Maier, S. F., 54, 166, 338, 340, 341, 342,
   345, 346
Maisiak, R., 72
Maki, W. S., 87, 241
Makous, W. L., 309, 310
Maleske, R. T., 87
Malott, M. K., 263
Malsbury, C. W., 245
Manning, A. A., 52
Marchant, H. G., 31, 84
Marler, P. A., 186
Marrazo, M. J., 331
Marsh, G., 269, 271, 279
Martin, G. K., 57
Martin, G. M., 169
Martin, L. K., 147, 148
Martin, P. A., 238
Martin, R. C., 332, 333, 335
Marwine, A. G., 300
Marx, M. H., 118, 296
Mason, W. A., 186
Masterson, F. A., 98, 102
Matthews, T. J., 44
Mauldin, J. E., 289
Mayer, J., 165
Medin, D. L., 238, 243, 286
Mednick, S. A., 258
Meinrath, M., 123
Mellgren, R. L., 118, 119, 121, 208, 221,
   222, 306
Melvin, K. B., 335
Melzack, R., 75
Merrill, H. K., 81, 82
Meyer, C., 262
Mickley, G. A., 168
Mikhail, A. A., 336
Mikulka, P. J., 89
Miles, R. C., 286
Millenson, J. R., 302
Miller, H. L., 126
Miller, K., 331
Miller, N. E., 55–56, 123, 305

Miller, R. R., 236, 237
Mineka, S., 308, 329, 330, 331
Mis, F. W., 69, 89
Misanin, J. R., 98, 105, 236, 249, 250, 251, 299, 334
Mistler-Lachman, J. L., 189
Mitchell, D., 168, 174, 319
Mochhauser, M., 153
Mock, L. A., 330
Modaresi, H. A., 102
Moe, J. C., 242
Moeser, S., 56
Molina, E. J., 343
Mollenauer, S. O., 302, 303
Moltz, H., 184
Montgomery, K. C., 317
Moore, B. A., 68
Moore, B. D., 215
Moore, J. W., 31, 54, 59, 84
Moot, S. A., 159, 202, 336
Morlock, H., 295
Morrison, G. R., 172
Morrow, M. W., 332, 333
Mostofsky, D., 258
Mountjoy, P. P., 263
Mowrer, O. H., 136–40, 143, 145
Moyer, K. E., 100
Mucha, R. F., 246
Mueller, P. W., 251
Mulick, J. A., 200
Murphy, A. L., 343
Murphy, W. M., 296
Murray, E. N., 123
Myers, A. K., 100
Myran, D. D., 328, 329

Nachman, M., 64, 65, 174, 175
Nageishi, Y., 56
Nagy, Z. M., 250, 251
Nation, J. R., 119, 121
Nettleton, N., 307
Neuringer, A. J., 111
Neville, J. W., 240
Nevin, J. A., 127
Newby, V., 183, 185
Newman, J. A., 250
Newman, J. R., 261
Norris, E. B., 69
Nottebohm, F., 186
Nowell, N. W., 105

Oakley, D. A., 54
Obrist, P. A., 57, 123
Olsen, P. L., 250
Olton, D.S., 115, 272, 319, 320
O'Neil, H. F., 335
O'Neill, W., 241
Osborne, B., 172
Osborne, S. R., 297
Ost, J.W. P., 63
Ostapoff, E. M., 263, 264
Ott, C. A., 236

Overmier, J. B., 144, 152, 153, 154, 312, 313, 345

Pacitti, W. A., 201
Pack, K., 152
Page, H. A., 330
Palmer, J. A., 258, 260
Park, D. C., 121
Parker, L. A., 168
Parsons, P. J., 237, 252
Pavlik, W. B., 300
Pavlov, I. P., 22–28, 32, 51, 76
Paynter, W. E., 330
Pearce, J. M., 44, 66, 100, 152, 179, 196
Peckham, R. H., 215
Pennes, E. S., 227
Perin, C. T., 103
Perkins, C. C., 262
Perry, R. L., 168
Peters, D. P., 122
Peters, R. J., 98
Peterson, G., 199
Peterson, G. B., 274, 275
Peterson, N., 263
Phillips, K., 123
Pies, R. W., 166
Pinel, J. P. J., 177, 245, 256
Platt, J. R., 63
Plotkin, H. C., 54
Powell, G. D., 339
Powell, R. W., 98, 203, 238
Premack, D., 118, 188
Prewitt, E. P., 75
Prochaska, J., 327
Prokasy, W. F., 57, 69, 258
Pubols, B. H., 282
Pulliam, H. R., 97
Purdy, J. E., 284
Purtle, R. B., 267
Pyke, G. H., 97

Rabin, J. S., 120, 122
Rachlin, H., 166
Rachlin, H. C., 126, 127, 166
Raich, M. S., 123, 124
Rajecki, D. W., 184
Ramirez, J. J., 57
Randall, P., 253
Randall, P. K., 104
Randich, A., 89
Rapaport, P. M., 346
Ratliff, A. R., 96, 106, 219, 220
Ratliff, R. G., 106, 219, 220
Ratner, A. M., 183, 184, 185
Rawson, R. A., 200, 201, 202
Ray, O. S., 246
Raymond, G. A., 98
Reese, E. P., 185
Reese, H. W., 286
Reid, L. S., 282
Reiss, S., 80
Remington, G., 344, 345
Renner, K. E., 102, 207, 306
Renner, K. W., 334, 335

Rescorla, R. A., 28, 40–42, 50, 54, 55, 56, 65–66, 70, 71, 72, 73, 74, 75, 80, 85, 88, 89, 147, 148, 149, 153, 154, 196, 197, 198, 202
Revusky, S. H., 168, 174
Reynierse, J. H., 202
Reynolds, G. S., 113
Reynolds, T. J., 243
Reynolds, W. F., 300
Riccio, D. C., 104, 250, 253, 330, 331
Richter, C. P., 298
Richter, M., 342
Riesen, A. H., 263
Riess, D., 66, 100, 147, 148
Rifkin, B., 79, 81, 82
Rigby, R. L., 187
Riley, A. L., 58, 150, 159, 166, 178
Riley, D. A., 241, 263
Riley, E. P., 249
Riley, J., 331
Rilling, M., 109, 110, 115
Rizley, R. C., 70, 71, 74, 75, 202
Robbins, T. W., 317
Robbins, D., 218
Roberts, W. A., 96–97, 204, 205, 238, 239, 240, 241, 242, 243
Robles, L., 238
Rodgers, W., 165
Rodgers, W. L., 165
Rohrbaugh, M., 330
Rohrbaugh, M. J., 253
Rosellini, R. A., 340, 341, 342
Rosenberg, L., 81
Ross, L. E., 52, 57, 67
Ross, S. M., 52, 57
Roth, S., 339
Roussel, J., 214
Rozin, P., 58, 60, 164, 165, 172, 173
Rubel, E. W., 263, 264
Rubinsky, H. J., 115
Rudolph, R. I., 263
Rudy, J. W., 81, 87, 221
Rumbaugh, D., 188
Rusiniak, K. W., 26, 58
Russo, D., 342
Ryan, F. J., 335
Ryerson, W. N., 176

Sadler, E. W., 27, 67
Sahakian, B. J., 317
Sainsbury, R. S., 273
St. Claire-Smith, R., 44
Sakai, D. J., 240
Sakellaris, P. C., 248
Salafia, W. R., 57, 69
Sallery, R. D., 336
Samuelson, R. J., 319, 320
Sandell, J. H., 81
Sanders, R., 118
Sara, S. J., 238
Sawabini, F. L., 102, 207
Scavio, M. J., 152
Schaub, R. E., 59
Scheuer, C., 56, 177

Schiff, D., 177, 282
Schiff, R., 327, 328
Schmaltz, L. W., 177
Schlosberg, H., 27
Schneider, B. A., 109, 111
Schneiderman, N., 52
Schnur, P., 79, 80
Schoenfeld, W. N., 108, 137, 138, 140
Schonfeld, L. I., 123, 124
Schrier, A. M., 285, 289
Schulenburg, C. J., 250
Schusterman, R. J., 289
Schwartz, B., 115, 176
Schwartz, G. E., 123, 124
Schwartz, G. F., 124
Schwarzkopf, K. H., 153, 154
Scobie, S. R., 147
Searle, J. L., 185
Sebeok, T. A., 186
Seelbach, S. E., 98
Seger, K. A., 56
Seidel, R. J., 73
Seligman, M. E. P., 56, 96, 159, 308, 338, 339, 340, 341, 342, 345
Senkowski, P. C., 215
Seraganian, P., 272
Seybert, J. A., 221
Sgro, J. A., 207, 215
Shanab, M. E., 118, 120, 207
Shapiro, D., 124
Sheafor, P. J., 53, 63
Sheffield, F. D., 298
Shepp, B. E., 285
Sheppard, L., 326
Sherman, J. E., 54
Shettleworth, S. J., 106, 159, 160, 161, 162, 163
Shimp, C. P., 114, 127
Shipley, R. H., 195, 330
Shnidman, S., 123
Shulman, R. I., 289
Sibly, R. M., 304
Sidman, M., 100, 138–39, 261
Siegel, S., 52, 63, 75, 78, 174, 307, 311
Silberberg, A., 127
Silvestri, R., 253
Singer, D., 183
Singh, D., 111
Singh, P. J., 248
Skeen, L. C., 335
Skinner, B. F., 11, 113
Skucy, J. C., 198, 202
Slotnick, B. M., 288
Sluckin, W., 183
Smith, M. C., 57, 63
Smith, N., 327
Smith, N. F., 98, 105, 197, 201, 250, 299, 334
Smith, R. D., 326
Solomon, R. L., 52, 148, 149, 312, 313, 315, 325, 326, 332, 339
Sparer, R., 98
Spear, N. E., 220, 237, 238, 245, 246, 247, 248, 249, 250, 252
Spence, K. W., 63, 69, 103, 277

Sperling, S. E., 206, 282
Spevack, A. A., 236
Spiker, C. C., 278
Springer, A. D., 236, 237
Starr, M. D., 316, 317
Starr, R., 140, 141
Staveley, H. E., 98
Stein, G. W., 106, 300
Steinsultz, G., 269
Steranka, L. R., 246
Stettner, L. J., 184
Stevenson-Hinde, J., 159
Stikes, E. R., 250
Still, A. W., 318
Stimmel, D. T., 215
Stoffer, G. R., 286
Stokes, L. W., 142, 159, 202, 299
Stratton, J. W., 183
Stratton, V. N., 185
Strongman, K. T., 315
Suboski, M. D., 236
Suissa, A., 344
Suiter, R. D., 84
Surridge, C. T., 227
Sutherland, N. S., 257, 279
Sutton, C. O., 177
Sytsma, D., 204, 205, 221, 222, 223
Szakmary, G. A., 70

Tapp, J. T., 106, 299
Tarpy, R. M., 102, 103, 109, 207
Tauber, L., 120
Taylor, G. T., 318
Taylor, P., 97
Teghtsoonian, R., 298
Terlecki, L. J., 177
Terrace, H. S., 69, 269
Terry, W. S., 69, 87
Testa, T. J., 62, 340
Teyler, T. J., 57, 63
Theios, J., 100, 101, 221, 282, 300
Thomas, D. R., 238, 262, 265
Thomas, E., 63, 69, 218
Thompson, R., 234, 235
Thompson, R. F., 73
Thorndike, E. L., 22, 32–35
Thornton, J. W., 339
Toffey, S., 185
Tolman, E. C., 14, 15
Tombaugh, J. W., 103, 207
Tombaugh, T. N., 103, 207, 208
Tomie, A., 343
Tortora, D. F., 329
Tracy, W. K., 263
Trankina, F., 303
Trapold, M. A., 98, 103, 104, 274, 275
Traupmann, K. L., 206
Treichler, F. R., 299
Treit, D., 177
Turner, C., 88
Turner, L. H., 312, 313
Turney, T. H., 271, 272
Tursky, B., 124

Uhl, C. N., 199, 200
Ulm, R. R., 336
Ulrich, R. E., 336

Vardaris, R. M., 67, 218
Verplanck, W. S., 303
Victor, I., 277
Viney, W., 209
Vogel, V. A., 215
von Glasersfeld, E., 188

Wade, M., 265
Wagner, A. R., 63, 67, 80, 85, 86, 87, 88–89, 204, 215, 218, 219
Waller, T. G., 246, 280, 281, 284, 344
Wang, G. H., 298
Warden, C. J., 304
Warren, J. M., 281, 288
Wasserman, E. A., 343
Waters, R. W., 227
Webber, M. I., 97
Weisinger, R. S., 308
Weisman, R. G., 147, 148, 149, 258, 260
Weiss, A. B., 115
Weiss, B., 115
Weiss, J. M., 344
Weiss, K. M., 315
Welker, R. L., 61–62, 212, 213
Wells, C., 165
Wepman, B., 342
Westbrook, R. F., 172
Weyant, R. G., 262
Whaley, F. L., 69
Wheatley, K. L., 61–62
Wheeler, R. L., 274, 275
White, C. T., 27
Whitlow, J. W., 87
Whitworth, T. L., 127
Wickens, D. D., 111
Wielkiewicz, R. M., 153, 275, 276
Wike, E. L., 205, 206, 208, 306
Wike, S. S., 208
Wilcoxon, H. C., 169, 170, 171, 172
Wilkie, D. M., 120
Williams, B. A., 44, 116
Williams, D. R., 115
Williams, J. L., 340, 341, 342
Williams, S. B., 206
Willis, R. D., 67
Willner, J. A., 84
Wilson, G. T., 330
Wilson, J. J., 220
Wilson, N. E., 166, 167
Wilson, R. S., 52
Wilson, W. H., 187
Wolfe, J. B., 102
Wolkoff, F. D., 319
Wollman, M. A., 166
Wong, P. T. P., 107, 300, 301
Woods, P. J., 98, 300
Woods, S. C., 308, 309, 310, 311

Wooten, C. L., 43
Worsham, R. W., 239, 240, 241, 244
Wrather, D. M., 118, 119, 121
Wright, J. H., 306
Wynne, J. D., 74
Wynne, L. C., 325, 326

Yelen, D., 215
Yorke, C. H., 165
Young, G. A., 302
Young, L. D., 124
Younger, M. S., 142, 202

Zahorik, D. M., 58, 165, 166
Zakubczak, L. F., 298
Zamble, E., 153, 306
Zaretsky, H. H., 300
Zentall, T. R., 243
Ziff, D. R., 227
Zimmer-Hart, C. L., 196
Zimmerman, R. R., 286
Zimmerman, R. S., 299
Ziriax, J. M., 127
Zuelzer, K., 308

Subject Index

Action patterns, 160–63
Adaptation, 4–6
Age-retention phenomenon, 249–53;
  theories of, 252–53
Aggression, 203–4;
  after punishment, 336–37
Analyzers, 279–81
Anticipation method of measurement,
  26–27
Attention theory, 279–84
Aversive conditioning, 25–26
Aversive unconditioned stimulus, 25–26
Avoidance relearning decrement
  phenomenon, 245–47
Avoidance training:
  and conditioning, 37–38, 100–102;
  defense reaction learning, 178–80;
  and extinction, 324–26;
  and punishment, 331–37;
  response extinction, 202–4;
  Sidman, 38, 138–39, 147, 203;
  two-factor theory, 136–43;
  and vicious circle, 332–35

Backward conditioning, 54
Belongingness, 166–67, 171
Biofeedback, 122–24
Biological drives, 297–300
Black box theory, 11–12
Blocking phenomenon, 84–90
Blood sugar levels, 309–11

Chain schedule, 116
Classical conditioning, 22–32;
  aversive conditioning, 25–26;
  conditioned excitation, 22–23;
  measurement, 26–28;
  terms of, 23–25
Color aversion, 169–72

Communication, 186–90
Competing response theory, 330–31,
  344–46
Conditioned emotion, 143–46
Conditioned excitation, 22–23, 28–29
Conditioned inhibition, 28–32;
  measurement of, 29–32
Conditioned response:
  definition of, 24–25;
  measurement of, 26–28;
  nature of, 95–96
Conditioned stimulus:
  compound, 83–90;
  definition of, 24;
  and emotions, 143–46;
  intensity of, 59–60;
  offset in avoidance, 137–43;
  quality of, 60–63;
  /US interval, 56–59
Consolidation theory, 233–35
Context conditioning, 90
Contrast phenomenon, 117–22
Counterconditioning, 201

Delay:
  conditioning, 51–52;
  learning, 172–76;
  of reward and extinction, 207–9
Delayed-matching-to-sample technique
  (DMTS), 239–44
Diet regulation, 163–66
Differential-reinforcement-of-other-
  behaviors (DRO), 199
Discrimination;
  definition of, 257;
  extradimensional shift, 285–86;
  intradimensional shift, 285–86;
  observational learning, 277;
  overlearning reversal effect, 282–84;

400

problem difficulty, 270–72;
and stimulus, 272–74;
theories of, 278–81
Discriminative stimulus, 34–35
Disinhibition, 76
Drive theory, 295–96
DRL schedule, 114–15
Drug studies, 227

Easy-to-hard effect, 270
Effort, 209
Electroconvulsive shock (ECS), and memory, 233–38
Emotions, 143–46;
in generalization testing, 268–69
Escape training, 37–38, 98–100;
two-factor theory, 136–37
Evolution, 3–7
Excitatory conditioning, 22–23, 28–29
Expectancy, 7–11;
in discrimination, 275–77
Experiments, in interaction, 146–55
Exploration, 317–18
Extinction:
avoidance training, 202–4;
and drive levels, 300–301;
general theory, 209–11;
and interference, 214–17;
and partial reinforcement, 218–28;
and punishment, 331–37;
resistance to, 225;
response effort, 209;
reward aftereffects, 211–12;
reward delay, 207–9;
stimulus learning, 194–98;
theories of, 209–11
Eyelid conditioning, 57

Fear:
and acquired motivation, 305–6;
avoidance extinction, 325–26, 329–30;
conditioned emotion, 143–51;
and memory, 245, 249–50;
response extinction, 215–16
FI-FR schedule (Fixed interval-fixed ratio schedule), 115–16
Fixed interval schedule, 109–11
Fixed ratio schedule, 111–12
Flooding, 326–31;
theories of, 329–31
Food deprivation:
and drive theory, 297–300;
and instrumental learning, 300–302
Food preferences, acquired, 163–66
Forgetting, 233
Forward conditioning, 51–52
Frustration:
drug studies, 227;
reaction, 214;
theory, 214–17, 222–27

Galvanic skin response (GSR), 27
Generalization:
definition of, 256–57;
degree of acquisition, 26;
and early experience, 263–65;
gradients, 257–60;
inhibitory-excitatory interactions, 267–69;
and motivation, 261
prior discrimination, 265–67;
and test interval, 262–63

Hull-Spence theory, 278–79
Hunger:
conditioned appetitive drive states, 305–9;
and drive theory, 297–305
Hyperglycemia, conditioned, 311
Hypoglycemia, conditioned, 311

Imprinting, 180–86, 316–17
Incentive theory, 296–97
Inhibition:
conditioned, 36–38;
of delay, 75–77;
in discrimination, 267–69
Instrumental conditioning, 22, 33–39;
and deprivation levels, 300–302;
major paradigms, 95;
procedures, 35;
and punishment, 35–36;
terms, 34–35
Interaction between classical and instrumental conditioning, 132–36;
experiments in, 146–55
Interference, 214–17, 226;
in memory, 233, 241–42
Intermediate-term memory, 244–49
Intertrial interval, in classical conditioning, 69

Language, animal, 186–90
Latent inhibition, 77–81
Learned helplessness phenomenon, 338–48;
principles of, 339–43;
reversibility of, 341–42;
theories of, 343–48
Learned laziness phenomenon, 343
Learned safety theory, 173–75
Learning abilities:
animal language, 186–90;
color aversions, 169–72;
defense reaction learning, 176–80;
diet regulation, 163–66;
food preferences, 163–66;
imprinting, 180–86;
sign language, 187–88;
taste aversions, 166–76
Learning sets, 286–89
Long-term memory, 249–53

Matching law, 125–28
Measurement, of conditioned response, 26–28
Memory, 64, 231–53;
general theories, 233–38;
stages of, 231
Memory-retrieval theory, 247
Misbehavior, 159
Mixed schedule, 116
Motivation:
acquired, 305–11;
deprivation and, 300–302;
and discrimination, 294;
drive measurements, 303–5;
drive theory, 296–97;
and fear, 333–35;
and incentive theory, 295–96;
and irrelevant drives, 302;
nonregulatory systems, 317–20;
stimulation theories, 312–17
Motor competition theory, 245–46
Mowrer's two-factor theory, 136–43
Multiple schedule, 116

Naturalistic response pattern, 160–63
Negative contrast, 118
Neophobia, 174
Noise intensity, 59–60, 78

Observational learning, in discrimination, 277
Omission training, 37;
response extinction, 199–202
Opponent process theory, 312–17

Partial reinforcement effect, 106–7, 218–28;
definition of, 218;
interference position, 226;
and reward magnitude, 219–22;
in stimulus learning, 218–19;
theory of, 222–28
Passive avoidance, 38
Pavlovian conditioning, 22–28, 32;
and incentive theory, 297
Peak shift, 267–69
Performance:
and drive theory, 295–96;
vs. learning, 13–15
Positive contrast, 118
Pseudoconditioned CRs, 52–53
Psychic secretions, 22–23
Punishment, 35–36, 98;
and aggression, 336–37;
avoidance extinction, 331–37;
and vicious circle, 333–35

Random control procedure, 55
Reinforcement:
contingency, 95–96;
delay, 102–5;
schedules of, 107–17
Reinforcer, 34

Reinstatement, 253
Relearning, 245–49
Response, 15–18;
blocking, 326–31;
expectancy, 22, 42–45;
extinction, 198–204
avoidance training, 202–4;
omission training, 199–202;
reward magnitude, 198–99
Retardation-of-acquisition technique, 31
Retention, 232;
interval, 244–48
Retrieval:
stage of memory, 232;
theory, 235–38
Retroactive interference, 241
Retrograde amnesia phenomenon, 236–37
Reward training:
aftereffect, 211–12;
discrimination, 274–76;
instrumental conditioning, 33–34;
magnitude, 96–98;
response extinction, 198–99, 204–6, 219–22

Salivary conditioning, 57
Schedules of reinforcement:
chain schedule, 116;
complex schedules, 115–16;
DRL schedules, 114–15;
FI-FR schedule, 116;
fixed interval schedule, 108, 109–11;
fixed ratio schedule, 108, 111;
mixed schedule, 116;
multiple schedule, 116;
tandem schedule, 116;
variable interval schedule, 108, 112–13;
variable ratio schedule, 108, 113
Second-order conditioning, 70–73
Sensitization, 53
Sensory preconditioning, 73–75
Sensory reinforcement, 105
Short-term memory, 238–44
Side-tracking, 274
Sidman avoidance, 38, 138–39, 147, 203
Sign language, 187–88
Skeletal conditioning, 57
Species-specific defense reactions (SSDR), 177–80;
avoidance extinction, 325;
and learned helplessness, 340
Species-typical behavior potential, 9–10
Spontaneous alternation, 318–20
Stimulus, 7–11, 15–18;
aftereffect, 211;
expectancy, 21;
extinction, 194–98;
frustration, 214, 223–25;
learning, and partial reinforcement effect, 218–19;
response learning, 132–36

Summation measurement method, 29
Survival, 4–7

Taste aversion:
  acquired, 166–76;
  learning, 57–59, 60–61
Temporal discrimination hypothesis,
  243–44
Thigmotactic behavior, 178–79
Thirst:
  conditioned, 308;
  and drive theory, 297–305
Trace conditioning, 51–52
Trace decay notion, 243
Trauma, early, and memory, 249

Ulcers, 336
Unconditioned response:
  definition of, 24;
  in instrumental conditioning, 34

Unconditioned stimulus:
  definition of, 23;
  duration, 66–67;
  in instrumental conditioning, 34;
  intensity of, 63–66, 96–102;
  and learned helplessness, 337;
  predictability, 67;
  preexposure effect, 89–90;
  quality, 67–69, 105–6

Variable interval schedule, 112–
  13
Variable ratio schedule, 113
Visceral conditioning, 57
Visceral responses, 123–24
Vicious circle phenomenon, 332

Wagner-Rescorla model, 88–89,
  128
Warm-up decrement phenomenon, 238